Science, the Never-Ending Quest

Other books by the same author

Science in Everyday Things

Mathematics in Everyday Things

Electronics in Everyday Things

Science,

THE NEVER-ENDING QUEST

by William C. Vergara

WITH 155 DRAWINGS BY THE AUTHOR

HARPER & ROW, PUBLISHERS

1817

NEW YORK,
EVANSTON,
AND LONDON

FIRST EDITION

LIBRARY OF CONGRESS CATALOG CARD NUMBER: 64-25142

To Patricia and Bob

Contents

Preface

In this book, I have tried to discuss and explain what seem to be the high points in the growth of the exact sciences. But I hope the reader will find more than a cold collection of facts. The story of science covers an exciting phase of human history, full of the thrill of discovery, and overflowing with the toil of exceptional minds. While the book is not strictly a history of science, I have tried to convey some of that feeling in the pages that follow.

We can understand science, of course, without caring a pin for its history. But we are poorer for it. Learning scientific principles the way their discoverers did gives us an insight into our groping for the truth. The great James Clerk Maxwell wrote that ". . . science is always most completely assimilated in the nascent state . . ."

We can best understand science by re-living its doubts along with its bold achievements. This means discussing its failures as well as its successes, its violent controversies as well as its accepted doctrine. The historical approach also helps us to recognize and accept the tentative nature of all scientific knowledge. Time has little respect for the theories of men.

In writing the book, there was a continual tugging between a chronological approach and one based on the separate scientific disciplines. Each method has its strong and weak points and I have tried to find a reasonable compromise between the two. The reader will find some chronological backtracking, and some crisscrossing of scientific disciplines. I hope he will overlook these inconveniences.

The book is written for the general reader having little background in science, so it deals primarily with basic principles and skips much of the detail surrounding them. For the same reason, emphasis is placed on clarity as opposed to scrupulously precise scientific language. Because it tries to entertain as well as to teach, interesting sidelights are also drawn into the text.

The general reader is often surprised to learn that science has great difficulty explaining its most fundamental concepts. Most of us probably expect scientific explanation to resemble a proof from geometry —starting with a few self-evident axioms and proceeding logically to an important conclusion. On the contrary, scientific research usually takes the other way around. It begins with a set of observations and works backwards to try and deduce something more fundamental about nature. More often than not, the basic "axioms" so deduced are anything but obvious. They are hard to accept because they usually have no apparent basis in our everyday world. Scientists use these fundamental concepts—not because they are intuitively satisfying—but because they work. The only justification for any scientific principle lies in the accuracy with which it can predict the curious ways of nature.

If this book should arouse an interest in science or in its history it will have served its purpose. But it has merely scratched the surface in most scientific areas. A bibliography at the end will help the general reader select authoritative books along his line of interest. The listed books by Newton, Galileo, Hertz, and Faraday are scientific classics that show in detail how great discoveries are made. Using the simplest of apparatus by today's standards, these men performed only the most humble experiments. But they watched these experiments with the eye of genius, and described them with a beautiful clarity. We are doubly fortunate when great scientists are also gifted expositors.

In a book such as this, covering so many scientific disciplines, errors and obscurities have a way of creeping in. I have tried to weed them out wherever they could be found. In that regard, I want to thank my colleagues Dr. T. J. Carroll, Mr. H. M. Greenhouse, and Dr. O. N. Rask for reading portions of the manuscript and offering many valued suggestions for improvement. Their help has been to the book's advantage.

WILLIAM C. VERGARA
Towson, Maryland
October 1964.

Science, the Never-Ending Quest

The Coming of the Calendar

1

According to legend a certain scholar once said, "If the extent of the Atlantic Ocean were not an obstacle, we might easily sail from Spain to India along the same parallel (of latitude)." This was neither Columbus speaking, nor Magellan, nor Amerigo Vespucci. The novel idea belonged to Eratosthenes, librarian of the Museum of Alexandria, who lived about 250 years before the Christian era. Eratosthenes, like most educated men of the day, knew that the world was round, not flat. He had even measured the earth's circumference to an accuracy better than 4 per cent. Yet almost two thousand years were to pass before man would put his idea to the test. This delay emphasizes the interdependence of science and the everyday life of mankind. Science seems to thrive only in soil made fertile by a society that makes use of new-found knowledge. Scientific achievement and the needs of society have been so closely associated throughout history that it is often impossible to determine which is cause and which is effect. Eratosthenes' concept of a western route to India lay dormant because the time wasn't right.

When a branch of science becomes a plaything of a leisure class, it must soon stagnate. The ancient Greeks raised geometry to an elegance unsurpassed to this day, yet they stubbed their mathematical toes on everyday problems any fifth grader can solve today. The Greeks were completely uninterested in solving the practical problems

→ NO NEED AS THEY UTILIZED SLAVE LABOR!

of the common man. Their science of mathematics became a game and it eventually lost touch with reality.

The story of science is closely associated with man's changing social needs. The printing press led to the development of eyeglasses even though their basic principle, refraction, had been observed 1,500 years earlier. Eyeglasses led to the telescope and microscope. The telescope soon revealed the moons of Jupiter which offered hope in solving the knotty problem of longitudinal determination. The microscope made visible a vast new world of microorganisms whose very existence defied belief. The cannon encouraged scientific inquiry on two fronts: in chemistry to help solve the problems of making gunpowder, and in the study of motion to understand a projectile's trajectory. For these reasons, the chapters that follow will tell the story of scientific achievements against the backdrop of man's changing problems and needs.

The Herdsman and Farmer

Thousands of years before Genesis was written, man learned to herd and to farm. But the success of these new activities depended upon a recognition of the passage of time. Over countless generations, man began to note subtle regularities in the movement of heavenly bodies. He noticed that regular changes in the heavens seem, somehow, to be related to the recurring seasons. Standing in his field, the farmer noticed that the sun always set on the same horizon, yet the exact point at which it set changed from season to season. The same observation held true for the rising sun, but on the opposite horizon. He noticed that his shadow changed in size and direction as he toiled in the field. As the morning progressed, the shadow became shorter and shorter until the sun reached its highest point in the sky. The shadow then lengthened as the sun fell toward the western horizon. His day was divided into two seemingly equal parts by the shortest length of shadow. It was a simple step, then, to erect a pole or column which would divide the day into smaller and more convenient parts by the angle of the shadow.

Continuing his observations by night, our prehistoric farmer noticed that the North Star always hangs above the same spot on the horizon. And strangely enough, the direction of the North Star always coincides precisely with the direction of the noonday shadow. The other stars, like the sun, rotate from east to west about the North Star during the night. As the shadow pole indicated the noon hour, so did the constellations divide the night into two. Tending his flock by night, a shepherd must have noticed certain stars that rise in the east at dusk and set in the west at dawn. Midnight would occur when such a star was directly above the North Star. Like the sun, each star rotates just fifteen degrees per hour in a counterclockwise direction about the North Star. With a little practice, he learned to tell time at night by noting the position of a suitable star.

Seeking to tie these facts together, early man concluded that the stars were fastened to a great celestial sphere. Since this sphere encloses the earth, we see only one half of it at a time. The stars become visible only when the sun is below the horizon. The axis about which the celestial sphere rotates extends from the North Star to some point on the earth—the North Pole. It rotates in a counterclockwise direction about this axis, completing one revolution in a day and a night.

The Moon and Our Month

One of the first impressions that early man must have had of the moon is the regularity with which it goes through its phases. A full moon is repeated, for example, about every thirty days. Further observation discloses that the moon rises a little later each day. It seems to slip backward each day against the backdrop of fixed stars on the celestial sphere. After about a month, the moon has returned to its original place in the celestial sphere. It was convenient, therefore, to reckon time by the lunar month which, in turn, could be divided into weeks of seven days' duration, each corresponding roughly to a quarter phase of the moon. Many backward people still keep track of time by the lunar month, especially those with no organized agriculture.

The Calendar Priests

For a farm economy to prosper, it was necessary that moon reckoning of time give way to the more useful yearly method. The ancient

Babylonians had a year consisting of twelve 30-day months, and by 2500 B.C. the Egyptians had already established a year of 365 days. This surprising accuracy was achieved by keen observation of the sun's shadow and of the rising and setting of stars.

We are all familiar with the fact that the sun is higher in the sky in summer than in winter. Less obvious, perhaps, is the changing length of the noonday shadow as the seasons progress. The shadow is shortest in summer when the sun is more nearly overhead, and longest in winter when the noonday sun slants toward the south. By measuring these changing lengths, ancient man found two days that were related to the year: the longest day of the year (June 21) when the noon shadow is shortest; and the shortest day of the year (December 21) when it is longest.

These important days were connected with another solar peculiarity. On or about December 21 (winter solstice) the sun rises and sets at its northernmost points on the horizon, while on June 21 or 22 (summer solstice) it has reached its furthest excursion to the south. Midway between these two dates are the vernal equinox (on or about March 21) and the autumnal equinox (on or about September 23) when day and night are of equal length. On these days the sun rises and sets precisely halfway between the points on the horizon corresponding to the solstices.

It was possible, therefore, for ancient man to determine the beginning of the year by noting the direction of the rising or setting sun. Pairs of stone pillars or monuments were often used for this purpose. The stone remains at Stonehenge (c. 1500 B.C.) were used by early Britons to determine the solstices and equinoxes with considerable accuracy.

Early man also made use of the stars to fix dates that were important in his everyday life. Observation shows that every star rises a little earlier each night. In midwinter, in northern latitudes, the three bright stars of Orion are visible most of the night. They rise just after sunset and set just before sunrise. By early spring, however, these stars are already directly overhead at sunset. By midsummer they have already passed below the western horizon when the sun sets and so are invisible in the summer sky. This behavior of the stars made it possible to fix the year's length and to pinpoint specific days of interest. At Stonehenge, for example, an early astronomer could sight along two accurately located stones in the twilight of August 6 and see a bright star (Aldebaran) on the western horizon. When he looked on earlier or later dates, the star would be above or below the line of sight of the stones. By selecting the correct angle and the appropriate bright star, he could determine days of importance in his religious or agricultural activities.

Although the calendar grew out of the needs of the common man —the planting and reaping of crops, for example—it soon ceased to be a common possession. In most societies a caste developed which was entrusted with the timekeeping necessary for a successful farming economy. Because nature might have to be bribed, these timekeepers were often the calendar-priests of the society. This was probably the most influential and respected job in the community, so it was only natural that such individuals should keep astronomical knowledge secret. As this knowledge developed further, the prosperous calendar business came to be farther and farther removed from the common man. Until comparatively recent times, the calendar was considered the property of its priestly custodians.

Julius Caesar and the Missing Hours

Although the Egyptian year of 365 days was an excellent example of applied astronomy, it was still about six hours too short. In the course of centuries, the months managed to get considerably out of step with the seasons. The ancient Babylonians solved the problem by adding an extra month to the year every so often, much as we add an extra day every fourth year.

Nevertheless, Julius Caesar's inauguration found the Roman calendar in hopeless confusion. So, after consulting the Alexandrian astronomer Sosigenes, Caesar established the Julian calendar in 45 B.C. With only minor adjustments, his calendar has served well for two thousand years.

The Julian calendar broke completely with all lunar considerations and fixed the true year at 365¼ days. For practical reasons, three years of 365 days were instituted, followed by a fourth year of 366 days. The extra day was obtained originally by repeating February 24, but it later took the form of our February 29. Caesar also changed the beginning of the year from March to January 1. This explains why September (meaning the seventh month) is actually the ninth month. To commemorate his work, Julius renamed the month Quintilis, calling it July after himself. Not to be outdone, his successor—Augustus—appropriated Sextilis, naming it August.

Pope Gregory and the Extra Minutes

In spite of all precautions, Caesar's year turned out to be about eleven minutes too long. By A.D. 1582, the accumulated error had reached ten days and the spring equinox had slipped back from March 21 to March 11. In order to correct the discrepancy, Pope Gregory ordered the elimination of ten days from the calendar. The day following October 4, 1582 became October 15 instead of October 5. That solved the immediate problem, but it was clear that something ought to be done to prevent a recurrence. The Julian calendar ran slower than the true year by about 3 days in 400 years. Pope Gregory decreed, therefore, that only one *century year* (that is 1600, 1700, etc.) in four should be a leap year. (Previously, every century year had been a leap year because all were divisible by four.) This was accomplished by specifying that, henceforth, only those century years divisible by 400 would be leap years. Pope Gregory's refinement speeded up the Julian calendar just enough to keep it in step with the sun—at least, until some future civilization becomes interested in seconds.

The development of the calendar was the first truly great scientific achievement of the human race. In learning to fix the year as the interval between successive repetitions of an astronomical event, man had acquired two fundamental scientific tools: observation and measurement. He learned that apparently random events in nature may sometimes be connected by subtle relations. It had taken countless thousands of years for man to notice and then to solve the great riddle of the heavens. Piece by piece he fitted the puzzle together by a process we would call the scientific method. Although his calendar science was intermixed with magical lore and priestcraft, it led to the fabulous geometry of the Greeks which became a milestone of scientific thought.

Learning to Count

2

Whether we live in a modern skyscraper or among the aborigines of Australia, our lives are continually affected by numbers. Objects and events about us seem to cry out for numerical treatment. How many are there? In which direction does it lie? When did it happen? How big is it? Before a science can evolve from human experience, questions like these must have numbers for their answers. And man had learned to answer such questions tens of thousands of years before he built the pyramids.

There is a common misconception that mathematics was born about 500 B.C., when the Greek aristocracy had the leisure time necessary to develop geometry. Actually, the Greeks inherited a world full of arithmetic and geometry. They were visited by tax collectors calculating taxes due. They saw artisans making intricate tiles and decorating vases with interrelated geometrical patterns. They watched slaves building temples with plumb line, water level and square. All about them were men measuring angles between stars, calculating interest charges, measuring heights by the length of shadows, predicting the dates when eclipses might occur, and taking bearings from the North Star. Ancient Greek geometry, which repels so many intelligent students, was an outgrowth of leisure time *and* a well-developed civilization—a civilization built by men who preferred activity to leisure.

Can Animals Count?

To understand why scientific progress is so completely associated with mathematics, we have to travel back mentally through great periods of time. The Golden Age in Greece is much too recent for our purpose, as are the civilizations of Egypt and Babylon. We must go back, in fact, to a period long before man's thoughtful ancestors evolved from a lower form of life. The beginnings of the number notion predates the existence of man. Ridiculous? Well, let's examine the facts.

Naturalists tell us that an amazing number sense exists in birds and insects. Consider birds, for example. If a bird's nest holds four eggs, we can probably take one of them without arousing suspicion. But if two are removed, the bird generally notices the theft and deserts the nest. Crows are even better mathematicians. The English naturalist Sir John Lubbock tells the story of a man who wanted to shoot a crow. To trick the bird, he sent two men to a watch house and had one of them leave while the other remained in hiding to do the deed. But the crow stayed clear of the hiding place until the second man had left. Then he repeated the procedure with three men, two of whom left the watch house. Still the crow refused to be tricked. He found it necessary to send five men to the watch house in order to upset the crow's counting. Can we assume from this that a crow can count to four? There are many other examples in nature that imply the existence of a number sense in lower forms of life. Is it scientific to ascribe all of these observations to instinct?

Stone Age Mathematics

Although we can't pinpoint the exact beginning of mathematics, we know that it is very old indeed. Back when prehistoric Europeans were still using rough tools and implements made of chipped stone, they had already learned to count beyond fifty with the help of notched tally sticks. A prehistoric computing machine of that kind survives from paleolithic times showing notches cut on it in groups of five. It's probably valid to assume that early man had little need to count large numbers of things until he began to keep herds and flocks. Even a prehistoric herdsman would want to count his animals from time to time to be sure that an overambitious neighbor had not taken up cattle rustling. The paleolithic tally stick shows that early man had learned to count in groups of five long before he found it advantageous to smooth and polish the stone implements he kept around the cave. His use of groups of *five* is undoubtedly associated with finger counting, a practice that can be found even today in some primitive societies. In our number system, which came to us from the Hindus, we count things in groups of tens. When the numbers get large, we use groups of tens of tens (or hundreds), and tens of tens of tens (or thousands). That is what we mean when we call *ten* the *base* of our decimal number system. The word *decimal,* of course, comes from the Latin word *decimus* meaning tenth. Our paleolithic herdsman used a *quinary* system whose base is *five.*

The Babylonian Number System

Number systems based on two, twenty and sixty have also been in vogue in the past. The Syriac system had numerals for one and two which were repeated in Roman numeral fashion to build up larger numbers. (Figure 1.) The Egyptians had symbols for 1, 10, 100 and so on which were added as shown in Figure 1 to form composite numbers. The Greeks and Alexandrians used letters of their alphabet for numbers in an additive system based on ten. The Chinese used a "matchstick" number notation also based on ten. All of these number systems have one thing in common: They depend upon the *addition* of different elementary number symbols to form larger composite numbers. The contribution of each symbol to the larger number depends only upon its configuration and not upon its position in the composite number. Roman numerals, which are familiar to all of us even today, use the same basic principle, although the location of a symbol may sometimes indicate subtraction: IV=4, but VI=6.

A much earlier and more sophisticated notation was that of the Babylonians—a sexagesimal system based on groups of sixty. Counting by sixties (and sixties of sixties, etc.) may seem cumbersome in our decimally oriented age, yet most of us would feel more at home with the Babylonian arithmetic and algebra of 2000 B.C. than with the clumsy Roman numerals used by Europeans during the Dark Ages. The Babylonians had a method of building up numbers that was remarkably similar in principle to our own. To illustrate, consider our number 6,243. The three is located in the first position to the left of

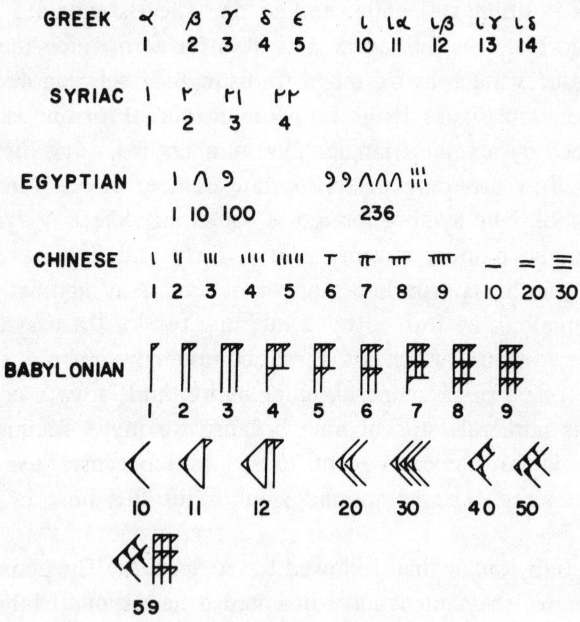

Fig. 1. Five ancient number systems.

the decimal point and stands for the number of *units* of something, say sheep. The four, being in the second position, stands for the number of *tens* of sheep. The two stands for the number of *hundreds* (or tens of tens) and the six stands for the number of *thousands* (or tens of tens of tens). In other words, the numerical value of any number symbol depends upon its *position* in the composite number. The combination of the numeral and its position really tells us *how many times* one, or ten, or one hundred, or one thousand the numeral stands for. The Babylonians did precisely the same thing, merely substituting powers of sixty for powers of ten. Their units position corresponds to ours, the first place to the left of an unwritten decimal point. It can hold any numeral from zero to fifty-nine. The next place to the left is their "sixties" position which corresponds to our "tens" position. Each symbol in the sixties position must be multiplied by sixty to gets its actual value. The next position is their thirty-six hundreds position (60×60), and so on. The Babylonians were also the first to hit upon the use of a symbol for zero which they used in later centuries much as we might to distinguish between 42 and 402.

As shown in Figure 1, the Babylonian symbol for one is a vertical line capped by a small triangle. The numbers from one through nine are formed by grouping an appropriate number of one's into a compact symbol. The symbol for ten is something like a V lying on its side, and the numbers twenty, thirty, forty and fifty have symbols derived from an appropriate number of tens. Any number from one to fifty-nine can be formed by combining two of these symbols into a compact group. Each of these compound symbols should be thought of as a single numeral, much as we think of two, or seven, or nine. Our numerals stop at nine because we use a decimal system. The Babylonian numerals go up to fifty-nine because their system is based on sixty. The compound symbol for fifty-nine is shown in Figure 1.

If the Babylonians had followed the logic of the Egyptians, Greeks and Romans, they might have invented a new symbol for sixty and proceeded to build up larger numbers by the principle of addition. But for some unknown reason they did something much more profound: they invented the important *principle of position*.

Figure 2 shows how the Babylonians wrote numbers larger than fifty-nine. The first number at the left of the diagram consists of a Babylonian one followed by a ten. This is obviously not meant to be eleven, for then the two symbols would be reversed, the one following the ten. It is translated as 1,10 (which we can read, "one comma ten") where the 10 stands for a Babylonian ten, the 1 for a Babylonian one, and the comma emphasizes that we are dealing with only two Babylonian symbols translated into our Hindu-Arabic number language. The remaining symbols on the top line are translated 1,20, 1,30, 1,40, 1,50, and 2,0 or 2. A study of Babylonian mathematical texts shows that the number before the comma in our translations above stands for the number of sixties which must be added to the subsequent number of units. The number 1,10 really means $1 \times 60 + 10$, or 70 in our system. The first position on the right is the units position, the next to the left is the sixties position. With these two positions, the Babylonians could build up all of the numbers from 1 to

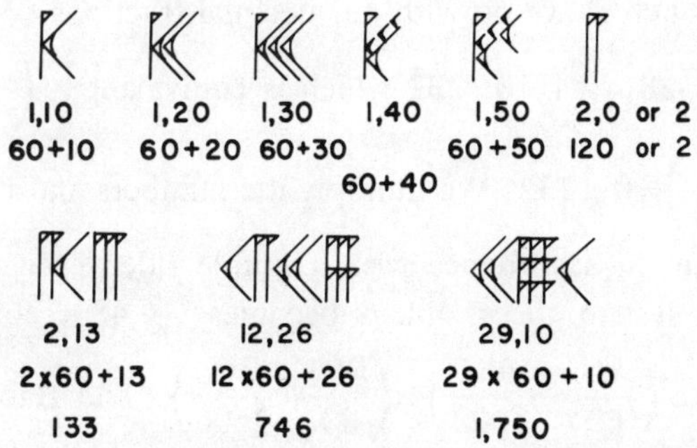

Fig. 2. The Babylonian principle of position illustrated by nine Babylonian numbers. Each of these contains Babylonian number symbols in two positions: the units position (on the right) and the sixties position (on the left). The Babylonian symbol in either position is often a compound symbol, consisting of two or more simpler symbols. The comma in the decimal equivalents of the Babylonian numbers emphasizes that we are dealing with two sexagesimal positions in translation.

59,59 or 3,599 in our system ($59 \times 60 + 59 = 3{,}599$). By adding a third position in which each number stands for 60×60 or 3,600, they could write numbers up to 59,59,59 or 215,999.

The Babylonian place value notation just described dates back to about 1800 B.C., a time before their development of zero. The lack of zero makes it possible to misread 1 20 as $1{,}20 = 80$ when actually 1,0,20 ($1 \times 3600 + 0 \times 60 + 20$) $= 3{,}620$ is meant. This ambiguity is often overcome in Babylonian texts by a large spacing between the 1 and 20 when the second sexagesimal place is missing. In succeeding centuries a special symbol for zero was used to indicate a zero within a composite number. The zero was never used, however, at the end of a number. This means that the Babylonian numeral for 1 might stand for one, sixty, or thirty-six hundred, and the actual value of the symbol must be made clear by the context.

The Babylonian place value system is particularly useful in calculating with fractions. An additive number system, such as our Roman numerals, runs into complications when fractions must be manipulated arithmetically. To illustrate, suppose you want to multiply $\left(\frac{\text{II}}{\text{X}}+\frac{\text{IV}}{\text{L}}\right)$ by $\left(\frac{\text{III}}{\text{X}}+\frac{\text{VII}}{\text{L}}\right)$. Problems of that kind were extremely difficult to solve in ancient times. The Babylonians got around the difficulties much as we do with our decimal fractions. We express the problem as $.28\times.44=0.1232$ which is equivalent to $\left(\frac{2}{10}+\frac{8}{100}\right)\times\left(\frac{4}{10}+\frac{4}{100}\right)=0.1232$. We multiply the numbers and then locate the decimal point in accordance with a simple rule. Translated into the Babylonian system the problem becomes .16,48×.26,24, which is equivalent to $\left(\frac{16}{60}+\frac{48}{3600}\right)\times\left(\frac{26}{60}+\frac{24}{3600}\right)$. The Babylonians then used their sexagesimal multiplication tables to get the answer, .7,23, 31,12. (They did not use a written decimal point. The absolute value of the number was determined at the end of the computation.)

Although the Greek astronomers used an additive number system (similar to that of the Egyptians and Romans) for whole numbers, they copied the sexagesimal system for fractions. Strangely, the astronomers of India, Arabia and Europe retained the sexagesimal fractions even though the decimal system eventually came into common use for whole numbers. This practice has persisted in astronomy to the present time. To illustrate, we write the latitude of a place as 120°36′41″ in which the degrees are given by a decimal number but the minutes and seconds are in units of sixty. (The Babylonians would have written 2.36,41 for the latitude.) When we want to be more accurate, we write 120°36′41.6″ where the ".6" is a decimal fraction. (Again, the Babylonian equivalent would have been 2.36,41,36 because $\frac{6}{10}=\frac{36}{60}$.)

Our final figure, therefore, is an amazing mixture of decimal and sexagesimal notation. After 4000 years of scientific evolution and migration, astronomical notation has manged to arrive at a completely absurd number system. Because time and navigation depend upon astronomical measurements, the same difficult conventions have grown up in these fields. All of these complexities are the bequest of ancient Babylonian science.

Counting in America, 500 B.C.

Although the Mayan civilization on the Yucatan Peninsula of Central America played no part in the development of modern science, its base-20 notation was the most highly developed number system of its time. Figure 3 shows one of the systems used by the Mayas. It employs symbols for 0, 1, 2, 3, 4 and 5 which are stacked vertically to form the compound number symbols up to 19. Larger composite numbers employ zero and the principal of vertical position. Their composite number for 20, for example, uses two digits arranged vertically—one in each of the two positions of the number. The lowest or units position has a zero in it. The second position has a one in it which stands for $1 \times 20 = 20$. The entire number is translated from top to bottom as $1 \times 20 + 0 = 20$. Similarly, the Mayan 400 is $1 \times 20^2 + 0 \times 20 + 0 = 400$ and their 8,000 is $1 \times 20^3 + 0 \times 20^2 + 0 \times 20 + 0 = 8{,}000$. Figure 3 also shows how they wrote a typical composite number 13,0,7 which is equal to our 5,207. The Mayas used a slightly different number system for their calendar. The units and twenties positions are identical to those described above, but each numeral in the third position was worth only 360 instead of 400. The inconsistency in the third position was probably related to their year which consisted of eighteen months of twenty days ($=360$ days) plus five extra days at the end of the year.

Both the Mayan and Babylonian civilizations developed remarkable number systems, although separated by half a world in distance and culture. Of the two, the Mayan system was superior because of the consistent use of the number zero. In our system we can use the digits 4 and 3 to mean 43, 403 or 4,000,300 merely by adding zeros and changing the location of the digits. The Romans might have written IV,III to mean 43 or IV,O,III to mean 403, but nobody seems to have thought of doing it. The concept of zero as a number was one of the most elusive in the history of science, yet one of the most important. It helped change the most difficult calculations into simple operations that any child can perform. Our simple rules of addition,

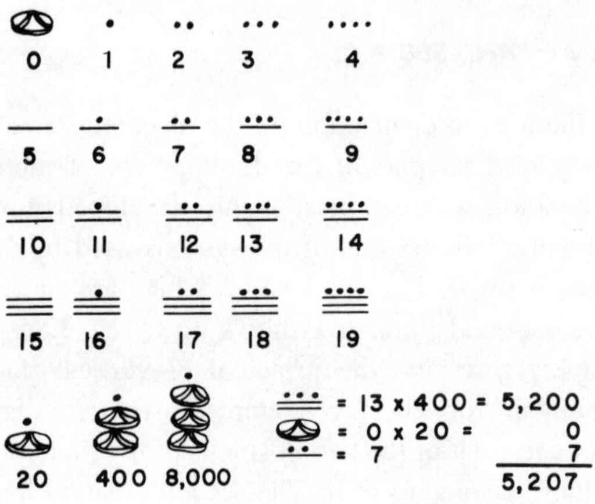

Fig. 3. The Mayan number system used the vertical principle of position based on units of twenty.

subtraction, multiplication, and division could not evolve until zero was elevated to the full status of a number. When zero was added to the Babylonian principle of position, we obtained a number system that stretches out to infinity in either direction—yet uses only a handful of number symbols.

Mathematics in Babylon

The use of the principle of position together with the number zero was a second great triumph of ancient science, equal in importance to the development of the calendar. Unfortunately, these concepts were not extended by the Greek civilization that followed the Babylonians and had to be kept alive or rediscovered by the Hindus.

It was India that gave us zero and the number symbols that have come down to us through the Arabs.

The Babylonian mathematics that grew up in the Mesopotamian valley was much more highly developed than that of Egypt or other contemporary societies. Cuneiform tablets dated about 4000 years ago show that Babylonians were familiar with solutions to problems in geometry that are usually credited to the Greek mathematicians Thales and Pythagoras, who lived some 1,500 years later. Thales, for example, is credited with the discovery that the area of a circle is divided into two equal parts by a diameter. Although it would never have occurred to a Babylonian that such a statement required a "proof," they were surely well acquainted with it as a useful fact. Similarly, the "Pythagorean" theorem ($3^2+4^2=5^2$) and similar Pythagorean numbers were used repeatedly by Babylonians in the solution of mathematical problems. They had formulas—some exact, others approximate—for the volumes of various solids. They also knew the formulas for the circumference of the equilateral triangle, the square, and the regular pentagon, hexagon, and heptagon, in terms of the radius of the circumscribed circle. One text gives π, the ratio of the circumference of a circle to its diameter as: $3.7,30=3+\frac{7}{60}+\frac{30}{3600}=3\frac{1}{8}$. The correct figure is 3.1416. The Babylonians also had a well-developed form of algebra, and tables of squares and cubes which were used to solve rather difficult equations.

Nevertheless, our knowledge of Babylonian mathematics is pitifully incomplete. There are probably close to a half million cuneiform texts in museums around the world, yet a mere handful have been carefully studied by Babylonian scholars. Otto Neugebauer tells us that this ancient civilization achieved a "level of mathematical development which can, in many aspects, be compared with the mathematics, say of the early Renaissance." Although Babylonian algebraists never learned to substitute letters for numbers in an equation, they did use the idea of a "general formula" to solve problems of the same class. Their formulas were in the form of written sentences. Their repeated use of numerical examples merely served to illustrate the general procedures described in the tables.

Greek Mathematics

Although the Babylonians were familiar with many geometrical theorems, there is no evidence of their having "proved" the truth of any them. For that reason, Thales of Miletus (fl. 600 B.C.), a Greek, is usually thought of as the first great mathematician. We are told that he was the first to state half a dozen geometrical theorems in general form and give their proofs. From this beginning, Pythagoras of Samos (c. 580–500 B.C.) established Greek mathematics as a science and built up a more complete logical structure. The Pythagorean mathematics had four branches: arithmetic, or numbers at rest; music, or numbers in motion; geometry, or magnitudes at rest; and astronomy, or magnitudes in motion.

Pythagoras and other early Greek mathematicians realized, as we do, that knowledge is most valuable when shared with others. He often gave lectures to large audiences on numbers and geometrical figures. But an attitude of secrecy soon developed among his students. The Pythagoreans who followed formed secret brotherhoods and enveloped mathematics in a sea of mystery and religious trappings.

Mathematical discoveries could be communicated only under oath. Numbers were given moral or human qualities. One stood for reason, because it was the basis of all other numbers. Two stood for opinion and four for justice. Five stood for marriage because it was the sum of two, the first female number, and three, the first male number. Six was a perfect number because it was the sum of its divisors one, two, and three ($1+2+3=6$). We are told by St. Augustine that "God created all thing in six days, because this number is perfect." About 400 B.C., Hippasus, a member of one of the Pythagorean brotherhoods, was said to have been killed for announcing the discovery of a regular solid having twelve identical surfaces.

By the time the brotherhoods relaxed their secrecy sufficiently to allow books to be written, their mathematics had lost touch with the everyday work and problems of the world. Worse still, they had divorced the fields of arithmetic and geometry. The excellent start that the Babylonians had made toward modern algebra gave way in Greece to a preoccupation with geometry. About the year 300 B.C.,

Euclid assembled all of the geometrical knowledge of the Greeks into a great book, *The Elements.* His contribution was not the solution of new problems in geometry; it was in organizing all known mathematical methods into systems in which known facts could be combined to discover and prove new geometrical truths. Beginning with simple definitions called axioms, Euclid combined them into statements called theorems or propositions which he proved by logic. Algebra, trigonometry, graphs and the calculus all depend on Euclid's geometry. Only ten or fifteen of the two hundred theorems of Euclid's *Elements* are required, however, to enable us to understand and use this new kind of mathematics. The others are merely elegant but complicated ways of doing things that can better be accomplished by newer methods. Nevertheless, geometry became the focal point of all Greek learning. This led to a concentration of Greek effort on three interesting problems: squaring of the circle, duplication of the cube, and trisection of the angle. All three turned out to be insoluble within the limitations of their numberless geometry.

It remained the task of a vigorous new society to put numbers back into mathematics. In 332 B.C, Alexander the Great conquered Egypt and celebrated the victory by establishing a great city in his name where the Nile meets the Mediterranean. When Alexander died in 323 B.C., Egypt became an independent nation under Ptolemy, one of Alexander's generals. Ptolemy's most important act was to establish the first center of secular learning in the history of mankind—the library of Alexandria—a combined museum, library and university.

The library accumulated most of the learning of the ancient world —mathematics, medicine, astronomy and navigation—and attracted some of the finest minds of the civilized world. For the next three hundred years—before the Christians burned the first library—Alexandria saw the most astonishing display of intellectual achievement in the history of mankind. The key to these achievements is the fact that mathematics had once again come into contact with the everyday problems of the world. From this union there would evolve a new kind of mathematics called trigonometry with which the Alexandrians would measure the earth and the distance to the sun and moon.

The Hindus and Arabic Numerals

While the Greeks and Alexandrians made great strides in science and Mathematics, their failure to build on the Babylonian arithemtic made their task more difficult. Luckily for succeeding generations, the Babylonian concept of zero and the principle of position came down to us through another path. A handful of ancient manuscripts tell us of the lofty heights that the Hindus attained in mathematics. Unfortunately, their path to mathematical greatness is lost in prehistory. It's quite possible, however, that Babylonian achievements filtered through trade channels to India and were accepted and advanced by Hindu mathematicians.

Although our numerals, 0, 1, 2, 3, 4, 5, 6, 7, 8, 9, are usually referred to as Arabic, the term is really a misnomer. They actually came to us—after many changes—from the Hindus. The Arabs were primarily intermediaries who carried Hindu and Greek mathematics to Europe. Perhaps the greatest mathematical contribution of the Hindus was the elevation of their *sunya*, or zero, to the status of a full-fledged number. The Babylonian zero was essentially a mark used to avoid ambiguity in writing numbers, and it never entered into their arithmetical operations. The Hindus used zero much as we do today. The caste system eventually stifled Indian science, however, through the same kind of secrecy employed by the Pythagoreans. Mathematical knowledge came to be the property of a few, and new discoveries were not readily communicated to all. It was in the Arabic world that scientific learning was kept alive during the anti-intellectual period of Christianity—the aptly named Dark Ages.

Following the death of Mohammed, in 632, the Moslems began to overrun the ruins of the Roman Empire—Egypt, North Africa, part of Spain and Sicily. Later, their conquests reached as far east as India, which explains their contact with Hindu culture. The subsequent translation of Hindu mathematical texts and tables into Arabic probably introduced the Hindu numerals and number system into that language. At about the same time, the califs had the ancient Greek texts translated into Arabic. The works of Plato, Ptolemy, Euclid, Aristotle, and a host of other classical writers were circulated from Baghdad to the Moorish universities that sprang up in various countries, notably Spain, during the ninth and tenth centuries. The Arabs improved upon the mathematics of the Alexandrians and Hindus and applied the elegantly simple number system of the Hindus to their task. Perhaps the greatest Arabian mathematician was Al-Khwarismi who lived in the ninth century. His book *ilm al-jabr wa'l muqabalah* had a great influence on the growth of mathematics. In fact, the *al-jabr* in the title eventually changed into our word, *algebra*. The full title can be freely translated, *On Transposing and Collecting Terms*. *Transposition* is the word we use to describe the shifting of terms from one side of an equation to another as, for instance, in going from $4x+3=8$ to $4x=8-3$. By *collecting terms*, we transform the equation to $4x=5$. Much of the algebraic and arithmetic knowledge of Babylon, Alexandria and India came into Europe through translations of Al-Khwarismi's work. Our word *algorism*, which refers to a sequence of mathematical steps such as long division, is a corruption of Al-Khwarismi's name.

Mathematics Returns to Europe

Mathematics returned to Europe from the Arabs chiefly through two routes: the Moorish universities of Spain and Sicilian trade in the Mediterranean. A Sicilian coin dated 1134 is the first known example of the official use in Christendom of the so-called Arabic numerals. By the thirteenth century, Italian merchants were using them in commercial calculations. Despite their obvious advantages, Arabic numerals encountered resistance both from the Church and from civil authority. An edict in 1259 forbade the bankers of Florence to use the "infidel" symbols. In 1348 the ecclesiastical authorities of the University of Padua decreed that the prices of books should be given in "plain" letters (i.e., Roman "numerals") rather than in "ciphers."

The mathematical learning of the Arabs also found its way into Europe because of the social need for timekeeping. The Christian religion had taken over the priestly task of calendar making. As custodians of the calendar, the monks were interested in astronomy and so, indirectly, in mathematics. Adelord of Bath disguised himself as a Moslem (about 1120) and studied in the Moorish university at Cordoba. He subsequently translated the works of Euclid and Al-Khwarismi as well as many Arabic astronomical tables. In a short period of time, hundreds of other Arabic texts were translated into Latin. In addition to its priestly applications, this learning soon found its way into the commercial activities of Europe. An astonishing proportion of the books published following the invention of the printing press were commercial arithmetics.

From its modest Stone Age beginnings in Europe, mathematics had taken thousands of years to move across three continents before returning again, immeasurably enriched by its travels. The language of science was ready and waiting when the printing press opened the modern European period of scientific enlightenment. Without mathematics, we might still believe, like Aristotle, that a heavy object necessarily falls faster than a light one.

The Size of the Earth

3

Man's first great scientific achievement, the calendar, grew out of the needs of agriculture. The next human activity to encourage scientific investigation was world trade. From this stimulus would come modern methods of land survey and navigation.

Overland transportation of goods in ancient times involved enormous difficulties. Roads had to be built without the help of modern earth-moving equipment, and animals or slaves were needed to haul heavy loads over great distances. The sea lanes offered an extremely attractive alternative. They dispensed with the need for roads, and later substituted the forces of nature for the muscles of man or beast. By the year 2000 B.C., the Phoenicians had set up many colonies in the Mediterranean world, including the great city-state of Carthage in the ninth century B.C. The Carthaginians subsequently came to regard the Atlantic as their private sea, and their ships sailed north regularly toward the tin mines in Cornwall, England. By the sixth century B.C., Cathaginian navigators had sailed below the equator along the western coast of Africa. At about that time, the Greeks began to absorb the ancient knowledge of the stars by which the Phoenicians navigated. Out of this star lore of the ancient world, the concept of latitude would soon evolve.

The latitude of any point on the earth's surface is merely the angle between the North Star and the horizon at that point. For a person standing at the North Pole, the North Star is directly overhead and the latitude is 90°. The equator, on the other hand, has a latitude of 0°. The idea of latitude probably arose from the ancient practice of navigating by the North Star. Sailing north, an observer would see the North Star located higher and higher in the northern sky on each succeeding night. Sailing south, the opposite effect would take place. By measuring the angular elevation of the North Star, a mariner could get an accurate measure of his distance north or south of the starting point. For centuries to come, however, longitude was to remain a matter of dead reckoning.

By the end of the seventh century B.C., numerous Greek settlements dotted the western Mediterranean. The most famous of these was at the first site of civilization in France, Marseilles. From this port, a Greek mariner, Pytheas, made one of the great voyages of antiquity in 300 B.C. Eluding the Carthaginians, who were still jealous of the Atlantic, he slipped through the Pillars of Hercules at Gibraltar and sailed north. After visiting the Tin Isles, he made accurate measurements of latitude which were used by later geographers to plot maps of northern Europe. He discovered that the North Star was not located exactly at the pole (it was about 3½° off in those days) and determined the latitude of Marseilles to within a few minutes of the correct figure.

The Greeks, Pro and Con

Perhaps the greatest Greek contribution to science was the realization that science need not be associated with priestly magic and ritual. As they acquired knowledge from other cultures, they retained the real science and discarded much of the priestly superstition that came with it. Nevertheless, Greek science soon became enmeshed in an ideology that was almost as destructive to science as priestcraft. Science and mathematics came to be regarded more and more as purely intellectual activities completely divorced from the observed world. Mathematics, which ought to have been the servant of science, became its master instead. In time, mathematics itself lost touch with the work of the world. The taking of measurements and the performance of experiments were ridiculed as inaccurate sources of knowledge. Even Plato (*Republic*, Book VII) used his great influence to foster such unscientific ideas:

It makes no difference whether a person stares stupidly at the sky, or looks with half-closed eyes upon the ground, so long as he is trying to study any sensible object, I deny that he can ever be said to have learned anything, because no objects of sense admit the scientific treatment.

Aristotle, the student of Plato and the tutor of Alexander the Great, dealt a blow to the exact sciences that would be felt for almost two thousand years. The methods of observation that he used so well in biology were completely ignored in his study of physics and astronomy. And yet, his influence was so great that his writings became articles of faith and were universally accepted simply because the great Aristotle had written them. According to Aristotle, all things were made of five elements: water, air, fire, earth, and an unchanging element of the heavens. Fire went up, earth went down, water went above earth, and air went above water but below fire. The four elements of the earth went up or down, but the heavenly element moved in a circular motion. The circle—being the perfect figure—was obviously the only motion that could be ascribed to the perfect, unchanging element. True scientists like Galileo, Copernicus and Kepler would later engage contemporary followers of Aristotle in bitter scientific controversies to overthrow such Aristotelian nonsense.

The knowledge of antiquity was rescued from this antiscientific climate by the military success of another Greek, Alexander the Great. We have already seen how Alexander's conquests resulted in the great library at Alexandria. This cosmopolitan city rapidly became the world's center of shipping, opening new contacts with the world's affairs.

Longitude

The concept of latitude had evolved slowly from the use of the North Star in navigation. The Alexandrian astronomers added the idea of *longitude*, which gives the location of a place in an east-west direction from some arbitrary starting point. Knowing both longitude and latitude accurately fixes a mariner's position and makes it possible to draw accurate maps of the world.

To understand how the longitude of a place can be measured, imagine two observers—one in Greenwich, England, and the other in New York City. Imagine, further, that the sun is directly overhead in Greenwich, so the *local time* there is 12 noon. At the same instant, it is 7 A.M in New York City, or just five hours earlier. The longitude at Greenwich is called 0° by modern convention and our problem is to determine the longitude of New York City. We know that the earth makes one complete rotation, or 360°, in twenty-four hours. In five hours, therefore, it will rotate $\frac{5}{24} \times 360°$, or 75°. Since the sun will be directly over New York City at the end of five hours, its longitude is 75° west of Greenwich. Taking a second example, if local noon occurs one hour apart at two places, their difference in longitude is 15°, because the sun "moves" just 15° in one hour.

The Alexandrians knew that a navigator could measure longitude by carrying an accurate clock aboard ship. The clock could be set to the local time at Alexandria and would tell the navigator Alexandrian local time throughtout his voyage. He could determine his longitude at any other place merely by reading the clock's time when the sun was directly overhead. His longitude relative to Alexandria would be the product of 15° and the difference in time (in hours) between 12 noon and the Alexandrian local time given by the clock.

Unfortunately, there were no portable clocks available to the Alexandrians that were accurate enough to maintain a continuous record of time over a long journey. Alexandrians were able to measure longitude, nevertheless, with the help of eclipses. Hipparchus prepared tables of future lunar eclipses for that purpose. An illustration will help to make their use clear. Suppose the tables predict a lunar eclipse at 9 P.M. at Alexandria. An observer at another place, say eastern Sicily, could wait for the eclipse to occur and read his local time from an hourglass or water clock when the eclipse began. (His local "clock" would measure time from the preceding noon.) If he saw the moon entering the earth's shadow at 8 P.M., he would know that local Sicilian time is one hour ahead of Alexandrian time. His longitude would then be $1 \times 15° = 15°$ west of Alexandria.

The practical difficulties of longitude determination were so great, however, that longitude failed to enter the practice of navigation until much later. Estimates of long distances depended upon the guesses of sea captains and travelers. Nevertheless, the groundwork was laid well in Alexandria and would be available when the chronometer supplied the needed accuracy of time measurement.

In addition to the use of latitude and longitude for terrestrial cartography, the principal geographical discoveries of the Alexandrians were: the approximate size of the earth, the first estimates of the distances to the sun and moon, and the use of latitude and longitude to map the stars.

Measuring the Earth

The first calculation of the earth's circumference was made by Eratosthenes (c. 250 B.C.), the librarian of Alexandria. His calculation depended upon the angle between the sun's shadow and the vertical at two places, one due north of the other. On June 21, the sun was directly overhead at noon at Aswan near the first cataract of the Nile. Its rays, therefore, made an angle of 0° with a vertical pole and cast no shadow. At the same instant, the sun's rays made an angle of 7½° with the vertical at Alexandria, about 5,000 stadia (500 miles) to the north. Eratosthenes' reasoning can be explained with the help of the accompanying diagram. Because of the sun's great distance, he knew that its rays are parallel everywhere on the earth's surface. The rays through the two places, therefore, must be parallel. He also knew that the ray through Aswan could be extended to pass through the center of the earth. From his knowledge of geometry, he knew that the two angles marked 7½° must be equal. The solution was then clear to him. The angle at the center of the earth between Aswan and Alexandria (which we would call their difference in latitude) was equal to 7½°, or just 1/48th of 360°. Since 360° corresponds to the entire circle, the earth must have a circumference of 48×500, or 24,000 miles. His estimate was within a few per cent of the true figure.

Another Alexandrian, Poseidonius, calculated the earth's circumference from the difference in latitude between Alexandria and the island of Rhodes, roughly 350 miles to the north. Poseidonius had to rely on mariners' estimates for his distance between the two places and these turned out to be much too low. As a result, his circumference has come down to us through the great astronomer, Ptolemy

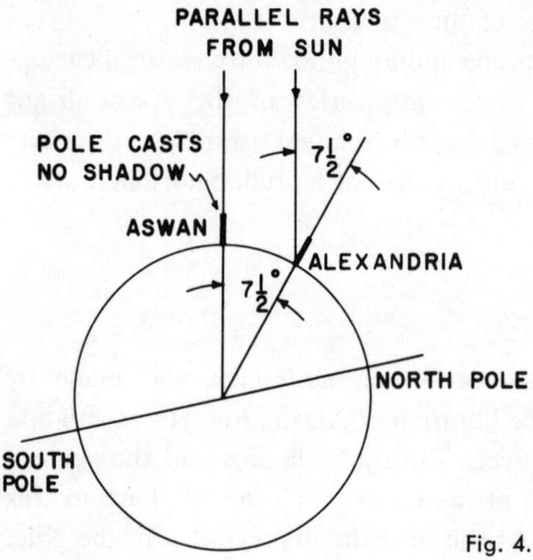

Fig. 4. Eratosthenes measures the earth.

(not the general), as 18,000 miles. It was this figure that gave Columbus hope of reaching India. Had he known the true distance he might well never have sailed.

The Distance to the Moon

Although geometry was the greatest achievement of Greek science, it turned out to be a clumsy tool for solving numerical problems. The Alexandrians, notably Hipparchus, soon developed a new kind of mathematics—trigonometry—that was eminently simpler to use in solving the new problems of their world. With this new mathematics, the Alexandrians would make the first scientific estimates of the size of the solar system.

The key to trigonometry is the so-called right triangle. As shown in Figure 5, a right triangle is merely a three-sided figure containing a right angle (90°). Hipparchus knew from geometry that the length of the sides of such triangles bear a definite relationship to each other —a relationship that depends upon the size of the two unspecified angles. This feature can be illustrated with the help of the numerical example in the accompanying diagram. One of the angles has a magnitude of 30° and the lengths of the sides are represented by the letters *a, b* and *c*. Hipparchus proved that the ratio of *a* to *c*—that is, the fraction *a/c*— is always equal to 0.5 when the angle shown is 30°. If *a* is 100 feet long, *c* is 200 feet long. For such a triangle, *a* is always precisely half as long as *c*. If angle *A* happens to be 45°, the fraction

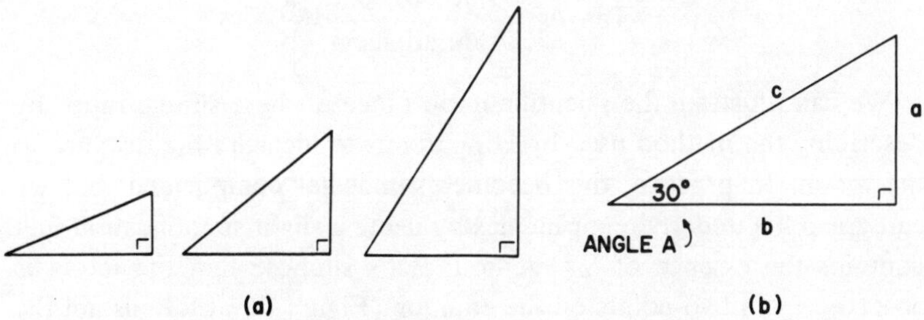

Fig. 5. Triangles give birth to a new kind of mathematics.

a/c is equal to a different number, 0.707. If the angle equals 62°, the number is 0.883.

The ablity to predict the ratio *a/c* if we know the angle is a simple fact, but one that astronomers and surveyors have long rejoiced over. It lets them draw a small right angle in a notebook using a measured number of degrees for angle *A* and know that the ratio of the sides in their little triangle is the same as for a large triangular piece of land or sky. And furthermore, each time the same triangle is encountered, the ratio will always be the same. Then why not measure the ratios in many different triangles and keep them on file for future use? That is just what Hipparchus did as he studied the heavens. His use of a right triangle simplified things, because that made it necessary to specify only one of the remaining angles. Any triangle contains a total of 180° in its three angles. If one angle is 90° (the right angle), and another 30°, the third is understood to be 60°.

Tradition has given names to the ratios of the sides in a triangle, that is to a/c, b/c and a/b. The ratio of the *side opposite* the given angle, a, to the *hypotenuse*, c, (the longest side) is called the *sine* of the given angle. For an angle of 30°, the relationship is written sin 30° = 0.5. The remaining ratios are called the "cosine" and "tangent" of the given angle and all three are defined for the angle A as follows:

$$\text{Sin A}^\circ = \frac{\text{side opposite}}{\text{hypotenuse}} = \frac{a}{c}$$

$$\text{Cos A}^\circ = \frac{\text{side adjacent}}{\text{hypotenuse}} = \frac{b}{c}$$

$$\text{Tan A}^\circ = \frac{\text{side opposite}}{\text{side adjacent}} = \frac{a}{b}$$

We can illustrate the scientific importance of these simple ratios by describing the method used by Hipparchus to measure the distance to the moon. In practice, this becomes somewhat complicated, but we can grasp the underlying principles by using a slight simplification that contains the essence of his method. Let's suppose that the moon is observed from two points on the equator (Figure 6) at the instant the moon is directly over one of these points, *D,* as shown in the diagram. At the same instant, an observer at *C* sees the moon rise on the horizon. A theorem from geometry tells us that angle *C* is a right angle, so the trigonometric ratios of Hipparchus can be applied to the triangle *ACB*. Using the cosine ratio we can write $\text{Cos A} = \frac{AC}{AB}$. Because he knew the location of points *C* and *D,* Hipparchus could easily calculate the angle *A*. A reasonable figure is 89°3′. He also knew that the distance *AC* was equal to the earth's radius, about 4,000 miles, and that cos 89°3′ = 0.01658. Using this general method, he calculated the moon's distance to be about seventy-two earth radii, a tolerable approximation for the time. The true distance is 239,000 miles, or about sixty radii.

His estimate of the distance to the sun was based on a study of eclipses and gave a figure of 13,000 earth radii, or about 51 million miles. Although this figure was much smaller than the correct distance, 92 million miles, it emphasized the fact that the sun was much more distant from the earth than the moon. And since both bodies seem to have the same size when viewed from the earth, it was clear that the sun was a vastly larger body than the moon.

With this information in hand, it's a simple matter to calculate the diameter of the moon. The only new measurement needed is the angular size of the moon, that is, the size of the angle between the two opposite edges of the moon. This angle is almost exactly half a degree. Using the trigonometric tables, the diameter turns out to be 2,160 miles, or about three-elevenths that of the earth.

Although observational errors crept into the early measurements

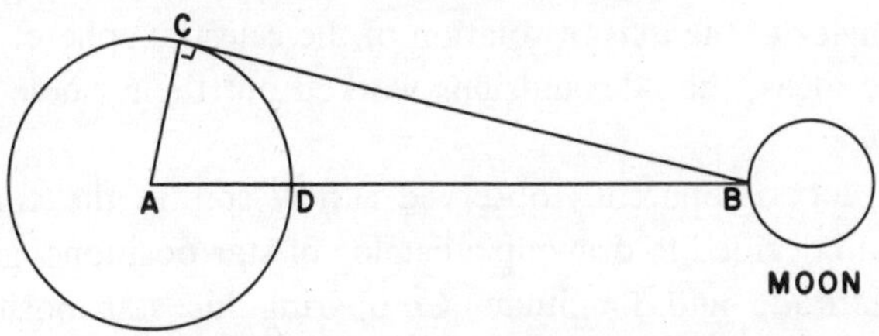

Fig. 6. Hipparchus measures the distance to the moon.

of the solar system, the Alexandrians managed to deduce the correct relative sizes and distances of the sun and moon. The world now knew that the moon was smaller than the earth, and that the sun was vastly larger and more distant than the moon.

The Wobbling Earth

The mathematical progress of the Alexandrians led to their crowning achievement in the field of astronomy—the Ptolemaic system. Ptolemy's *Almagest* gives a full and accurate description of the motions of the heavenly bodies, and predicts the motion of the planets to an incredible accuracy of about one-twelfth of a degree. Not until the seventeenth century would adjustments to his mathematical theory be made in order to allow for the increased accuracy of observation.

Because of its success, the *Almagest* displaced most of the earlier astronomical books, so it's hard to be sure how much of it was borrowed from Ptolemy's predecessor's, notably Hipparchus. Its great value, however, was in its superb presentation of the heavenly scheme as a whole.

The Alexandrian observer thought of himself as being at the center of a vast heavenly sphere containing the fixed stars. He could see only half of the sphere at a time because the other hemisphere was blocked by the opaque earth. The horizon formed a *great circle* on the heavenly sphere. The celestial *poles* were points on the sphere pierced by the axis about which the heavens seemed to turn. On this sphere he marked out a *meridian* which traces a great arc from one pole to the other and includes the observer's *zenith* (the point on the celestial sphere directly above the observer). The *equator* was a great circle at right angles to the axis of rotation of the celestial sphere. Beginning with these ideas, the Alexandrians worked out their whole system of astronomy.

In 134 B.C. Hipparchus observed a new star in the constellation Scorpio and decided to draw up a catalog of star positions, giving their celestial latitude and longitude. Comparing his star positions with those taken about 150 years earlier, he found that all of the stars had moved somewhat relative to certain fixed places in the heavens. We now know that this apparent motion is caused by a slight wobbling of the earth's axis, much as a spinning top wobbles when it slows down. Because this effect causes the equinoxes to fall a little earlier each year, it is called the "precession of the equinoxes." A complete cycle of precession takes 26,000 years. Hipparchus also devised methods of predicting lunar eclipses to an accuracy of an hour or two. Solar eclipses were predicted less accurately, however.

Cycles and Epicycles

While the motion of the stars was easily accounted for, the planets presented a much more difficult problem. Instead of moving steadily across the heavens, the planets exhibit extraordinary peculiarities. Sometimes they speed up, and at other times they slow down and even move backwards for a while against the background of the stars. Figure 7 shows the apparent path of Mars through the heavens in the winter of 1928–1929. Mars has continually repeated this kind of motion through the centuries. It takes nearly two years to move about 40° to the east; then it moves west for about seventy days over an arc of about 16°; finally it begins its direct eastward motion again.

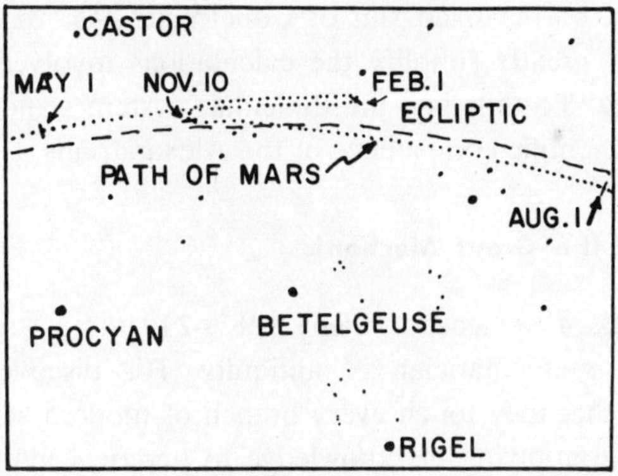

Fig. 7. The path of Mars through the heavens, 1928–1929.

This looped motion comes about because of the motion of the earth—the observer's platform. The apparent motion of the planets is the resultant of two motions, those of the earth and planet as both revolve about the sun.

Ptolemy, who believed the earth was stationary, combined two clever ideas to explain the motions of the planets. The first was to have the planets rotate about a point displaced from the center of the earth; the second was the use of *epicycles* to explain the strange

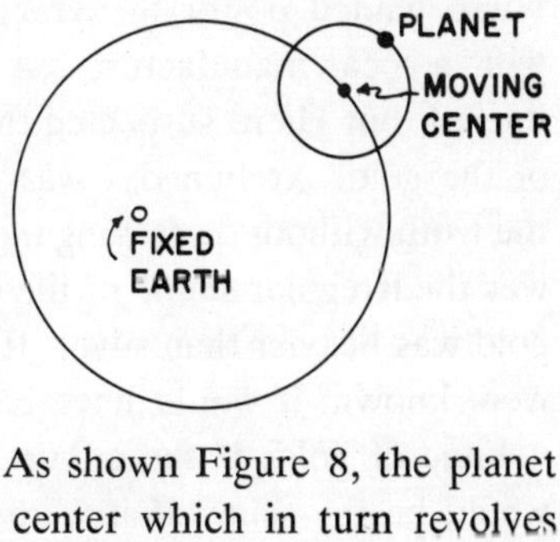

Fig. 8. According to Ptolemy, each planet moved about a hypothetical moving center which in turn rotated about a point near the center of the fixed earth.

back and forth motion of the planets. As shown Figure 8, the planet revolves about a hypothetical moving center which in turn revolves about the earth. This combined motion together with the eccentric circle concept discussed above closely approximates the observed motion of the planet. While the notion of a fixed earth is completely wrong, we should keep in mind that it makes no difference to the accuracy of astronomical predictions whether we use Ptolemy's stationary earth, or the fixed sun of Copernicus. As we shall see, the latter did not greatly simplify the calculations involved, or improve their accuracy. To this day, the Ptolemaic system stands as a monument to the scientific competence of the Alexandrians.

Archimedes, the Great Mechanic

Archimedes of Syracuse in Sicily (287–212 B.C.) was the greatest scientist and mathematician of antiquity. His discoveries were so fundamental that they touch every branch of modern science. But he was not above applying his knowledge to practical matters. In addition to inventing the pulley and designing ingenious engines of war for the defense of Syracuse against the Romans, his name is connected with the Archimedean screw, a kind of water pump. As shown in Figure 9, water can be raised to a higher level merely by rotating the screw. The use of the screw as a fastener or to apply mechanical force was unknown before Archimedes and was probably suggested by his device. More than any other man, Archimedes can be thought of as the father of the science of mechanics which deals with forces and motion.

A story tells us of his success in another practical matter. When Hiero gained power in Syracuse, he contracted for a golden crown with a local manufacturer. A crown of the correct weight was delivered, but Hiero suspected that silver had been substituted for some of the gold. Archimedes was brought in as a consultant to discover the truth without destroying the crown. The major difficulty, of course, was the irregular shape of the crown. Scientists of the time knew that gold was heavier than silver. If the volume of the crown's solid matter were known, it would have been a simple matter to weigh an equal volume of gold. If the crown were made of pure gold, both weights would be the same. If it were made of a mixture of gold and silver, the crown would weigh less than an equal volume of gold.

The problem was on Archimedes' mind when he happened to go to the bath. On getting in, he noticed that the more his body was immersed, the more water ran over the top. Shouting for joy, he rushed home, exclaiming, "Eureka! Eureka!" ("I've found it! I've found it!") What he had found was an ingenious way of measuring the volume of an irregularly shaped body. It would lead to his principle of buoyancy, which explains why some bodies float while others sink.

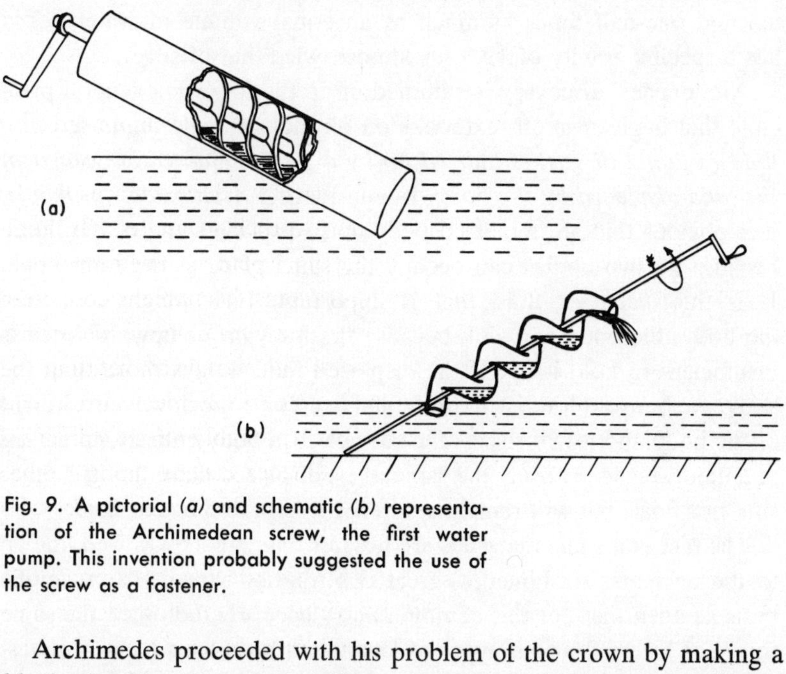

Fig. 9. A pictorial *(a)* and schematic *(b)* representation of the Archimedean screw, the first water pump. This invention probably suggested the use of the screw as a fastener.

Archimedes proceeded with his problem of the crown by making a block of gold and another of silver, each equal in weight to the crown. Next he filled a jar to the brim and dropped in the gold block, catching the water that ran over. He then repeated the process for the block of silver and found that more water had been displaced by the silver block than by the gold. This is because a pound of silver occupies a larger volume than a pound of gold. The same operation was now repeated with the crown. Less water ran over than for the silver block, but more than for the gold. This showed that the crown was made neither of gold nor silver, but a mixture of the two.

Archimedes went on to expand his ideas in his book *On Floating Bodies*. This was the first scientific discussion of what we now call "specific gravity"—the ratio of the weight of a given volume of a substance to the weight of an equal volume of water.

$$\text{Specific Gravity} = \frac{\text{Weight of a body}}{\text{Weight of an equal volume of pure water}}$$

Silver, for example, has a specific gravity of 10.5 because it weighs ten and one-half times as much as an equal volume of water. Gold has a specific gravity of 19.3, or almost twice that of silver.

Archimedes' discovery is summed up in the following general principle that is given in all textbooks on physics, *A body immersed in a fluid (a liquid or gas) is buoyed up by a force equal to the weight of the fluid displaced by the body*. In considering *Archimedes' principle*, it is obvious that an immersed body must displace some of the liquid because no two bodies can occupy the same place at the same time. It is this *displaced* fluid that is important. If it weighs less than the body, the body will sink because the buoyant or upward force is insufficient to hold it up. If the displaced fluid weighs more than the body, the buoyant force will more than overcome the downward weight of the body and it will rise to the surface. If a body entirely immersed in a fluid weighs precisely the same as its displaced fluid, it will neither sink nor float, but will remain wherever it is placed in the fluid.

The fact that some materials are heavier than others was well known to the ancients. Archimedes' great contribution was in the scientific basis he unearthed for this common knowledge. He did much the same thing for levers. Various types of levers had been used from remotest antiquity; even an intelligent ape will use a stick as a lever. But it is one thing to use the lever, and quite another to explain its exact mathematical principles and to follow them to their theoretical conclusions. Archimedes once said, "Give me but a place to stand, and I can move the world." This was no idle boast. He subsequently built a compound lever and used it to move a loaded ship with only the slightest effort.

On Plane Equilibrium is one of the earliest of Archimedes' books to have survived. Following the pattern of Euclid's *Elements*, the book opens with his famous "postulate": "Equal weights at equal distances are in equilibrium; equal weights at unequal distances are not in equilibrium but incline toward the weight at the greater distance." This is the principle behind the balance or scale used for weighing things, and it eventually led him to discover the center of gravity of various geometrical figures. As its name implies, the center of gravity of a body is the point at which all of its weight seems to be located.

A simple lever is nothing more than a long bar such as a common seesaw balanced at the center, called the "fulcrum." A beam balance is a lever balanced at the center (fulcrum), and having a pan suspended from either end. When a mass of unknown weight is put on one pan, the lever is thrown out of balance, or equilibrium, and tips over. If

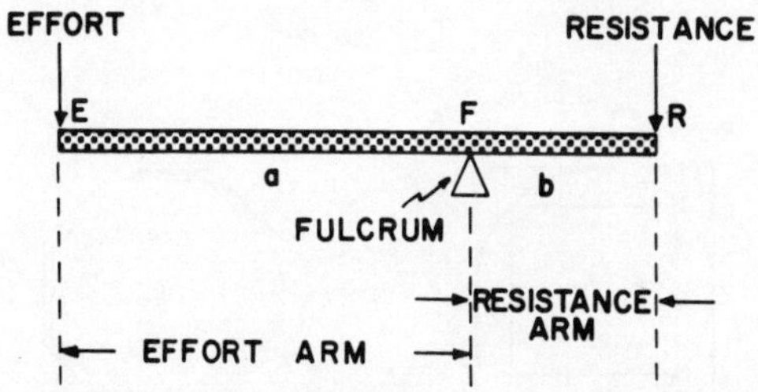

Fig. 10. The law of the lever. The product of the effort and the effort arm is equal to the product of the resistance and the resistance arm.

masses of known weight are added to the other pan so that the lever just balances, the unknown mass weighs the same as the known mass. As shown in Figure 10, the two weights are called the "effort" and the "resistance." The distance from the fulcrum to the point where the effort is applied is called the "effort arm," and the distance from the fulcrum to the resistance is called the "resistance arm." The law of the lever tells us that, *The effort times the effort arm equals the resistance times the resistance arm.* Using *E* and *R* for the effort and resistance forces, and *a* and *b* for the corresponding distances, this law can be expressed by the equation: $Ea = Rb$. A ten pound weight placed two feet from the fulcrum will balance a five pound weight placed four feet from the fulcrum because $(10)2 = (5)4$.

In the field of mathematics, Archimedes extended and made systematic use of an older idea called "limits." His methods are of utmost importance because they form a main foundation of modern mathematics. To illustrate, consider a square inscribed within a circle as shown in part (*a*) of Figure 11. It is obvious that the area of the square is less than the area of the circle. Similarly, the area of the circumscribed square of part (*b*) is greater than the area of the circle. The Greeks knew how to calculate the area of straight sided

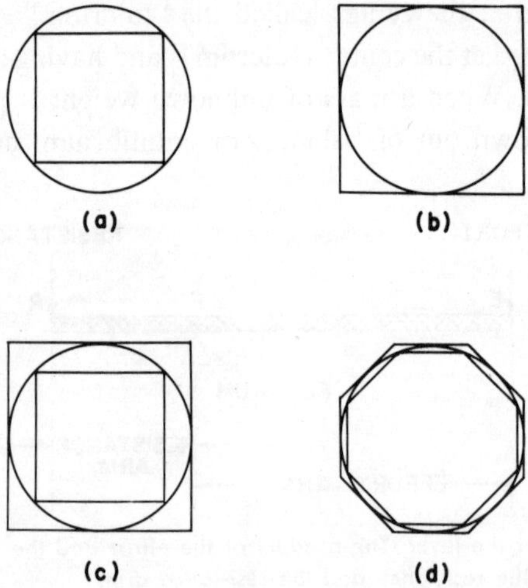

Fig. 11. Archimedes determined the area of a circle.

figures, but not the area of a curve such as a circle. This method told Archimedes, however, that the area of the circle in part (*c*) lies somewhere between the areas of the two squares.

It is a simple matter to double the number of sides and make eight-sided figures still touching the circle as in (*d*). This gives a closer approximation of the circle's area. Archimedes knew he could go on doubling the number of sides to 16, 32, 64, 128, 256, or to any higher number. The greater the number of sides, the more closely will the polygons approximate the area of the circle. In the *limit,* when its sides are no larger than points, each polygon changes into a circle. Archimedes realized that this limit can never be reached, but that it can be approached as closely as we wish. Using the method of limits, he found π, the ratio of the circumference to the diameter of a circle, to lie between $3\frac{10}{71}$ and $3\frac{10}{70}$. The average of these numbers is 3.1419 compared to the true figure for π, 3.1416.

Archimedes may be rightly called the Einstein of antiquity. He was so far ahead of his times that it almost seems as though he could read the great minds of the Renaissance. But the great Archimedes, his mind lost in the geometric diagrams traced in the sand in front of him, was brutally murdered in 212 B.C. by a conquering Roman soldier. He used his last words to request that his circles be left undisturbed.

Curves and Alchemy

Before leaving the Alexandrians, mention should be made of their work with curved lines that comes down to us in the *Conic Sections* of Apollonius (fl. 220 B.C.). He gave us the terms "ellipse," "parabola," and "hyperbola" and showed that all three can be obtained from the same cone. He cut a cone at different angles in order to obtain the different curves (Figure 12). Subsequent work would show that the circle is really a special case of the ellipse. This early knowl-

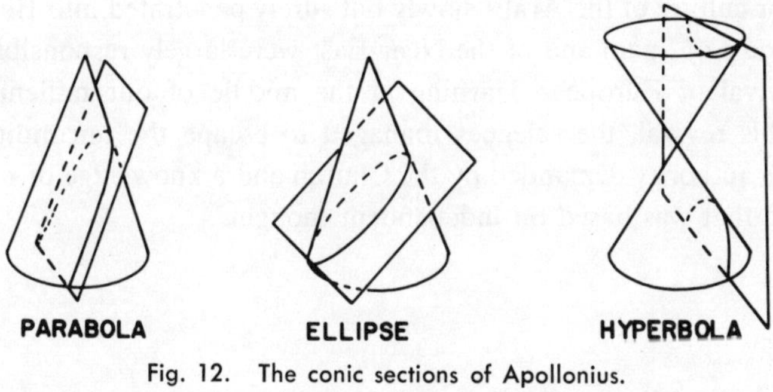

Fig. 12. The conic sections of Apollonius.

edge of the conic sections was to be of great value when Kepler would prove that the planetary orbits are elliptical rather than circular.

Little is known of Alexandrian chemistry prior to A.D. 100. Nevertheless, the writings that have survived from that period tell us that

Alexandrian alchemy was very sophisticated and clearly the result of a long period of evolution. These early writings carry marks of Christian, Jewish, Greek, Egyptian and possibly Persian and Chinese elements. They carry ideas that have persisted to the very dawn of modern chemistry. Some contain processes that are so practical that they are certainly not a part of alchemy as that term is now understood. There is little doubt that the early Arabic alchemists had access to a rich source of Alexandrian knowledge in this field.

Alexandrian science, which had risen to such great heights in the early centuries of our era, soon eroded to nothing as Christianity became the state religion of the Roman Empire. The magnificent pagan library at Alexandria, which contained practically all of the world's knowledge, was burned twice—first by Christian fanatics in the fourth century, then by Moslem fanatics in the seventh. Because of this vandalism, great numbers of ancient manuscripts disappeared and would have been lost forever were it not for their Arabic translations. By the year A.D. 1000, paganism was dead, but so was science. Whatever was left of culture and knowledge in the world belonged to the Arabs. Then came the Crusades.

The acquisition of knowledge was certainly not a motive of the Crusaders' program. Yet, almost in spite of themselves, this is exactly what the Crusaders accomplished. During three centuries of war, the superior culture of the Arabs slowly but surely penetrated into Europe. The Arabs of Spain and of the Near East were largely responsible for the revival of European learning at the middle of our millennium. With this revival, the sciences managed to escape the unconditional faith in authority demanded by the Church and a knowledge of nature evolved that was based on independent thought.

Through the Looking Glass

4

Arriving at an understanding of human vision was one of the most difficult, yet vital, assignments that science has had to face. How does the human eye perceive the form, color and size of everyday objects and surroundings? When this question was answered satisfactorily in the seventeenth century, it triggered a fundamental change in the attitude of scientists toward the methods of science. The human mind lost its ancient role as the ultimate source of all truth about nature. It suddenly became an intelligent spectator capable of probing her secrets. When Kepler confirmed Galileo's discovery of the moons circling Jupiter, and denied the general belief that they were optical illusions, a revolution of vast extent took place. Man, who had been the lord of the universe, was finally dethroned. Formerly, the aim of science had been to explain how the human mind came to know the outside world. Now its purpose was to know the structure and laws of the external world, independently of the observer. But most important of all, now that scientists had come to understand the eye, they were willing to believe the amazing things it told them.

Acrobatic Skin-Images

Solving the riddle of vision involved a long scientific effort extending across two thousand years of speculation. The ideas of the ancients ran something like this. All of our senses depend upon physical *contact* with something from the external world. The soul locked within our mind can't go out to touch external objects, so these must come to the soul and touch it by passing through the senses. A likeness called "eidolon" covers an object, moves about on its surface, and can detach itself and become transported to the soul through the pupil of the eye. The eidolon, or "species" as it was called in medieval times, brings with it the form, color and all the other qualities of the body from which it comes. Sight, like all the other senses, is merely a sense of touch. It was known that direct contact is responsible for touch and taste. For smell, we have a vapor coming to our nose. For sound, we have a vibration coming through the air. For sight, we have species or images, a kind of skin that carries the form of the object to our eyes. But how can the skin of a mountain fit into the tiny pupil of the eye, an opening barely the size of a large raindrop? To answer that question the images were given another amazing property. They were supposed to contract uniformly along the way until they became small enough to enter the pupil. Because different pupils might be located at any point whatever, the images had to diminish at different rates for each observer. Despite the acrobatic nature of these ideas, the images carried with them no information concerning the size and distance of an object. Neither could the theory explain why we can't see in the dark, or why images are reversed when we look in a mirror. The theory raised more questions than it answered.

The Probing Eye

An opposing theory was based on the fact that a blind man can learn something about the form of an object by touching it with a stick. By analogy, the eye was thought to send out straight sticks—called "visual rays"—that examine the world and bring back knowledge of its forms and colors. The visual ray theory was about as absurd as the skin-image theory, and left about as many questions unanswered. Why do the rays fail to work in the dark? Why do they become confused for objects very close to our eyes? How can they possibly reach to the sun and the stars? Proponents of the two theories argued bitterly through the ages, each attacking the ideas of the other; but when it came to constructive ideas that was another story.

Refraction, the Bending of Light

The first step in the right direction was taken by an Arabian mathematician of the eleventh century, Ibn al-Haitham, who became known to Europeans as Alhazan. To understand the gigantic step he took, it is necessary to go back to Alexandrian times when Ptolemy, the astronomer, looked into a subject called "refraction"—the bending of light rays that usually takes place as they pass from one medium into another. As a result of his research, Ptolemy was able to explain

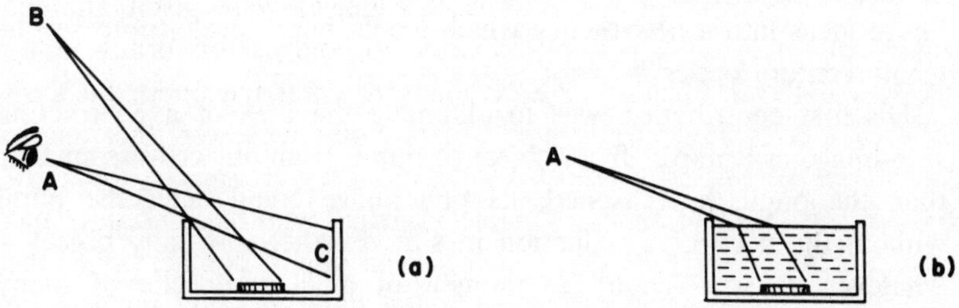

Fig. 13. The ancient disappearing coin trick.

why we see the sun for a short time after it is actually below the level of the horizon. Figure 13 illustrates how refraction accounts for the ancient disappearing coin trick. When the jar is empty, the eye at point *A* can see no deeper than the level at *C*. To see the coin, the eye must be raised to point *B*. But after the jar is filled with water, the coin is visible even when the eye is at point *A*. This is because light is bent or *refracted* when it passes obliquely from a medium of one density (air in the coin trick) to a medium of different density (water). No refraction takes place, however, if a light ray enters a second medium perpendicular to its surface. Measurements made by Ptolemy on the refraction of light are the first recorded body of numerical data in the history of experimental science.

Ptolemy explained the visibility of the sun when slightly below the horizon on the basis of refraction in the atmosphere. Imagine a light ray coming into the atmosphere from near the horizon. As it moves from outer space into our atmosphere, it enters progressively more dense layers of air. This bends the ray downward somewhat. If the sun is a little below the horizon, a ray that would ordinarily pass far over our heads is bent down by atmospheric refraction and enters our eye. This makes it possible to see the sun even though it is physically located about one-half degree below the observer's horizon.

Alhazan was familiar with refraction, of course, and also with the structure of the eye which had been described by Galen, another Alexandrian, in the second century of our era. He knew, therefore, that the important parts of the eye should be considered spherical in shape. With these wellknown facts in mind, he modified the skin-image ideas into a new theory which would later develop into seventeenth century optics.

His first contribution was to eliminate the idea of a contracting skin-image as it moves from object to pupil. If an object were smaller than the pupil, he reasoned, its tiny image could enter the pupil without the need of a reduction in size. Furthermore, any object—whatever its size—could be thought of as a composite of many minute picture-units or points. Each of these points emits tiny skin-images in all directions which can enter a pupil wherever they meet it. Alhazan went on to explain how all of these point-images could reconstruct a complete picture within the eye (Figure 14). A point-image strikes the outer spherical coating of the eye (the cornea) in one of two ways: its straight line path is either perpendicular to the spherical cornea, or it is not perpendicular. If it is perpendicular, no refraction takes place and the image continues straight on into the eye's interior. If the path of the image is not perpendicular, the image is deviated by refraction. According to Alhazan, an image deflected by refraction must lose its power to stimulate the "sensorium" at the back of the eye. The sensorium is affected, therefore, only by those point-images that strike the cornea perpendicularly. For each point on the object, there is just one point-image that can strike the cornea perpendicularly and reach the sensorium. These point-images enter the eyeball, cross at the center, and redistribute themselves over an extended area of the sensorium, preserving the exact order and

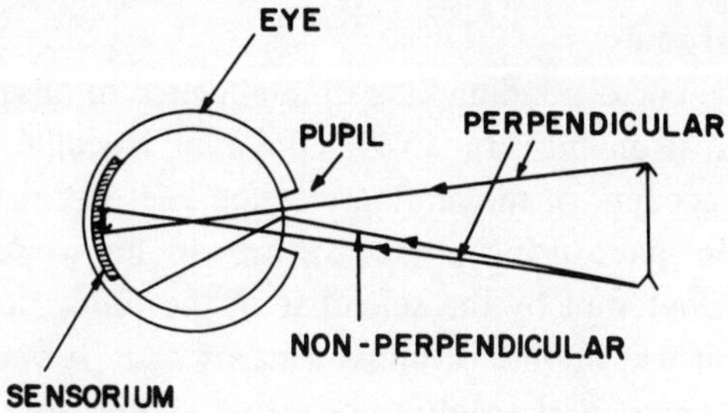

Fig. 14. Alhazan's concept of vision. Only perpendicular rays could reach the "sensorium" and stimulate the eye.

arrangement they had whey they left the object. Alhazan's new theory was not yet perfect, but it was an indispensable step toward the ultimate solution.

Lenses, Those Diabolical Bits of Glass

Although Alhazan's theory of vision was far more logical than the older ideas, it failed to gain acceptance in medieval Europe, probably because it was too difficult to understand. Instead, a monstrous mixture of the old *eidolon* and *visual ray* theories gained universal acceptance.

The scientific atrophy of the times is illustrated by the attitude of scientists of the Middle Ages toward eyeglasses. These small circular pieces of glass, which were somewhat thicker in the middle than at the edge, had been used since the thirteenth century to correct the far-sighted vision (presbyopia) so common in older people. Their polished, transparent surfaces possessed the mysterious power of making old people see well at close range. Instead of arousing the interest of scientists and mathematicians, eyeglasses aroused their scorn. Although eyeglasses are transparent, they argued, they cause refraction and distortion of the image. It is neither logical nor sensible to bend the rays that come out of the eye to explore the world, and thereby falsify the result. The aim of vision is to know the truth. Eyeglasses make figures look bigger or smaller, nearer or farther away, often distorted, inverted or colored. In view, of this deception, they cannot possibly make the truth known and should not be used for serious purposes.

There was a corresponding lack of confidence in observations made with optical instruments. In 1523, Giovanni Rucellai, a Florentine, published an account of the anatomy of the bee. Because his observations had been made using a concave mirror, his work was received with amused contempt by the scientists of the time. No one followed his lead in the use of this primitive microscope. It was an accepted dictum, *Non potest fieri scientia per visum solum;* no scientific value can be attached to anything observed by sight alone. Visual observation is valid only when confirmed by touch.

Silly as theses arguments may sound, they convinced the *entire* scientific world to ignore lenses for almost four hundred years! Spectacles became increasingly important in the everyday world only because generations of modest craftsmen—who knew nothing of theories—found them to be a useful device. They ground lenses, mounted them, and made their living by selling them. It was not until the seventeenth century that the science of optics caught up with its applications in the everyday life of man.

The Miracle of Vision

In 1604, Johannes Kepler found the key to the riddle of vision. His thinking was as follows. An object consists of a great number of points, just as Alhazan had theorized. Each point emits light in all directions in the form of straight, or rectilinear rays which go on forever unless they meet an obstacle. As shown in Figure 15, a point is like a star that emits light in every direction. All the rays that enter the eye form a cone whose origin or vertex is the star and whose base is equal to the diameter of the pupil. Only these rays are important. They are refracted by the crystalline lens within the eye, and go on to make a new cone whose vertex is a point on the light-sensitive retina. If the position of the point of light changes, the point-image on the retina moves an equivalent amount. Each point on the retina corresponds to a specific point on the object being viewed. In this way, a complete picture is projected by refraction on the light-sensitive retina of the eye. Kepler also asserted that we estimate the distance

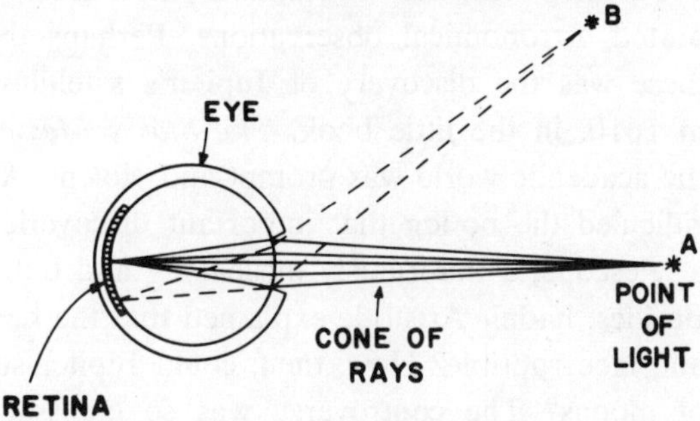

Fig. 15. Kepler explains vision.

to an object by *unconsciously* making use of the triangle whose base length is the distance between the eyes and whose sides are the rectilinear lines of vision drawn to the object. These ideas of Kepler are the basis of the modern science of geometrical optics.

Telescopes and Satellites

At about the same time, several Dutch spectacle makers began to produce a simple telescope that had been invented by an unknown Italian in 1590. But since these new instruments contained curved pieces of glass, they received the same abuse from the learned that had been heaped upon eyeglasses for hundreds of years. The telescope remained a toy in the hands of glassworkers—weak in magnification, condemned in philosophy, and misunderstood in operation. But this time the period of rejection lasted only a few years. In 1609 Galileo Galilei (1564-1642) heard about a marvelous new instrument that magnified images, and recognized its importance to science. He made his first telescope, which magnified objects three times, by fixing a convex lens in one end of a tube and a concave lens in the other end. He soon made others that magnified eight times, and finally one that magnified thirty-three times. Galileo—the last great man of history to be known by his given name—was the first scientist to believe firmly that what he saw in this marvelous new instrument was *true*. It might be nearer or farther away, bigger or smaller, inverted or upright—but it was a *true* representation of what he would see with the naked eye.

Fortified with this revolutionary new faith, Galileo made his celebrated astronomical observations. Perhaps the most sensational of these was the discovery of Jupiter's satellites which he announced in 1610, in the little book, *The Starry Messenger*. The response of the academic world was prompt and violent. A great chorus of voices ridiculed the notion that important discoveries could be made with a telescope, a notoriously misleading and untrustworthy contraption. Besides, hadn't Aristotle explained that the heavens were unchanging and incorruptible? How, then, could Jupiter suddenly acquire a bevy of moons? The controversy was so colossal that the followers of Aristotle refused even to look through the telescope for fear of being deluded by its diabolical distortions. At that time Galileo was a professor at the University of Padua. The following excerpt from a letter which he wrote to Kepler is striking evidence of the amazing extent to which faith in authority can produce blindness to truth:

> I wish, my dear Kepler, that we could have a good laugh together at the extraordinary stupidity of the mob. What do you think of the foremost philosophers of this university? In spite of my often repeated efforts and invitations, they have refused with the obstinacy of a glutted adder, to look at the planets or the Moon or my glass (telescope)! . . . Why must I wait so long before I can laugh with you? Kindest Kepler, what peals of laughter you would give forth if you heard with what arguments the foremost philosopher of the university opposed me, in the presence of the Grand Duke, at Pisa, laboring with his logic-chopping argumentations as though they were magical incantations wherewith to banish and spirit away the new planets (the moons of Jupiter) out of the sky!

All of the scientific world showered abuse on Galileo—all, that is, except Kepler. In August 1610, Kepler obtained a telescope in expectation of proving Galileo wrong. He saw the moons for himself, however, and soon came to believe that Galileo was right. Two of the greatest scientists of all time had agreed that the telescope was a trustworthy instrument and that Aristotle had been wrong. Although the world of science was still hostile, complete victory was now assured.

Within a matter of weeks Kepler worked out the theory of lenses and published it in his *Dioptrics* in January 1611. After more than three centuries of scientific abuse, lenses had acquired a mathematical theory and quickly assumed the respectability of scientific instruments. The story has been given in some detail because it describes one of the most significant revolutions in the development of science, a revolution—strangely enough—that is practically unknown. The development of optics as a science produced a radically new attitude of the scientist toward his observational instruments. In place of the Aristotelian skeptic who refused to look through the telescope, there was now a new kind of investigator who refined his instruments to the limit in order to learn more about nature. The information that he got might be fragmentary or even deceptive, but it offered the only means of studying these previously unseen phenomena. Lenses and mirrors opened a vast new horizon to the scrutiny of science, and the unscientific philosophy of Plato was finally laid to rest.

The Magic of Mirrors

When a ray of light strikes a mirror, or any other highly polished, flat metallic surface, it is reflected at exactly the same angle as it strikes the mirror (Figure 16). This gives us the important law of reflection which was known to the Alexandrians, The angle of reflection, *r*, equals the angle of incidence, *i*. The angles are measured conventionally between the rays and a hypothetical line drawn perpendicular to the mirror as shown in the diagram. This kind of reflection is called "regular," or "specular." *Diffuse* reflection takes place when light rays strike a rough surface, because the rays are scattered in random directions.

Kepler used the law of reflection to explain why a reflected object seems to be located behind a mirror. Referring to Figure 17, light rays from a point such as *A* are projected on the mirror, which sends them on to the eye according to the law of reflection. The bundle of rays that we see forms a widening or *diverging* cone as the rays move from point *A* to the mirror and on to the eye. Now let's consider the bundle of rays from the eye's point of view. If each ray is prolonged backward, behind the mirror, each must pass through a point *X* in

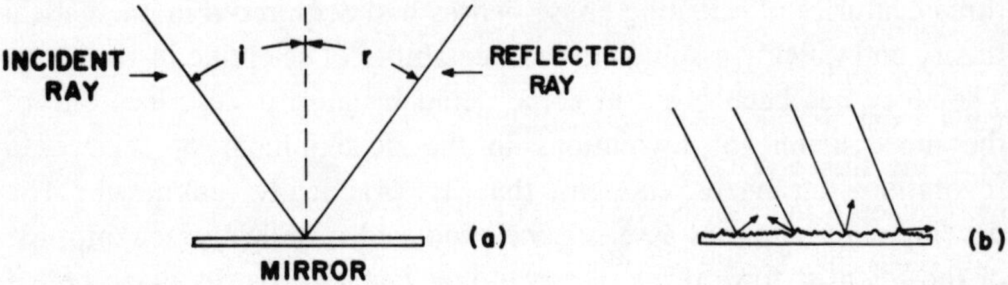

Fig. 16. The law of reflection. When a light ray strikes a plane mirror, the angle of incidence, *i,* equals the angle of reflection, *r.* Diffuse reflection takes place when light strikes a rough surface.

the diagram. The rays form a narrowing or *converging* cone from eye to mirror and we extend that *real* cone by an *imaginary* cone whose vertex is located at point *X*. We see an image of point *A* and locate it mentally at point *X* because our mental processes cannot take into

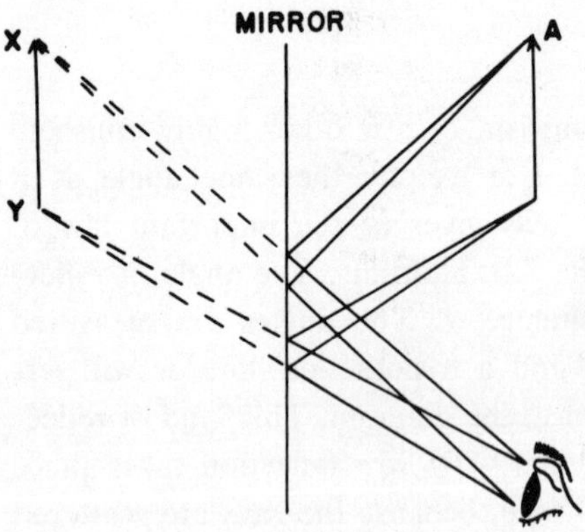

Fig. 17. Reflection by a plane mirror. The eye projects the rays it receives back behind the mirror, the place at which the object seems to be located. Because no real rays exist behind the mirror, we speak of a "virtual image" at that location.

account the fact that the light rays were changed in direction by the mirror. Kepler explains this effect in *Dioptrics,* as follows:

> The position (direction) of an object is estimated from the direction in which the ray of sight *originally* emerges from the eye, no matter how this direction may be altered by refraction in its path between the eye

and object. For the eye cannot detect what happens to the rays outside itself in intervening media, but assumes that these proceed in the original direction.

Ptolemy was also familiar with this view, *Because* X *is the point where imaginary rays intersect when prolonged backward, we speak of a virtual image at point* X. The image is vital because no light rays really exist at the place where the image seems to be located. By repeating this process for every point on the object, a virtual image is produced at *XY* as shown in Figure 17.

We can understand the nature of a *real image* from a discussion of curved mirrors. The mirror in Figure 18(*a*) is a small portion of a sphere. Light rays from a source at point *S* strike the mirror and are

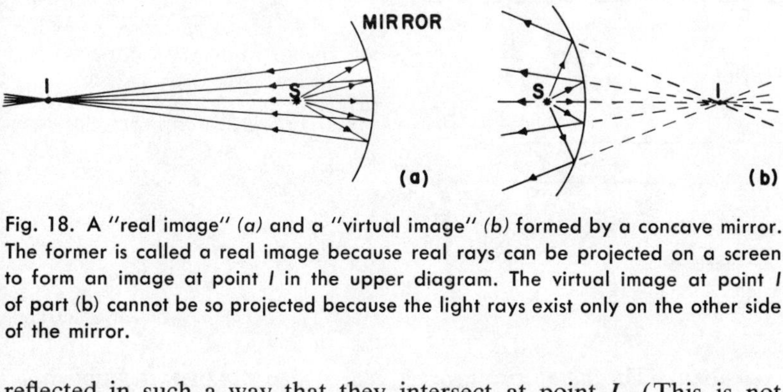

Fig. 18. A "real image" *(a)* and a "virtual image" *(b)* formed by a concave mirror. The former is called a real image because real rays can be projected on a screen to form an image at point *I* in the upper diagram. The virtual image at point *I* of part (b) cannot be so projected because the light rays exist only on the other side of the mirror.

reflected in such a way that they intersect at point *I*. (This is not exactly true, but the approximation is good if the curvature of the mirror is small.) The image at point *I* is called a "real image" because real rays intersect at point *I*. In part (*b*) of the diagram, the rays from point *S* are *divergent* and do not intersect to the left of the mirror. But they can be extended behind the mirror by imaginary rays that intersect at point *I*. By analogy with what happens in a plane mirror, the image at *I* in in Figure 18(*b*) is called a "virtual image."

The center of the sphere of which the mirror is a part is called the *center of curvature,* and is shown at point *C* in Figure 19. Now imagine a number of rays, parallel to the principal axis, that strikes the mirror on the conclave side. All of these rays intersect at point *F*, the *principal focus* of the mirror. This is a very important point to keep in mind. *The principal focus of a curved mirror is halfway between the mirror and its center of curvature*. The distance from the mirror to its principal focus is called the *focal length* of the mirror. You can examine the kinds of images formed by a concave surface by looking in an ordinary concave shaving mirror. If you hold it near your face, it magnifies, but if you look at it from a distance you see your image upside down. The former location produces a virtual image because the object is located between the mirror and the principal focus. The latter produces a real image because the object is placed beyond the principal focus.

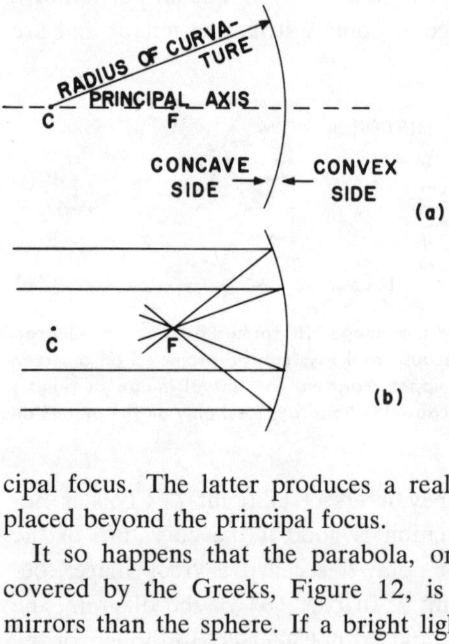

Fig. 19. Concave mirrors.

It so happens that the parabola, one of the conic sections discovered by the Greeks, Figure 12, is a better curve for reflecting mirrors than the sphere. If a bright light is placed at the focus of a mirror having a parabolic cross section (a paraboloid), all of the light rays emerge precisely parallel to the axis. Such a mirror casts a parallel beam of light over a great distance. Incidentally, the same curved surface is used to form beams of sound, heat and radio waves. The opposite effect is also true: all parallel rays entering the mirror are concentrated at the focus. This fact is made use of in the modern reflecting telescope.

Bent Rays

The principle of refraction led Kepler to the correct explanation of vision which in turn created the science of optics. In 1621, Snell discovered the precise mathematical relation that we call the law of

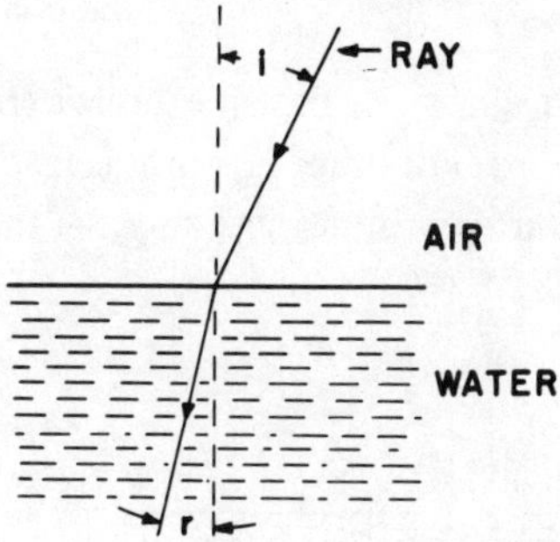

Fig. 20. Refraction of light.

refraction. Imagine a ray of light passing from air to water as shown in Figure 20. If *i* is the angle of incidence and *r* is the angle of refraction, Snell's law states that:

$$\frac{\text{sin of i}}{\text{sin of r}} = \text{n}$$

where *n* is a constant number that depends on the properties of the two media (in this diagram, air and water). A numerical example will help to illustrate the meaning and use of Snell's law. Suppose a beam of light enters the water in Figure 20 so that angle $i = 60°$. We then measure angle *r* and find it to be 40½°. Looking up these angles in a table of sines we find that sin 60° = 0.866 and sin 40.5° = 0.649. Substituting these numbers in Snell's law we have: $n = \frac{.866}{.649} = 1.33$: The constant $n = 1.33$ is called the "index of refraction" of water with respect to air. This number is called a "constant" because it remains unchanged for any angle of incidence *i* so long as air and water are involved. If glass is substituted for water, the index of refraction falls between 1.5 and 2.0 depending on the type of glass.

The Critical Angle

Snell's law tells us that a ray entering the water of Figure 20 perpendicular to the surface (angle $i=0°$) moves straight into the water with no bending:

$$\sin r = \frac{\sin i}{n} = \frac{0}{1.33} = 0.$$

If $\sin r = 0$, the sine tables show that the angle of refraction, r is also zero and there is no refractive bending. The law also points out another important angle—the *critcal angle*. Referring to Snell's law

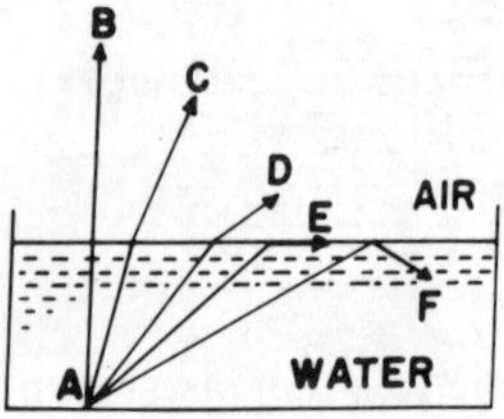

Fig. 21. Rays from A that strike the water-air boundary at an angle greater than the critical angle are totally reflected back into the water.

again, suppose the incident ray just grazes the surface so that the angle of incidence, i, is 90°. What would be the angle of refraction? The sine of 90° is equal to 1, so $\sin r = \frac{1}{n} = \frac{1}{1.33} = 0.75$: Referring again to a table of sines, we find the angle of refraction to be 49°. Figure 21 illustrates the importance of this angle.

Imagine a light source located at point *A* in the diagram. A vertical ray *AB* emerges without any refraction. Rays *AC* and *AD* are refracted in accordance with Snell's law because the angle of incidence is off from the vertical. Ray *AE* is at the *critical angle,* however, and just grazes the surface of the water. Rays such as *AF,* which have an angle of incidence greater than 49°, do not emerge from the water but are totally reflected back into it. The surface of the water acts like a mirror, and the law of reflection takes over.

The critical angle for most optical glass is between 30° and 40°, a fact that is made use of in ordinary optical prisms. As shown in Figure 22, light entering a glass prism perpendicularly, as at face *A*, undergoes no refractive bending. When it strikes face *B* at 45° it is totally reflected because 45° is greater than the critical angle. It proceeds in this way through both prisms until it reaches the eye.

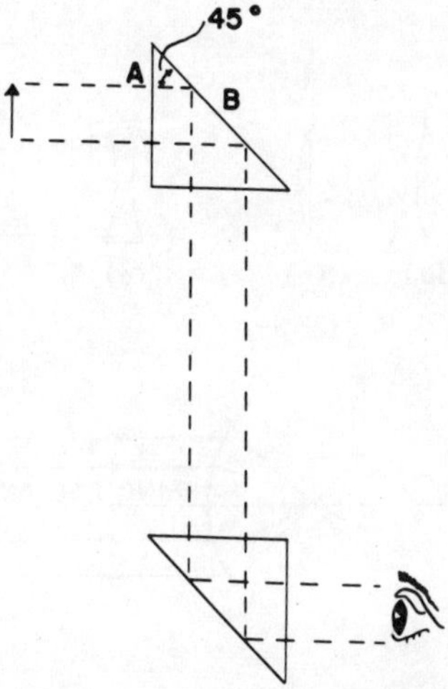

Fig. 22. A periscope made of glass prisms.

Prisms are widely used as mirrors in periscopes, binoculars and telescopes.

Lenses

A lens is a piece of transparent material—usually glass—whose polished surfaces are portions of a sphere. Figure 23 shows several common types of lenses. Of these, the double convex lens is most widely used in optical instruments. Referring to the diagram, all rays that pass through the optical center go through without refraction. All others (parallel to the principal axis) go through with different amounts of refraction, according to their distance from the principal axis. All of the latter pass through the focus of the lens, *F*. It is worth remembering that all rays are bent toward the thicker part of the lens.

Figure 24(*a*) shows how an ordinary magnifying glass makes objects appear larger. Starting with the tip of the arrow, point *A*, a cone of rays is seen striking the lens, being refracted, and then diverging toward the eye. The brain then projects this cone to the left of the lens until it converges at point *B*. (Once again, the brain has no way of knowing that the rays were bent by the lens.) Repeating the procedure for all other points, a virtual image is produced at *BC*

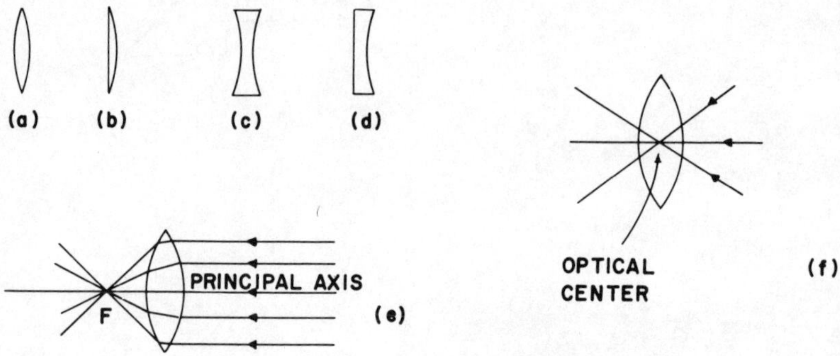

Fig. 23. Several common lenses.

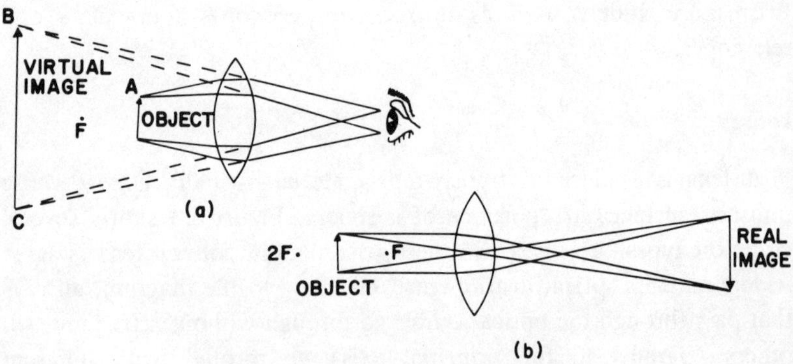

Fig. 24. Magnification by a convex lens.

which is larger than the object. A virtual image is produced whenever the object is placed between the focus, *F*, and the convex lens.

If an object is placed between the focus, *F*, and twice that distance, 2*F*, as shown in Figure 24(b), a real image is formed that is inverted and larger than the object. If the object is beyond 2*F*, the image is still real and inverted, but smaller than the object. It is important to keep in mind that real images can be projected on a screen because all of the rays from a point on the object *converge* or meet at a place in space *outside of our eye*. Virtual images are caused by rays that *diverge* or spread apart until they reach our eye. Only then does the eye cause them to converge and meet at a point on the retina.

Color, the Nuisance of Optics

The early part of the seventeenth century saw the ancient mysteries of vision and refraction dissolve into a precise geometrical and mathematical theory—mainly because of Kepler's genius. But there were still problems. One of these was the nature of Kepler's *lumen,* or "light" as we would say; the other was color.

Kepler had avoided talking about color. According to scientific opinion of the time, light was white and colorless, and sunlight was pure and perfect. When sunlight fell upon a body, it became "soiled," and came away "clothed" in color. Unfortunately, this simple little theory left a mass of unanswered questions. The most important of these was the reason for the color fringes that showed up in telescopes.

When telescopes of high magnification are manufactured of simple lenses, the image is always surrounded by a colored fringe that blurs its outline. The higher the magnification, the more troublesome is the distortion caused by the colored fringe. The same effect had long been noticed, of course, in cut glass prisms used for ornamental purposes. A phenomenon that had once been accepted thankfully for its beauty, now became a scientific nuisance. Men like Sir Isaac Newton (1642–1727), who were interested in improving the telescope, now found it necessary to investigate color. It was clear that to make good telescopes they would have to get rid of the colored fringes.

Newton's *Opticks*

It was still not well understood in the seventeenth century that scientific theories must have their basis in experimental facts. It must have been very difficult to make the reader take that point of view because Newton often went out of his way to emphasize its importance. He begins his great book *OPTICKS* by saying:

These Principles I consider not as occult Qualities, supposed to result from the specifick Forms of Things, but as general Laws of Nature, by which the Things themselves are form'd: their Truth appearing to us by Phaenomena, though their Causes be not yet discover'd. For these are manifest Qualities and their Causes only are occult. . . . To tell us that every Species of Things is endow'd with an occult specific Quality by which it acts and produces manifest Effects is to tell us nothing.

With these beliefs firmly held, Newton attacked the problem of colored fringes in the only way that he could—by finding out what causes them. Only when their cause was understood might it be possible to eliminate them. Newton did his work in a dark room into

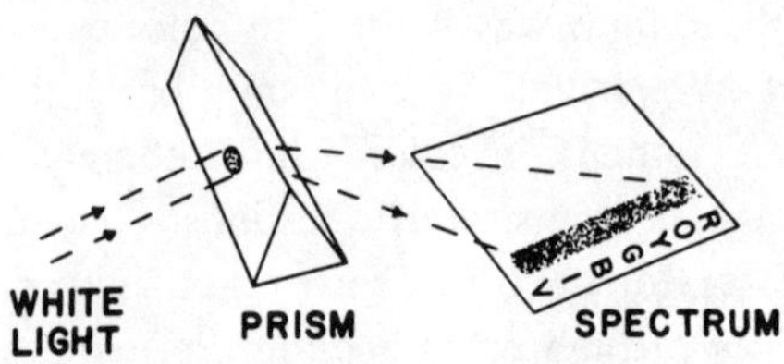

Fig. 25. Newton's experiment with a prism. White light is refracted by the prism and spread into a spectrum of all colors.

which a small beam of sunlight entered through a round hole. He placed a prism in the beam as shown in Figure 25, and he noticed that the white light was spread out into a colored strip. The colors of the strip were red, orange, yellow, green, blue, indigo and violet, each blending smoothly into the next. This pattern of colors is called the "visible spectrum." Newton explained the existence of the spectrum as follows: (1) white light contains a mixture of light of all colors, and (2) each color is bent by a different amount in passing through the prism. Red is bent least, violet most, and all the other colors in between.

Newton proved that white light is a mixture of all of the colors by bringing them back together again with a second prism. When all the colors of the spectrum were made to overlap, they produced light that was "entirely and perfectly white." He then performed what he called the *experimentum crucis,* the crucial test. He produced a spectrum as before, by passing a beam through a prism. Using a board with a a small hole in it, he intercepted all of the light except for one color, say red, which passed through the hole (Figure 26). This red beam was then passed through a second prism. He found that the red beam was not changed in color after passing through the second prism. It was bent once again, however, and the amount of bending was exactly the same as for red light passing through the first prism. If a blue beam was passed through the hole in the board and on

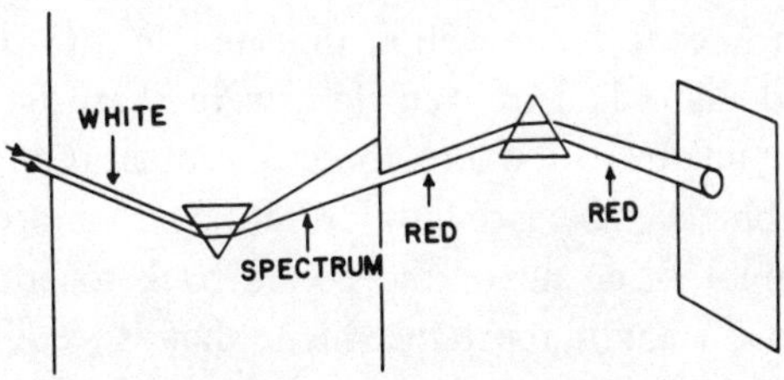

Fig. 26. Newton's crucial experiment. A color separated out of a spectrum remains that same color although refracted by a second prism.

through the second prism, it was bent more than the red beam but its color remained the same shade of blue. In this way he showed that each color underwent a different amount of bending when it passed through a prism. The amount of bending was always characteristic of the color that he used. Newton explained this observation by saying that the "refrangibility" was different for each color. We say glass has a slightly different index of refraction for each color.

Newton insisted that the colored light rays were initially in the white beam. Once a colored beam—orange, for example—was separated out, it stayed orange no matter how many prisms it was passed through. It could be made broader or narrower, fainter or brighter, but its color could not be changed. Light is indeed a mixture of different sorts of rays that appear white when combined in the normal way.

Newton's Telescope

At the completion of his experiments, Newton knew that a telescope made of ordinary convex or concave lenses could not produce an image free of color fringes. He illustrated that fact with a simple experiment—one that you can easily repeat in your own home. He painted a card with two colors—half of it blue and half of it red. Next, he wound a fine black thread round and round in order to form black lines on the red and blue background. Then he brightly illuminated the card and used a convex lens to focus the black lines onto a piece of white paper. He found that when those on the red background were sharp, those on the blue were out of focus and blurred; when those on the blue were sharp, those on the red were out of focus. It was impossible to get both sets of lines to focus well at the same lens-to-paper distance. He also noticed that when the red lines were sharp he had to move the white paper nearer to the lens to get the blue lines sharp. Just as for the prism, blue light was bent—or refracted—more than red. A simple lens cannot bring all of the colors back together again at a single point, so it cannot reproduce light that is "entirely and perfectly white."

These experiments confirmed Newton's belief that lenses—in any shape whatever—could never be used to make a telescope free of color fringes. He was right, of course, if you use only one kind of glass. Much later, it was discovered that different kinds of glass do not spread out a spectrum to the same extent. When this discovery was made, it became possible to combine lenses of different, suitably chosen glasses which get rid of colored fringes to a considerable degree.

On the basis of available information, Newton concluded that the best way to make a telescope was to use a spherical mirror as the main optical magnifier. A mirror reflects all light at the same angle and is not bothered by the color problems associated with refraction. He made and polished the mirror with his own hands, he said, because "if I had stayed for other people to make my tools and things for me I had never made anything of it." The concave mirror was made of a special alloy which he prepared of copper, tin and arsenic, called "speculum" metal, *speculum* being Latin for mirror. It is silvery in color and takes a high polish. The mirror was small in size, about an inch across, and was arranged to reflect rays toward a small flat metal mirror as shown in Figure 27. The small mirror was placed at a 45° angle to send the rays out the side of the tube through a small lens or eyepiece. Although the instrument was only six inches long, it magnified objects forty times.

News of Newton's new telescope soon reached the members of the Royal Society, or to use its full name, *The Royal Society of London for the Promotion of Natural Knowledge*. Founded in 1660, it is one of the most respected societies of the scientific world. At the urgent

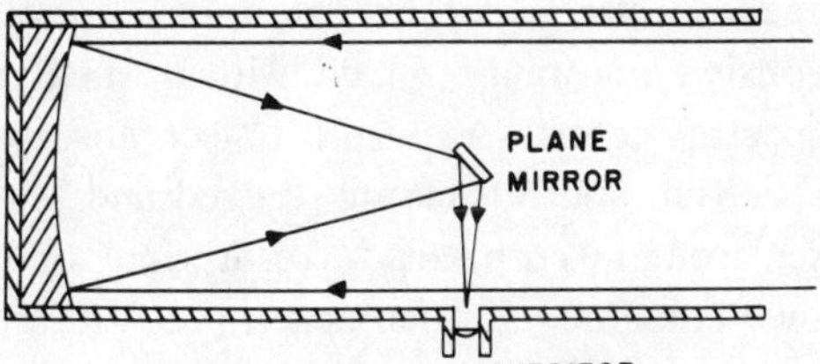

Fig. 27. Newton's reflecting telescope.

request of the Society, Newton presented its members with a second and better instrument containing a two-inch mirror which is still one of its most treasured possessions.

Color-Making Rays

Although Newton's great optical contributions were based on the measurement of physical things, he emphasized that the concept of color was entirely in the mind:

"The homogeneal light and rays which appear red, or rather make Objects appear so, I call rubrifick or red-making; those which make Objects appear yellow, green, blue and violet, I call yellow-making, green-making, blue-making, violet-making, and so of the rest." He tells the reader that if, at times, he speaks of colored rays, he must be understood to be speaking, "not philosophically (scientifically) and properly, but grossly. . . . For the rays, to speak properly, are not coloured. In them there is nothing else than a certain power and disposition to stir up a sensation of this or that colour."

In his *Opticks,* Newton discusses the ordinary colors that we see around us—the color of flowers, cloth and objects of every kind. He points out that the color we see is usually not one of the pure spectral colors that make up white light. A red book does not send a pure spectral red to our eyes, and brown is not a spectral color at all! Newton proved that the observed colors are produced by *mixtures* of pure spectral colors. Some objects look red because they absorb the blue and green part of the spectrum and reflect the red and some of the yellow and orange. If we illuminate a red object with spectrally pure blue light, it looks black because there is no red light for it to throw back. A brown object absorbs a large amount, but not all, of each spectral color except orange. The "brown" light that we see consists of orange mixed with a small amount of all of the other spectral colors. A purple object absorbs the middle colors of the spectrum and reflects mostly red and blue. Colors like these that we see around us are called "absorption" or "subtractive" colors, because they come about when spectral colors are subtracted from the white light that falls on a surface.

Colors by Artificial Light

Ordinary artificial light from an incandescent lamp has a slightly different spectral make-up from sunlight, being comparatively weaker at the blue and violet end of the spectrum. Consequently, colored fabrics, especially those rich in blue tones, appear different in color by incandescent light. Many fluorescent lamps have a different deficiency, being comparatively weaker than sunlight in spectral colors at the red end of the spectrum. Fluorescent light tends to emphasize the blue tones in a colored object.

It is now easy to understand why mixed paints give different colors from mixed lights. If we mix yellow and blue paints or powders, for example, we see green, but if we project yellow and blue light beams on a screen so they overlap, they produce white. (Actually, the white is rather pinkish; it takes three colors to produce a "true" white.) A mixture of yellow and blue paints gives green because the yellow paint absorbs red and blue and reflects yellow, green and orange; the blue paint absorbs red, orange and yellow, but reflects blue and green. The only spectral color that escapes the two absorptions is green—the color we see. But when blue light and yellow light overlap on a white screen, which reflects them both, we see a color that is nearly white. Color sensations of this kind, which are produced by blending spectral colors on a screen, are called "addition colors." Newton was the first to explain the difference between subtraction and addition colors, and to attempt a scientific study of the sensation of color perception in general.

Corpuscles and Waves

Newton thought of light as tiny particles, or corpuscles, which move at great speed. His *corpuscular theory* explained reflection by analogy with a ball bouncing off a smooth surface. He explained refraction by assuming that the particles move faster in more dense materials.* In order to explain the colors we see in thin plates and soap bubbles, he was forced to give the corpuscles a wave-like character. He pointed out, however, that light was not exactly like water waves. He noticed that when a beam of light passes through a screen with a small hole in it, the part that passes through does not appear to spread out. Such spreading does take place, however, when sea waves enter a harbor made of stone walls. The water waves that get through the opening in the wall do not form a straight "beam," but spread out sideways, forming a large triangle with its point at the entrance. Newton expressed himself as follows on the wave theory:

> If light consisted in motion (waves) *it would bend into the shadow,* for motion cannot be propagated in a fluid in right (straight) lines beyond an obstacle which stops part of the motion, but will bend and spread every way into the quiescent medium beyond the obstacle. . . . A bell may

* Referring to Figure 20, the velocity of the light corpuscles is the result of two simultaneous motions: a vertical motion (down, in the diagram) and a horizontal motion (to the left). Newton believed that the horizontal velocity would be unchanged when the corpuscles enter the water. The downward velocity would be increased, however, by the gravitational forces between the water and the corpuscles. When the two new velocities were added, the ray would be found closer to the vertical line than had been the case in air. This explanation was later found to be incorrect.

be heard beyond a hill which intercepts the sounding body. . . . but light is never known to bend into the shadow.

What Newton did not know was that light *does* spread out ever so slightly after passing through a hole. In addition, it was later found that the speed of light decreases when it enters a more dense material, in opposition to Newton's corpuscular theory. For these reasons, the corpuscular theory lost favor and light came to be thought of as a system of waves travelling through a mysterious medium called the ether. As we shall see, science was eventually forced to return to the general ideas of Newton, which gave light a dual nature consisting of both waves and particles.

The century of Kepler and Newton saw the emergence of optics as a full-fledged science. Through the instruments it provided, optics was to shed light upon many other areas of scientific enquiry. Astronomy, biology, medicine and physics—all would feel the influence of optics through the succeeding years. The development of seventeenth century optics was truly an epochal achievement in man's struggle to understand the nature of time and space.

The Riddle of the Heavens

5

The seventeenth century is often called the "century of genius." In addition to the birth of optics, it saw the riddle of the planetary motions dissolve into a beautifully simple scheme. But it would be wrong to assume that any one particular century has a monopoly on gifted people. The laws of probability see to it that no abrupt change takes place in the ability of our intellectual leaders. The remarkable scientific achievements of the seventeenth century can be attributed only to environment rather than to accident. Throughout the ages, each society has received the degree of scientific accomplishment that it has deserved.

The Printing Press

There was no single factor that led to the scientific awakening of the seventeenth century. It is probable that we will never know the complete story. Nevertheless, it is clear that the printing press played an important part. In Alexandrian times, many thousands of books were available in the great libraries, but they could be read only by those having the time and money necessary to travel to them. Private ownership of books was rare and costly, because of the colossal effort involved in manually copying a book onto costly parchment or papyrus. But all of that changed in the fifteenth century with the effective beginning of printing in Europe.

Printing is basically an old art. For example, stamps were used in medieval times to make inked impressions of decorative capital letters at the beginning of manuscripts. As early as 1300, man had learned to make woodcuts of pictures and short inscriptions. But the machine that contributed to the great intellecual revolution of the seventeenth century used movable type made of cast metal. However, printing machines could be used on a large scale only when it became possible to substitute paper for parchment. Although paper was used in China during the early Christian centuries, it did not become common in Europe until the fifteenth century.

The origin and date of the invention of modern printing is disputed, but its effective application in Europe began in 1447 at Mainz with the press of Johann Gutenberg (c. 1400–1466). Printing spread rapidly from that date onward. During the next century, many Greek and Arabic scientific classics were translated, printed and studied.

The Measurement of Longitude

During the same period, navigation attained a level far above the achievements of antiquity. The ships of that period could weather the great storms of the Atlantic and sail over great distances propelled only by the wind. Nevertheless, the measurement of longitude remained a difficult task. Accurate portable clocks were still not available and the motions of the moon, sun and planets were the only means by which a navigator could determine Greenwich time on a long voyage. Astronomical tables were prepared giving the times in the future when the planets would be in conjunction with the moon; that is, when they would have the same celestial longitude as the moon. This data was then used by the great navigators to determine their longitude on long voyages.

Amerigo Vespucci measured his longitude at a latitude of 10°N by observing a conjunction of Mars with the moon. He deduced from several observations that the conjunction took place at 6:30 P.M. local time; that is, just six and one-half hours after local noon. An almanac prepared by the astronomer Regiomontanus predicted that the conjunction would take place at midnight at Nuremburg. When it was 12 P.M. at Nuremburg, it was 6:30 P.M. at the ship. Amerigo's local time was just 5½ hours behind Nuremburg. His longitude, therefore, was $5\frac{1}{2} \times 15°$ or $82\frac{1}{2}°$ west of Nuremburg. Columbus used similar methods to measure longitude on his voyages to the New World.

Astrology and Science

The use of planetary motions in the determination of longitude fulfilled an important social need and helped to make astronomy the queen of the sciences. Of equal importance, perhaps, was the universal belief in astrology which taught that God ruled the earth through the motions of the celestial spheres. The stars influenced life on earth and the planets caused variations in this influence depending upon their changing relative positions.

Luckily, man had a way out. His gift of free will came directly from the Supreme Being. Only when he failed to offer resistance to the power of the stars and planets was he swept along by the whims of nature. Horoscopes were invaluable, therefore, as warnings of dangerous periods during which every precaution must be taken to avoid pitfalls. Astrology had reconciled itself with theology and became a dominant part of Christian thought. Much of this nonsense persists to the present time.

The changes that had been developing gradually in Europe erupted into a social and spiritual revolution in the sixteenth century. The growing power of businessmen, the rise of monarchies, the opposition to the Church, all combined to cause a break with the past. In addition, the navigators had opened up the entire earth, thereby helping to liberate human minds from old, narrow ideas. On the religious scene, new systems replaced the all powerful Church of Rome: Lutheranism, Calvinism and the reformed Catholic Church. Mingled with these theological changes was a rash of political and religious wars. The Church was unable to guarantee the quiet predictable life of the past, and its unchallenged authority began to wane. With earthly conditions so chaotic, man looked to the stars for refuge. Through astrology he could readily calculate the movement of the all-important heavenly bodies, and thereby find a degree of predictability in an uncertain world.

For those reasons, the sixteenth century found astronomy at the center of practical life. There it encountered the spirit of restless change that was growing in the minds of men. It is not surprising that people were no longer content to accept blindly the ideas of antiquity. Everything favored a period of scientific growth, and such a period followed—a sprinkling of new ideas in the sixteenth century, and an overwhelming torrent in the seventeenth.

Copernicus and the Moving Earth

The idea that the sun is the center of the universe did not originate with Copernicus (Figure 28). Aristarchus of Samos (c. 310–230 B.C.) taught at Alexandria that the earth and the planets revolve in circles about a stationary sun. It is also believed that Philolaus, the Pythagorean, held the same view. During medieval times, a number of theologians attacked Aristotle's doctrine that the earth is fixed. One of these was Oresme, Bishop of Lisieux (1323–1382) who was an advisor to Charles V of France. Another was Nicholas of Cusa (1401–1464), a fisherman's son who rose to the office of Cardinal of the Church. Still another was Domenico Novarra under whom Nicolas Copernicus (1473–1543) studied astronomy and mathematics at Bologna, Italy. The writings of the Pythagoreans taught that the ultimate truth about the universe must consist of simple and elegant geometrical relations, and Novarra felt that Ptolemy's astronomy—with its circles and epicycles—was much too complicated to be true. His doubts and criticisms probably influenced the thoughts of his young student, Copernicus, who kept them in mind long after he had returned home to a Canonry in the Cathedral of Frauenburg, Poland.

Putting the Earth in Its Place

About the year 1512, Copernicus wrote a summary of his ideas in a *Commentariolus* (short comment) which he sent to a few friends and astronomers. In it, he states seven principles which are the basis of his theory of the universe:

1. There is no single center for all heavenly bodies.
2. The earth's center is only the center of gravity and of the moon's orbit.
3. All planets circle the sun.
4. The distance from earth to sun is negligible compared to the distance to the stars.
5. The stars are fixed and appear to move across the sky because the earth rotates daily about its axis.
6. The sun is stationary and its apparent motion is caused by a rotation of the earth.
7. The back and forth motions of the planets are caused by the earth's motion in its orbit around the sun.

Copernicus had rejected the Ptolemaic system because of its complexity. We have seen how it placed the earth at the center of the universe with the sun moving around it in a circular orbit. Out beyond the sun were three additional orbits in which nothing moved but fictitious planets. The real planets—Mars, Jupiter and Saturn—moved around these fictitious planets in small circles called "epicycles." This was artificial enough, but Ptolemy's system had to be amended time and again in succeeding centuries as the planetary motions became known more accurately. By the time of Copernicus, Ptolemy's world system required about eighty circles to explain the observed planetary motions. Nevertheless, Ptolemy's system did work quite well and Copernicus realized that no one would accept his new theory until it could be used to predict the planetary movements with equal accuracy. This phase of his task took years of work, most of which was wasted.

We now know that the earth and the planets move about the sun in slightly oval curves or, to be precise, in ellipses. These ellipses are nearly, but not precisely, circular.

Looking from the North Star, we would see the earth and the other planets moving in a counterclockwise direction around the sun. Mercury, the planet closest to the sun, would move most rapidly, taking less than three earth-months to make one complete revolution. The other planets: Venus, earth, Mars, Jupiter, Saturn, Uranus and Neptune would have speeds that were progressively slower according to the distance of each planet from the sun. Pluto, the outermost planet, would be seen moving slowest of all, taking almost 250 earth-years to circle its orbit. All of the planetary orbits would lie almost in the same plane, and close observation would show them to be slightly elliptical. But Copernicus was so convinced that the circles of Aristotle were "natural" and "inevitable" that he found it impossible to conceive of noncircular motion for the planets. He added epicycles to his system along with other complications borrowed from Ptolemy's ancient scheme. He tried, in short, to get better agreement with measurement by using the same threadbare techniques that he had objected to in the older system. When his new world system was published in *De revolutionibus orbium coelestium* ("On the Revolutions of the Celestial Orbs") in 1543 it contained thirty-four circles—a little less than half the number needed by the Ptolemaic system. The complexity had been reduced, but it had not been eliminated.

The Insignificant Earth

Copernicus went on to answer the objections that had been raised against a moving earth for two thousand years. One was the terrific wind that would always be blowing from west to east. Copernicus merely explained that the air moved along with the earth. Another objection was that so rapid a rotation would cause everything to fly apart. Copernicus replied that a moving earth was the lesser of two evils. If the "sphere of fixed stars" were to rotate 360° in one day it would have to travel much faster than the earth because of its vastly greater diameter. So it would be even more likely than the earth to fly to pieces. Finally, the Pythagoreans had argued that any motion of the earth in an orbit would result in an apparent motion of the stars—much as a child on a merry-go-round sees nearby objects moving from left to right against the background of distant objects. But no such motion was observed. The stars remained fixed in the sky and the constellations did not change their shape. To this objection Copernicus answers that the stars are so far away that the minute motion of the earth in its orbit makes no real difference. "The earth and its orbit stand in the same proportion to the size of the universe as a point does to a clod. . . ." In short, we live on a speck of dust.

Man, who had always believed himself to be the crown of creation, suddenly found himself living on a minor planet that revolved around a far larger sun. Such a revolutionary idea ought to have made a great stir in the minds of thinking men. Strangely enough, it made almost

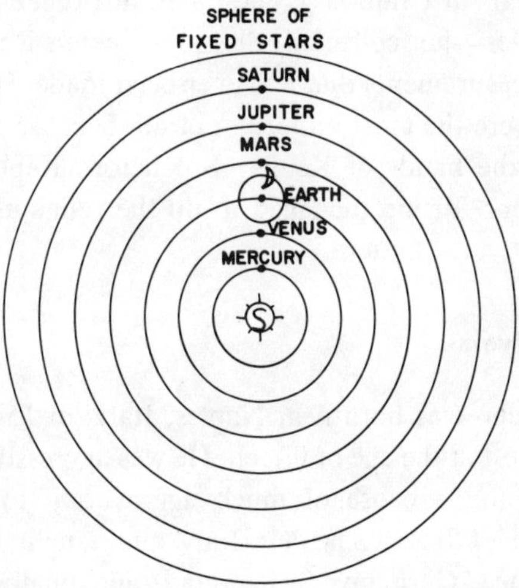

Fig. 28. The world system of Copernicus, 1543.

none at all. The Copernican world system was thought of, essentially, as a clever method of reducing eighty circles to thirty-four. Nevertheless, the little notice that was taken of it during the next generation was often friendly. Although the Lutherans, from Luther on down, had detested the book from the start, most of Christianity was at least noncommittal. It is even possible that Christianity might have absorbed the moving earth of Copernicus eventually, just as it had earlier accepted the spherical earth of Aristotle. But it was not to be that simple. A renegade monk by the name of Giordano Bruno (1574–1600) opened the eyes of the Church to the dangerous implications of the Copernican theory and a tactless Galileo closed the issue completely.

Tycho Brahe

The Dane Tycho Brahe (1546–1601) was born just three years after the death of Copernicus. Where Copernicus was a competent theorist, Tycho Brahe was a most patient and accurate astronomical observer, perhaps the greatest of all time. He was given an island observatory, Uraniborg, by Frederick II of Denmark and proceeded to equip it with instruments perfected with the best techniques of the time. For ten years at Uraniborg, and, later, at Prague—where he was joined by Kepler—he collected the most extensive and accurate astronomical measurements that had ever been made. His observations on the planets were the most important of all, because of the part they played later in the hands of Kepler. But before Kepler had time to complete his work, Bruno managed to tip the religious scales against the Copernican world system.

The Infinite Universe

Giordano Bruno was born near Naples, Italy, in 1547 and became a Dominican priest at the age of fifteen. He was aggressive, independent and intolerant, and a cause of much aggravation to his superiors. Under suspicion of heresy, he left Italy and taught at several universities in France, Germany, Switzerland, and finally in London in 1583. Here he published three little books, one of which—*Dell'infinito Universo e Mondi* ("On the Infinite Universe and Worlds")—caused all the trouble. Since God is infinite in all ways, Bruno felt that there could be nothing finite about his universe. He wrote, "There are endless particular worlds similar to this earth; with the Pythagoreans I regard it as a star, and similar to it are the moon, the planets, and other stars, which are infinite in number, and all these bodies are worlds." He also explained that each world has its own sun about which it moves. Expanding on the ideas of Copernicus, Bruno not only displaced the earth from the center of the universe, but also the sun because, "As the universe is infinite, no body can properly be said to be in the center of the universe or at the frontier thereof." All of the planets circling around all of the suns were equally important to God, and sometimes he thought they *were* God.

The Church had not shown any active disapproval of the Copernican ideas, but Bruno's theories were too revolutionary to be disregarded. The Church had placed heaven beyond the sphere that held the fixed stars. Bruno's new universe was infinite in size and left no room for heaven. In addition, Bruno preached that God had an infinite number of worlds to keep track of, emphasizing the insignificance of the earth. All of this was too antagonistic to the established doctrines of the Church to be passed over in silence.

In 1593, Bruno made the fantastic mistake of returning to Italy. His presence was soon discovered by the Inquisition—a religious tribunal that had jurisdiction in ecclesiastical matters—and he was placed under arrest. After being imprisoned for seven years, he was tried on a number of counts and found guilty. The judgment was punishment "with all possible clemency, and without shedding of blood." These mild words meant, in practice, that Bruno was roasted alive at the stake.

It is important to remember that Bruno's views were not based on observation. His contribution was a philosophy, not a scientific system. Nevertheless, some of his ideas lived on after his death and eventually found vindication in more rational times.

Kepler's Three Laws Simplify Astronomy

Johann Kepler was the first scientist to use mathematical methods in order to learn the laws of planetary motion. He was a confirmed believer in Copernicus and spent much of his life searching for a simple law that would explain the motions of the solar system. By 1618, using Tycho Brahe's precise astronomical observations, he had published the three laws that have become the cardinal principles of modern astronomy:

1. The planets revolve around the sun in ellipses with the sun at one focus of the ellipse.

2. Each planet moves at a variable speed such that a line from planet to sun sweeps out equal areas in equal periods of time.

3. The square of each planet's period of revolution around the sun is proportional to the cube of its distance from the sun.

The first law can be explained with the help of Figure 29. To draw an ellipse we begin by putting two tacks into a piece of paper at points *A* and *B*. A loop of string is then placed over the tacks as shown. If we

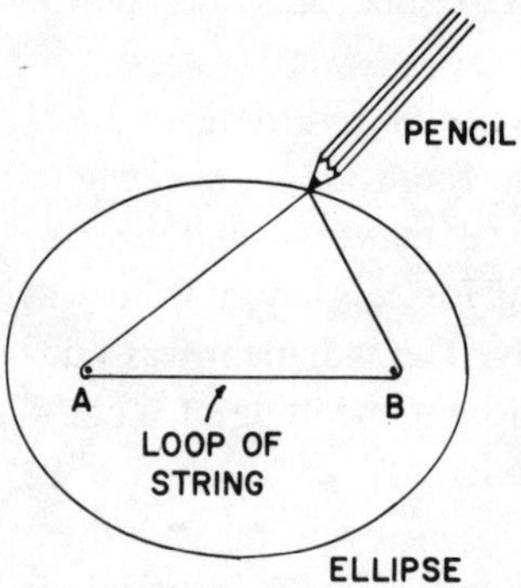

Fig. 29. Drawing an ellipse with tacks and a loop of string.

then run a pencil around the tacks, keeping the string taut, the line draws a perfect ellipse. The points *A* and *B* are called the *foci* of the ellipse. A circle, by the way, is merely a special ellipse in which *A* and *B* are at the same point. While the planetary orbits are really

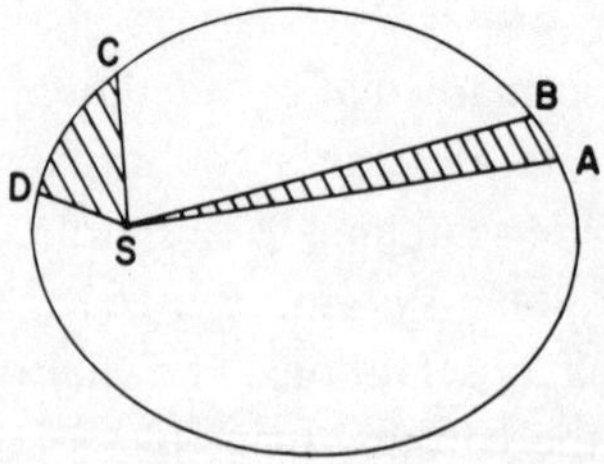

Fig. 30. Kepler's second law. A line drawn between planet and sun sweeps out equal areas in equal periods of time.

ellipses, their foci are extremely close together, which helps to explain why the orbits were so long thought to be circles.

Kepler's second law explains certain periodic changes noticed in the velocities of the planets. An elliptical orbit is shown in Figure 30 with the ellipticity greatly exaggerated for clarity. When the planet is at the far end of its orbit from the sun, it sweeps out an area *ABS* in a given length of time. When it is closer to the sun, however, it must move faster for the same length of time in order that area *CDS* will be equal to area *ABS*. Because the distance from *C* to *D* is the greater, the planet must speed up as it approaches the sun.

The third law tells us the relative times required by the planets to complete one revolution around the sun. If one planet *X* is four times as far from the sun as a second planet *Y*, then *X*'s year is eight times as long as *Y*'s. Taking an actual example, Jupiter is 5.2 times as far from the sun as the earth. The ratio of their solar distances is $\frac{5.2}{1}$ or 5.2. The cube of that ratio is $5.2 \times 5.2 \times 5.2$ or 140.6. But this product is equal to the square of the ratio of Jupiter's period of revolution (its year) to that of the earth. The square root of 140.6 is 11.9 which tells us that Jupiter's year is about twelve times as long as a year on earth.

Kepler's three laws have been confirmed by countless observations. We now know that they are not absolutely exact, but they are so nearly perfect that not the slightest error was found in them for over two hundred years. The eighty circles of Ptolemy, which Copernicus had reduced to thirty-four, had now been superseded by seven ellipses. The Pythagorean requirement of mathematical simplicity and elegance had finally been achieved.

Galileo and the Inquisition

Despite the accuracy and simplicity of Kepler's laws, the notion of a moving earth was to receive another blow. As soon as Galileo made his first telescope in 1609, he turned it toward the heavens and began making discovery after discovery with breakneck speed. Turning it toward the moon, he found mountains and other irregularities showing it to be a world like the earth. He next turned his telescope toward the constellations and found "a host of other stars, so numerous as to be almost beyond belief." Bruno had been right after all. "The galaxy is . . . a mass of innumerable stars planted together in clusters." Copernicus, too, had probably been right because Jupiter and its moons formed a miniature solar system patterned on his ideas. Looking at Venus, Galileo saw that it passed through phases just like those of the moon (Figure 31) which was a direct confirmation of the Copernican theory. The Ptolemaic system required that Venus should never show more than a semicircle of illuminated surface toward the earth. The Copernican system called for exactly the sequence of phases that Galileo had actually seen. This one single discovery proved Copernicus to be right to all who would believe the evidence of their eyes. At about the same time, Galileo discovered the rings

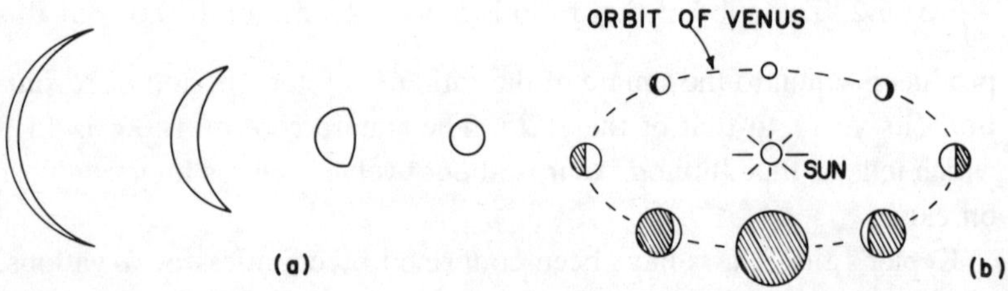

Fig. 31. The phases of Venus. Venus looks smallest when full because it is then farthest from us, on the other side of the sun from the earth.

of Saturn, but interpreted them incorrectly as "three spheres that almost touch one another." He also discovered moving black spots on the surface of the sun, although others noticed them at about the same time. These sunspots prove that the sun rotates on its axis, and its period of rotation can be determined by the motion of the spots.

Unfortunately, Galileo became involved in a controversy with a Jesuit rival named Christopher Scheiner who claimed to be the first discoverer of sunspots. The controversy soon spread far beyond its original topic, including the question of the habitability of the moon and the planets. Galileo made no secret of the fact that he believed firmly in the Copernican view. This, his critics pointed out, showed that he believed in the existence of other habitable worlds. This doctrine, which had been preached by the heretic Bruno, was contrary to Aristotelian and Christian teaching.

Galileo now found himself opposed by a variety of powerful interests. The Aristotelians had long fumed against him, and they were now joined by the powerful Jesuits. In addition, most pious churchmen were outraged by the thought that other worlds might share the Creator's favor. Behind all of this opposition was the great unspoken fear of Bruno's infinite universe with no place for a Christian heaven. In 1616, the Inquisition stepped in and Galileo was warned "to abandon these opinions and to abstain altogether from teaching or defending or even discussing them." Shortly thereafter, the work of Copernicus was placed on the index of prohibited books and withdrawn from circulation until it had been "corrected." The moving earth could be discussed only as a convenient "hypothesis," not as an absolute truth.

Science and Measurement

After a period of truce with the Inquisition, Galileo published *Il Saggiatore* ("The Assayer") in 1624—a work that contains an idea of great importance to modern science. In it he distinguishes between two kinds of qualities of an object: those which can be described numerically; and those which cannot. An object's velocity and position, for example, are *primary qualities* because we can measure them. *Secondary qualities* such as smell or beauty exist only in the mind of the observer and cannot be measured. This distinction may seem obvious in the twentieth century, but Galileo was writing for people who refused to look through his telescope in deference to Aristotle. Galileo was a prime mover in the modern belief that science is measurement.

The Dialogue

While Galileo worked on *The Assayer* and other non-controversial subjects, he tried repeatedly and in vain to have the new Pope, Urban VII, remove the ban on discussing the Copernican theory. At last his impatience got the better of him. He thought he had found a way to meet the pope's objections by giving his ideas in the form of a discussion among three men. The first would represent the old arguments of

Aristotle and the Church; the second would be a Copernican; and the other would be an impartial third party. His great skill as a writer made this book an immediate best seller, and his logic completely demolished Aristotle's ideas of motion and astronomy. It appeared in 1632 with the title, *Dialogo sopra i due massimi systemi del mondo* ("Dialogue on the Two Principal World Systems"). Then the lightning struck.

Although the book pretended to submit to theological authority, the intent to defy was painfully evident. The anti-Copernican character turned out to be a hopelessly stupid ignoramus, incapable of understanding the most logical arguments. Worse still, Galileo had promised to include certain Aristotelian arguments to be supplied by the pope, and these he put in the mouth of the riduculous anti-Copernican character. The sunspot-observing Jesuit Scheiner, who had become Galileo's arch enemy, used the book to make mischief. The stupid half-wit, he asserted, was actually intended to be a pen-picture of His Holiness, the pope.

Galileo, of course, was soon back in trouble with the Inquisition. His book was banned and, in 1633, he was tried, found guilty and compelled to recant all of the Copernican doctrines. Galileo spent the rest of his life under house arrest in Arcetri, near Florence.

A lesser man might have lived out his days in quiet comfort—recalling past glories as well as his cruel treatment at the hands of the Church. But not Galileo. As we shall see in the next chapter, his last and greatest book, *Discourses on Two New Sciences,* was published in Holland in 1638. By that time he was seventy-five years old and totally blind. On January 8, 1642, he died at Arcetri.

After Galileo's death, many writings appeared for and against the outlawed Copernican system, but opposition among scientists was steadily weakening. The ecclesiastical ban could impede the progress of science, but it could not stop it entirely. It merely made it difficult for Catholic scientists to exchange or publish new ideas. They did manage to circumvent the will of the Holy Office, however, by treating new scientific knowledge as mere "hypotheses." By the eighteenth century all important opposition ceased, and the new ideas became generally known to all intelligent people. It was not until 1835, however, that the works of Copernicus, Kepler and Galileo were all removed from the index of prohibited books. It had taken two centuries for truth to win out over obstinacy and ignorance. The attempt of men of the Church to use their great authority to impede the progress of science had been soundly defeated. The main scheme of the universe was now an open book and the final victory rested posthumously with Copernicus, Bruno, Galileo and Kepler.

Clocks and Cannon Balls

6

Scientists of the seventeenth century were faced with two everyday problems that defied solutions. One was the nagging problem of longitudinal determination aboard ship; the other was the motion of projectiles. The determination of longitude by modern navigators depends upon a comparison of local time at the ship, determined by the sun or stars, with the local time at Greenwich, England. In the seventeenth century, Greenwich time could only be inferred from relatively rare celestial phenomena such as eclipses or conjunctions. Besides being laborious, these methods required the talents of competent astronomers aboard ship, and were quite inaccurate at best. It would not be until 1765 that Harrison's chronometer would offer a means of keeping accurate track of Greenwich time aboard ship. At about the same time, Le Roy of Paris invented a better timepiece closely following the principle of the modern chronometer. But these inventions were beyond the capability of seventeenth century science. Much more would have to be learned about motion before accurate mechanical clocks could be built.

This lack of understanding of motion also had its detrimental effects on gunnery. Gunpowder had entered the arsenal of war and good marksmanship required an understanding of where the projectile would go after it was fired. For the first time in history, the measurement of long-range motion became a matter of practical importance. This, in turn, made it necessary to measure extremely short intervals of time—intervals corresponding to a small fraction of a projectile's time-of-flight. No clocks capable of such measurements were available.

The problems of time and motion measurement constituted a vicious circle, each depending upon a knowledge of the other for its solution. To make matters worse, the discovery of the true laws of motion—which would eventually lead to better clocks—was hindered by the "common-sense" ideas of Aristotle. Before any serious effect would be applied to a study of motion, it would have to occur to someone that the great Aristotle might be wrong. This possibility occurred almost simultaneously to two men working independently. The first was Simon Stevin (1548–1629) and the second was Galileo. Building upon the work of Archimedes, these two men laid a solid foundation for a new science called "mechanics." Stevin was concerned mainly with the mechanics of bodies at rest (statics), and Galileo mainly with the mechanics of bodies in motion (dynamics).

Falling Bodies

Aristotle had taught that heavier objects fall more quickly than lighter ones. He had not, of course, verified this belief by experiment, but it was not too unlikely a conjecture. Our everyday experiences with falling objects seem to confirm it. We see leaves falling from a tree, or a pencil falling from our hand, and the heavier object does fall faster. Our lives are full of falling objects that seem to agree with Aristotle's teaching. This makes the attempt to prove it wrong all the more daring.

In 1586 Stevin, or Stevinus as he is also known, dropped stones of different weight and found that all fall at the same speed. Although his experiment proved that Aristotle had been wrong, Stevin did not pursue the discovery. He is known primarily for two other contributions: the introduction of our modern decimal method of representing fractions, and the famous "parallelogram of forces."

Adding Forces

It is not unusual for an object to have more than one force acting upon it at the same time. A falling leaf, for example, is pulled down by gravity while the force of the wind pushes it to the side. In addition, air resistance produces an upward force that slows the fall of the leaf, Even if we know the magnitude of these forces, how can we determine where the leaf will fall. Or, in general terms, how can we determine the combined effect of several forces acting simultaneously on an object?

Stevin found the answer by a brilliant combination of logic and intuition. We can understand his ideas with the help of Figure 32.

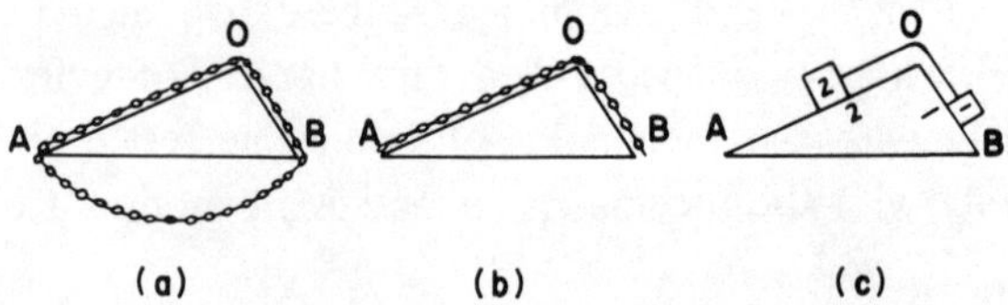

Fig. 32. Stevin's chain. The chain is in equilibrium for any triangular wedge.

Stevin imagined a triangular wedge, like *AOB* in part (*a*) of the diagram, with its longest side horizontal. A loop of chain surrounds the wedge as shown. Stevin knew that the chain would be stationary even if friction were completely absent. Any alternative would imply perpetual motion, an idea that had been discarded since the time of the Greeks. Stevin assumed next that he could remove the hanging portion of the chain, from *A* to *B,* without disturbing the equilibrium of the remainder as in part (*b*). He knew, of course, that he could replace the two remaining lengths of chain, *OA* and *OB,* by equivalent weights as shown in part (*c*) of the diagram. He also knew that the weight of a piece of chain is proportional to its length. The equivalent weights, therefore, would be in the same ratio as the sides of the triangle, *OA* and *OB*. If the sides have lengths of 1 and 2 feet—as shown in Figure 32(c)—the weights must be in the ratio of 1 to 2: 1 and 2 ounces, 1 and 2 pounds, or 1 and 2 of any other unit of weight.

While the example chosen uses a specific triangle, the same logic applies to any triangle whatever. As the angles and sides of the wedge change, the lengths of chain—and the resulting weights—change in a corresponding way to maintain equilibrium. It was clear to Stevin that the magnitudes of the weights were related geometrically to the shape of the triangle. From these ideas a mathematical proof, which we will omit, gives the method for determining the effect of any two forces acting simultaneously on the same object.

To illustrate Stevin's method, imagine two forces acting on a falling object such as a leaf. The weight of the leaf is a downward acting force, and the wind produces a force pushing it to the side. Suppose the leaf

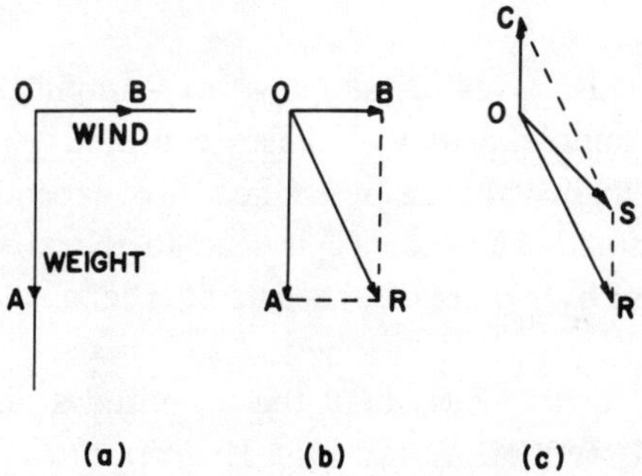

Fig. 33. The parallelogram of forces.

is at point O in Figure 33(a). Now draw two extended lines representing the *directions* of the forces, the weight vertically and the wind horizontally. An arrowhead is then placed at point A such that the length OA is proportional to the downward force or weight of the leaf. If the leaf weighs 2 grams, the length OA might be 2 inches, one inch corresponding to each gram of weight. The length OB is then measured off using the same proportionality between force and length. As shown in part (b) of the diagram, the parallelogram (in this example, a rectangle) is completed by drawing the dotted lines parallel to OA and OB. The length OR is now drawn along the diagonal from point O. The lines OA and OB represent the individual forces acting upon the leaf, and the line OR represents their combined effect. In other words, line *OR* represents a single force that can be used to replace the two original forces. The new force's direction is merely the direction of *OR* and its magnitude is proportional to the length of *OR*, using the same relationship between force and length that was used throughout: 1 inch = 1 gram.

Part (*c*) of the diagram shows how a third force, air resistance, can be taken into account. The line *OR* is redrawn from part (*b*) as shown. Line *OC* is added in the upward direction because the resistance of air impedes the downward motion of the leaf. The length of *OC* is made proportional to the magnitude of the air resistance, just as was done for the other forces. Completing the parallelogram, the line *OS* is found to be the effect of all three forces: the leaf's weight, the wind, and air resistance. Line *OS* is called the "resultant" of the three forces.

The example just given uses many over-simplifications. The air resistance, for example, actually changes constantly, as the leaf turns and tumbles in flight. The air resistance also depends on the leaf's speed at any instant. These concepts lead to the kinds of questions that are answered by *dynamics,* the part of mechanics that interested Galileo.

Stevin explained his ideas about the magnitudes and directions of forces in a book entitled *Statics and Hydrostatics,* which appeared in 1586. The book also told how he and Grotius had experimented with falling bodies. A few years later, Galileo performed the same experiments and came to the same conclusion as Stevin: something was wrong with Aristotelian mechanics.

The First Pendulum

Galileo's interest in falling objects began with a chandelier in the cathedral of Pisa. Once while attending Mass he noticed a chandelier that had been set swinging by an attendant who lit the candles. Galileo noticed that its successive swings became smaller and smaller in the course of time. He wondered if the time required for each swing also became shorter as time went on. He decided to find out by measuring the time of each swing. This he did by the ingenious method of counting his pulse beats. Probably to his surprise, he found that the candelabra did not slow down. Each swing took the same length of time, whether it moved through a wide arc or a small one. After church, he repeated the experiment with a stone tied to a piece of string and obtained the same result. He also discovered that the length of time required for one swing depends *only* upon the length of the string and *not upon the weight of the stone*. With the completion of these experiments, the modern pendulum was born. Today we speak of the time required for one complete back and forth swing of the pendulum as its *period,* and the size of its swing as the *amplitude*.

Back in his study, Galileo must have wondered about these amazing results. Why is the period of a pendulum independent of the weight of the stone? He reasoned that a pendulum, after all, is merely a special case of a freely falling body. If we drop a stone, it falls to the ground under the pull of gravity. But if we attach it to a string—and make a pendulum—it swings down in an arc instead. Both depend on the pull of gravity. Suppose we make two pendulums having the same length, but with different weights. The two weights swing in synchronism, as Galileo had demonstrated. Each weight takes exactly the same length of time to move from the top of its swing to the bottom. Then, stones of different weight must also fall with the same speed when dropped in free fall from a height. This conclusion agrees with the discovery mentioned earlier about the identical rate of fall of different weights.

The Laws of Fall

Galileo now set to work to discover the law that governs the rate of fall of objects. He suspected that the speed of a falling body might be proportional to the time that had elapsed since it had been set free. But this was little more than a guess, and proving it right or wrong would be a difficult task. It was quite hopeless to try to measure directly either the velocity or the time for a freely falling body. Today, we can make the measurements easily with high speed photography, but Galileo had to work with a ruler and a water clock. Confronted with these limitations, he decided to slow down his experiment. The laws of fall, after all, seemed to apply equally well to a pendulum and to a falling body. He was convinced that they would also apply to a ball rolling slowly down an inclined plane. So he set up a gently sloping plank, about 15 feet long, with a narrow groove cut down the center of it. A polished bronze ball rolled down the groove slowly enough to permit making accurate measurements of time and distance. Galileo explained his experimental method beautifully in *Discourses and Demonstrations Concerning Two New Sciences:*

Having placed the board in a sloping position, by lifting one end about one or two cubits above the other, we rolled the ball, as I was just saying, along the groove, noting, in a manner presently to be described, the time required to make the descent. . . . Having performed this operation and having assured ourselves of its reliability, we now rolled the ball only one-fourth the length of the groove; and having measured the time of its descent, we found it to be precisely one-half of the former. Next we tried other distances, comparing the time for the whole length with that for the half, or with that for two-thirds, or three-fourths, or indeed for any fraction. In such experiments, repeated a full hundred times, we always found that the distances traversed were to each other as the squares of the times, and this was true for all inclinations of the plane, that is, of the groove, along which we rolled the ball. We also observed that the times of descent, for various inclinations of the plane, bore to one another precisely that ratio which, as we shall see later, the author had predicted and demonstrated for them.

For the measurement of time, we used a large container of water placed in an elevated position. To the bottom of this container we soldered a pipe of small diameter giving a thin stream of water, which we collected in a small glass during the time of each descent, whether for the whole length of the groove or for a part of its length. The water thus collected was weighed after each descent on a very accurate balance. The differences and ratios of these weights gave us the differences and ratios of the times. . . .

Galileo's experiments proved that the distance traversed by the ball depends upon the square of the elapsed time. If it travelled one foot in one second, it would travel four feet in two seconds, nine feet in three seconds, etc. Since this square law was true for any inclination of the plane, he concluded that it would also be true for the limiting angle when the plane was vertical. The ball would merely fall alongside the plane, but its distance at any instant would be proportional to the square of the time of fall. He also proved mathematically that if the distance is proportional to the square of the time, then the velocity at any instant is proportional to the length of time since it was released. In modern terminology, these laws can be written as:

$$\text{velocity} = \text{acceleration} \times \text{time}$$

and

$$\text{distance} = \tfrac{1}{2}\ \text{acceleration} \times \text{time}^2$$

where the acceleration is a constant number that depends upon the pull of gravity.

Velocity, distance and time are familiar concepts in everyday life, but acceleration is not. *Acceleration* is a measure of the rate at which an object's velocity changes. Suppose you are riding in a car at a constant speed of 30 miles per hour. Your acceleration is zero, because your velocity is not changing. Now imagine that the car speeds up uniformly so that its velocity is increased by 10 miles per hour at the end of each succeeding second. After one second the car is moving 40 miles per hour, after two seconds 50 miles per hour, and so on. Under those conditions, your acceleration is 10 miles per hour per second because the car speeds up by 10 miles per hour in one second.

The acceleration of a falling body caused by the force of gravity (its weight) is 32.2 feet per second per second. Using this number for the acceleration due to gravity and neglecting air resistance, we can calculate the distance and velocity of a falling body after each second of fall:

Distance Fallen (feet)	Velocity (feet per second)	Time of Fall (seconds)
16.1	32.2	1
64.4	64.4	2
144.9	96.6	3
260.6	128.8	4
402.5	161.0	5

The velocities and distances given above show graphically why Galileo had to resort to a "diluted" force of gravity in order to make his measurements. Arriving at the true laws of falling bodies in the face of such difficulties was truly a great scientific achievement.

Inertia

Motion like that of a car moving at constant speed is called "uniform motion." Motion like that of a falling body is called "accelerated" motion. Galileo was the first scientist to explain the essential difference between the two—an explanation that led to the important concept of *inertia*. To do so, he elaborated logically on his experiments with falling bodies. He knew that the times of fall of a heavy and light object differ slightly because of air resistance. He described an experiment using two balls, one weighing about ten times as much as the other, one of lead and the other of oak, "both allowed to fall from an elevation of 150 or 200 cubits." He pointed out that they "reach the earth with slight difference in speed, showing that in both cases the retardation caused by the air is small." If the wooden ball were retarded to a greater degree, he argued, the lead ball ought to reach the ground a considerable distance ahead of the other. But this does not happen. "Indeed, the gain in distance of one over the other does not amount to a hundredth part of the entire fall."

He went on to show that the retardation caused by the air is usually negligible only because the speeds involved are relatively low. He was now leading up to the really important part of his argument. "For as to velocity, the greater this is, the greater will be the resistance offered by the air." To prove this point to yourself, merely compare the forces on your hand when you hold it out the window of a car as the speed changes. Galileo concluded that a falling body, if dropped from a great height, will encounter a steadily increasing force of resistance as it speeds up. Eventually, the resistance of the air will become equal to the weight pulling the body down. The two opposite forces will then offset each other exactly and the *falling body* will have *no force* at all acting upon it. The words in italics are the crux of this important matter. Although the net force acting on the body is zero, the body does not stop moving, but keeps moving at constant speed. Until the precise instant when the forces become equal, the body moves with accelerated motion. When the forces reach equality—and the net force on the body is zero—the accelerated motion changes into uniform motion. Galileo concluded, therefore, that if all the forces acting on a body balance out to zero, the body will nevertheless continue to move, and to move uniformly. This conclusion is contrary to Aristotelian doctrine which tells us that a body should stop moving when the net force acting on it is zero. As we shall see, Galileo had correctly stated a part of Newton's first law—the principle of inertia. He had discovered the important fact that it is equally "normal" for a body to move uniformly or to remain at rest. A body once set in uniform motion will continue in that state of motion until it is acted upon by an unbalanced force.

The Path of a Cannon Ball

In his *Discourses and Demonstrations Concerning Two New Sciences,* Galileo considered the concept of inertia in relation to the path of a projectile. He believed, correctly, that a projectile's motion is made up of two separate and independent motions: one uniform and the other accelerated. If we neglect gravity and air resistance for a moment, a projectile fired horizontally would continue to move in that direction indefinitely with uniform speed. Once it had left the gun barrel, there would be no force tending to change its velocity. If it were fired to the left from point *A* in Figure 34, it would move along the straight line *ABCDE*. These letters are equally spaced, so the projectile would take precisely the same length of time to move from any letter to the adjacent letter on its left. But in the real world, a projectile is also accelerated in the downward direction just as a freely falling body is accelerated. Nevertheless, gravity can neither increase no decrease the horizontal speed. Whatever the true path may be, the distance travelled horizontally depends *only* on the time that has elapsed since motion began. Suppose the projectile falls a distance equal to *BF* during the first second of flight as it moves from *A* to *B*. Its true location, then, is at point *F*. Because all of the horizontal spacings are equal, the projectile takes 2 seconds to move from *A* to *C*. But the distance it falls in that interval of time is proportional to the square of the elapsed time. The distance *CG*, therefore, is 2^2 or 4 times as great as *BF*. Similarly, the distance *DH* is 3^2 or 9 times *BF* and *EK* is 4^2 or 16 times *BF*. The projectile must follow a path having the shape *AFGHK*. Galileo proved that this curve is a parabola, one of the conic sections of the Greeks. He also proved that the

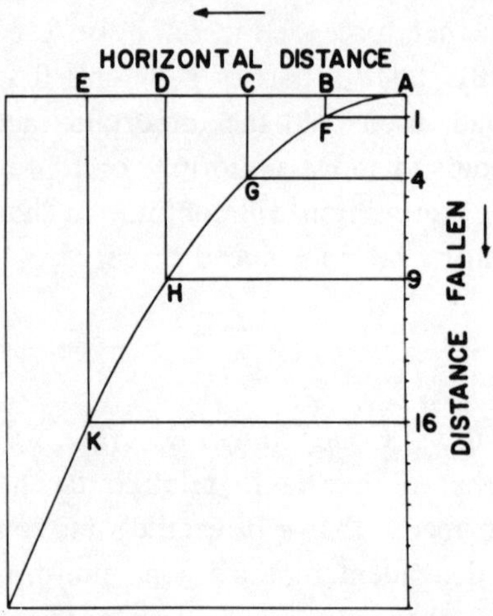

Fig. 34. Galileo's diagram illustrating the path of a projectile.

path of a projectile is parabolic, whether fired horizontally or at any other angle of elevation. He drew up a table giving the dimensions of the parabola for various angles and proved that the range is greatest when the angle of elevation of the gun barrel is 45° (Figure 35).

The Galilean Revolution

Galileo, more than any other man, introduced a new way of thought that led to modern science. Contributions had been made by others, of course—by Copernicus, by Tycho and by Kepler—but Galileo's share was overwhelming by comparison. He accomplished more than the mere acquisition of new knowledge; more even than a change in the basic structure of physics. It was rather a change in the kind of questions that a scientist should ask and the kinds of knowledge he should seek. Galileo's teachings shook the very foundations of philosophy—the knowledge of the world as a whole.

Galileo conceived of two completely separate bodies of knowledge. One relates to a physical world which is like a machine, obeying mathematical laws. The action of any of its parts can be calculated once the appropriate law is known. The other is a moral world about which science can have nothing to say. The two, which had been united in medieval times, became divided into two categories, *natural philosophy* and *moral philosophy*. This division of labor, so to speak, between the scientist and philosopher was not apparent in Galileo's

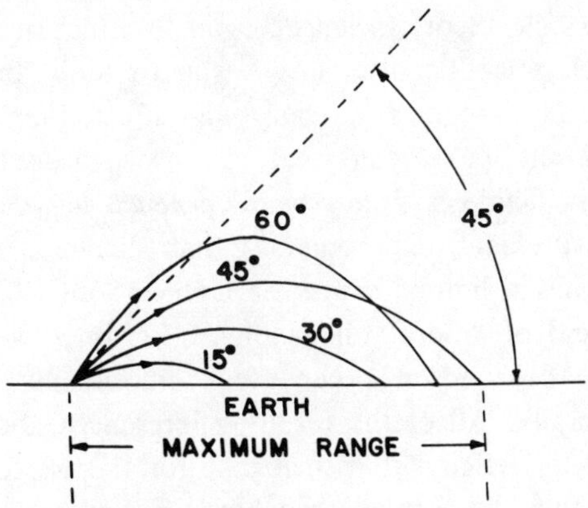

Fig. 35. The range of a projectile is greatest when fired at an angle of elevation of 45 degrees.

lifetime. If it had been, he might have had even greater difficulties with the authorities. Nevertheless, this new view of the world caused an enormous change in the scientist's approach to his work. Galileo's new world was one in which men might reasonably search for mechanical principles or mathematical laws. These principles and laws might apply to any part of the universe: to heavenly movements, to effects on earth, to the workings of the human body, or to the structure of a minute insect. With the development of Galilean mechanics and Keplerian astronomy, it even began to appear that all parts of the universe are mechanically interrelated.

From the time of Galileo, men of science have had to deal with an increasingly mechanical world. Faith in authority was no longer sufficient to explain the happenings in nature. Law must reign, and where moral law is inappropriate, physical law must take over. A *cause* must be found for every observed *effect*.

Cause and Effect

The concept of a mechanical universe guided by "causes" that produce "effects" is summed up in the word *determinism*. The new determinism that eventually gripped the scientific world led to a change in the scientific method. Instead of trying to explain all of nature with a single theory, science began to concentrate on limited objectives. Each scientific discipline began to look for relations in nature that can be thought of as cause and effect. Furthermore, only those effects would be studied that can be measured, or at least estimated. From Galileo's time onward, science has depended completely on measurement, and some say that science *is* measurement.

But science has a limited scope, as Galileo well knew. It has no place in the field of "moral philosophy." At times we look at the world scientifically, and at other times emotionally, socially, artistically or morally. All of the resulting inferences about the world affect the kinds of religion that men take for themselves. And there is no question that the Galilean revolution has played an important part in modifying predominant religious thought, especially on the part of men who accept the evidence of scientific experiment. Nevertheless, Galileo's private world was far from wholly mathematical. We know him as a pious man who was convinced that science and theology must be compatible because both convey truth. The great majority of modern scientists and theologians agree with him.

The Falling Moon

By 1618 Kepler had published the three great laws that described the motions of the planets. Then Galileo discovered the law that governed the fall of bodies, and came close to discovering the three laws of motion that we associate with Newton. But it was Newton who first clearly stated these laws and succeeded in connecting them logically with Kepler's laws. Newton was the first scientist who understood that the force that causes a stone to fall is the same kind of force that keeps the planets in their orbits. He began by considering the moon's motion. The moon can be compared to a metal ball swinging in a circle on a strong. The ball tends to fly off, and its pull can be felt on the string. If the string breaks, the ball keeps moving in a straight line in the direction in which it was moving when the break occurred (Figure 36). Newton compared the pull in the string to the pull of gravity between the earth and moon. He wondered whether this gravitational attraction might be sufficient to keep the moon from flying off into space.

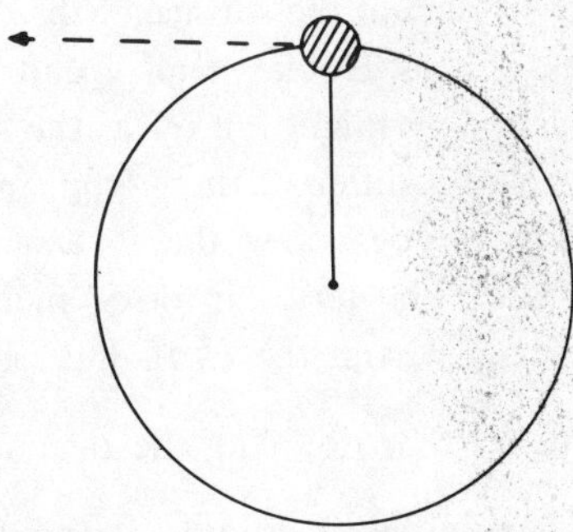

Fig. 36. A ball moving in a circular path tends to fly off on a tangent as shown by the dotted lines.

The inward pull that restrains the ball or the moon is called a *centripetal* force. It is equal in magnitude but opposite in direction to the outward pull or *centrifugal* force which results from the circular motion. The magnitude of the centripetal force necessary to keep any object moving in a circular path is given by $F=\frac{Wv^2}{gr}$ where F is the centripetal force in pounds, W is the object's weight in pounds, v is the velocity in feet per second, r is the radius in feet and g is the acceleration caused by gravity, 32.2 feet per second per second. It is probable that Christian Huygens, the great Dutch scientist, had correctly calculated the centripetal force before Newton, but he had not published his results and Newton was unaware of them.

Newton knew the size of the moon's orbit and the time it takes to make one revolution around the earth. Using these figures, he calculated the velocity of the moon. He also knew that the moon, like the ball in Figure 36, would fly off on a tangent if there were no centripetal force turning it in a curved path. (Although the moon's orbit is an ellipse, it is nearly circular.) But there *is* such a force, and the moon *is* constrained to follow her elliptic path. At the end of a second, for example, the moon—like a falling stone—has "fallen" a calculable distance toward the earth. The earth has attracted the moon just as it attracts any terrestrial object.

Convinced that the moon obeys the same laws as a moving body on earth, he searched his mind for the kind of a force that would account for the constant fall of the moon toward the earth. The force of gravity would surely be weaker at the great distance of the moon's orbit than at the surface of the earth. But how much weaker? Newton, as well as Hooke, Halley and others, had speculated that the force of gravity might fall off as the square of the distance. But saying this is a very different thing from proving it mathematically. The force of gravity does obey the *inverse square law,* as Newton would later prove. The term "inverse" means that the force gets smaller as the distance from the earth increases. The "inverse square" in the law means that doubling the distance cuts the force to $\frac{1}{2^2}$ or one-fourth of its initial magnitude. Tripling the distance reduces it to one-ninth of its initial magnitude, and so forth.

As a young man of twenty-four, Newton tested his ideas with a preliminary calculation. As shown in Figure 37, he calculated the distance that the moon must fall in one second in order to maintain its orbital motion. Using Galileo's law of fall, he also calculated the distance that a stone would fall in one second on the surface of the earth. Then he reasoned as follows: The moon is about 240,000 miles from the earth's center and the falling stone about 4000 miles. The ratio of their distances, therefore, is $\frac{240{,}000}{4{,}000}$ or 60. Because of the inverse square law, the pull on the moon would be $1/60^2$ or 1/3600 of that on the stone. The moon ought to fall just 1/3600 as far as a stone in one second. Comparing this to the moon's calculated distance of fall, he "found them answer pretty nearly." He actually found that the moon's "fall" was but seven-eighths of what he expected. His comparison was not perfect because he did not have the precise numbers for the moon's orbit and the diameter of the earth. But he had seized on the great concept of universal gravitation. He now suspected strongly that every bit of matter in the universe attracts every other bit and that the attraction varies directly with the product of the attracting masses, and inversely as the square of the distance

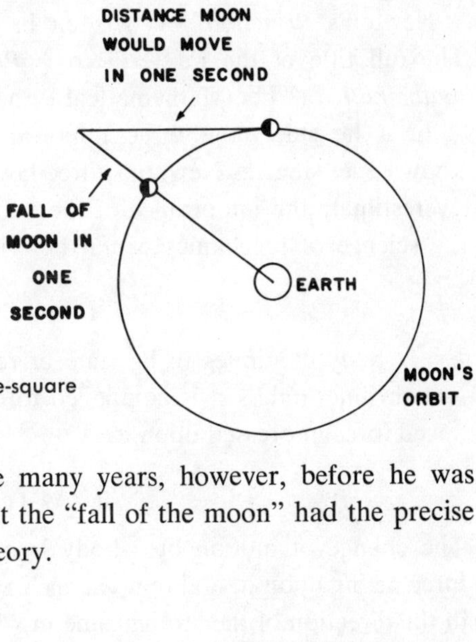

Fig. 37. Newton tests the inverse-square law.

separating them. It would be many years, however, before he was able to prove conclusively that the "fall of the moon" had the precise magnitude required by his theory.

Newton's Laws of Motion

Some eighteen years later, in 1684, the question of planetary motions was still occupying the thoughts of prominent fellows of the Royal Society. Early in that year, the great astronomer Edmund Halley—for whom a famous comet was named—discussed the matter with Sir Christopher Wren and Robert Hooke. Hooke was confident that the inverse square law was at the bottom of the matter, but he could produce no mathematical proof. In the following August, Halley went to Cambridge, where Newton was teaching, to get his views. Without mentioning Hooke's ideas, he asked Newton what would be the orbit of a planet supposing it was attracted to the sun by the inverse square law. Newton immediately answered, *an ellipse.* When asked how he knew it, he replied, "Why, I have calculated it." It was typical of Newton that he couldn't immediately find the right piece of paper with the proof on it, but he later sent Halley two different geometrical proofs of his discovery. Halley saw the tremendous importance of Newton's ideas and induced him to write a book about them. Many scientists agree that the book that followed is the greatest scientific work ever written.

Newton's *Principia* was written in Latin and appeared in 1687. The full title of the masterpiece is *Philosophiae naturalis principia mathematica* ("The Mathematical Principles of Natural Philosophy").

In it he states the three important principles which have been known ever since as Newton's three laws of motion. It is impossible to overestimate the importance of these laws or to imagine how difficult the science of mechanics would be without them.

LAW I

Every body continues in its state of rest, or of uniform motion in a straight line, unless it is compelled to change that state by an unbalanced force impressed upon it.

LAW II

The change of motion of a body is proportional to the unbalanced force acting upon it, and that change in motion (acceleration) is made in the direction of the straight line in which that force is impressed.

LAW III

For every action, there is an equal and opposite reaction.

Newton's first law is sometimes called the principle of inertia. We can understand what it means with the help of an example. Imagine that you are in a railroad car that has stopped at a station. If you toss a ball straight up in the air it will fall back into your hand. Now suppose the train is moving with uniform velocity. If you toss the ball straight up in the air it will once again fall straight down into your hand. Before the toss is made, the ball and every other object on the train is moving forward with the same uniform motion as the train. In accordance with the first law, the ball retains that forward motion until a new force is impressed upon it. Your vertical toss did not affect the horizontal motion of the ball, so it maintained enough forward motion to keep up with your moving hand and the train.

Newton illustrated the principal of inertia by the motions of projectiles. If a projectile were shot horizontally from the mountain top, as shown in Figure 38, it would move in a straight line *AF* were it not for gravity. But, in reality, its weight causes it to describe a

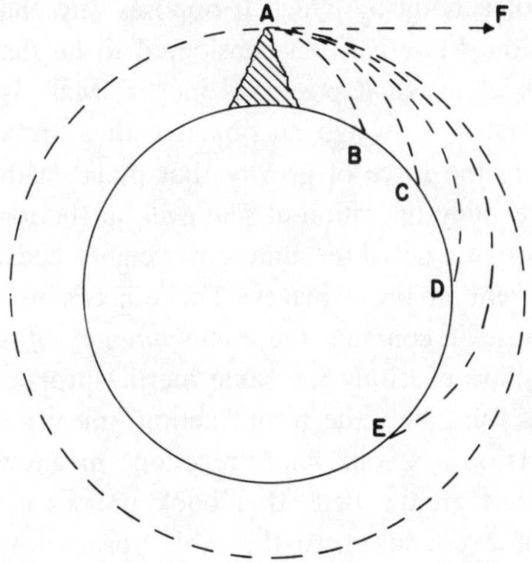

Fig. 38. Newton explains that a projectile becomes a satellite when fired horizontally from a high place with sufficient speed. In the absence of gravity the projectile would move along the line *AF*.

curved motion and strike the earth at *B*. The greater the velocity with which it is projected, the farther it goes before it falls to the earth. With increased velocity it may reach to *C* or *D*, or *E*. Finally, "exceeding the limits of the Earth, it should pass quite by without touching it." The motion of the projectile is maintained in its orbit because of its inertia (Law 1) and the pull of gravity.

The first law has an important implication. Whether an object is at rest or moving with uniform velocity along a straight line, its acceleration in either case is zero. The law means, therefore, that a body will not have an acceleration unless an unbalanced force acts upon it.

The second law tells us that the presence of an unbalanced force acting upon an object gives it an acceleration, or change in velocity, in the direction of the force. Of the three laws, the second is the only one that can be expressed as a mathematical equation:

$$\text{Force} = \text{mass} \times \text{acceleration}$$

The term "mass" is sometimes defined as a measure of the quantity of matter in an object. It can best be defined in terms of inertia. Inertia is a property of a body by which it opposes any change in its state of rest or motion. Mass is often considered to be that property of a body by virtue of which it possesses inertia. Mass is quite different from weight. When we weigh an object with a spring scale we are really measuring the force of gravity that pulls on the object. If we perform the weighing operation at sea level, at the top of a mountain or on the moon we get three different weights because the pull of gravity is different in those places. The object's mass is the same, however, because it contains the *same amount of matter* in those places and displays precisely the same inertial properties.

In Newton's third law, the term "action" means the force which one body exerts on a second, and "reaction" means the force which the second exerts on the first. If a book presses down on a table with a force of 2 pounds, then the table pushes back on the book with a force of 2 pounds. Notice that two objects must be involved before a force can be exerted. An axe striking a tree can exert only as much force on the tree as the tree is able to exert on the axe. Two objects are always involved in the application of a force and the action and reaction are never exerted on the same body. It is also important to remember that action and reaction can never balance each other out even though they are equal and opposite. In order for two equal and opposite forces to negate each other (cancel out), they must be exerted on the same object.

Newton's third law accounts for the operation of a rocket engine in the vacuum of outer space. The engine ejects gases at high speed toward the rear by exerting a force on them. An equal and opposite force pushes the engine in the opposite direction. Although these forces are equal and opposite they are quite different from the forces acting on a book at rest on a table. Gravity pulls the book down, and the table pushes it up. These two forces are equal and opposite, but both act on the same object. There is no unbalanced force acting on the book and it remains at rest in accordance with Newton's first law.

Universal Gravitation

Newton's calculations concerning the "fall of the moon" convinced him that the force of gravitation obeys an inverse square law. The force of attraction between two bodies is inversely proportional to the square of the distance between their centers. In mathematical terms, $F \propto \frac{1}{d^2}$ where F is the force and d is the distance. The sign $\propto$ means "proportional to." Although the inverse square law was known or suspected by others, the next step was entirely Newton's.

Because of Galileo's law of falling bodies Newton knew that a 2 pound mass and a 1 pound mass fall at the same rate. Furthermore, the acceleration of the masses is precisely the same: 32.2 feet per second per second. But Newton's second law, force = mass × acceleration, tells us that the forces on the two masses must be different in order that both achieve the same acceleration. Let's illustrate that fact by substituting numbers in the equation:

(1) Force on 1 lb. mass = 1 pound × 32.2 feet/sec/sec
(2) Force on 2 lb. mass = 2 pound × 32.2 feet/sec/sec

Clearly, the force of gravity on the 2 pound mass is twice as great as that on the 1 pound mass. In the general case, the force is *directly proportional* to the mass of the object. But what about the mass of the earth? Does the mass of the earth have anything to do with the force of gravity? The sun pulls on the planets which are smaller than the sun. The earth pulls on the moon which is smaller than the earth. And the earth's pull on ordinary objects is directly proportional to the mass on which it acts. Perhaps the pulling power of the sun and the planets and the earth is connected with their own masses as well as the mass of an object being attracted. If so, every piece of matter must exert a force on every other piece of matter in the entire universe. Newton concluded that the force of attraction between two bodies depends on both of their masses. In mathematical language, the force is directly proportional to the product of the two masses:

$$\text{Force of attraction} \propto Mm$$

where M and m are the two masses. Combining this law with the inverse square law Newton arrived at the general law of universal gravitation:

$$F = \frac{G\,Mm}{r^2}$$

where F is the gravitational attraction between two masses M and m whose centers are separated by a distance r. The G in the equation is called the "gravitational constant" and is quite different from the acceleration caused by gravity which is g. G is the same number in all circumstances—whether in calculating the force of attraction between two pebbles on the beach, between a stone and the earth, or between the sun and Saturn. It is called a "universal" constant because it seems to apply everywhere in the universe. Another universal constant is c, the speed of light.

The law of universal gravitation was a master-stroke of a great genius. With a single simple law Newton could explain the forces that keep the planets in elliptical orbits around the sun, hold satellites in elliptical orbits around the planets, cause objects to fall as observed, and hold objects on the earth. The discovery of universal gravitation was possibly the most supreme scientific achievement of the human mind.

Measuring Seconds with a Pendulum

A few examples will underscore the magnitude of Newton's accomplishment in discovering the law of gravitation. In Newton's day it was known that the period of time required for Galileo's pendulum to swing back and forth was not exactly constant for small and large amplitudes of swing. But the error was negligible if the amplitude was kept small. The equation for the period T (in seconds) of a common pendulum swinging through a short arc is

$$T = 2\pi\sqrt{\frac{l}{g}}$$

where l is the length in feet between the point of support to the center of the weight, and g is the acceleration of free fall—approximately 32.2 feet/sec/sec. If we substitute 2 seconds for T in the pendulum equation, we find the length to be about 3 feet 3 inches. A pendulum of that length takes 2 seconds to complete its back and forth motion, or just 1 second to swing from one side to the other. For that reason, it was called a "seconds pendulum" in Newton's day.

The Pendulum Clock

Shortly after Galileo's death, his pupil Mersenne (1644) determined the length of a seconds pendulum with great care. A little later, in 1645, Huygens made the first practical pendulum clock with an escapement capable of withstanding the rocking of a ship. It appeared, at last, that the age-old problem of carrying Greenwich time aboard ship had been solved. Perhaps longitude determination would soon cease to be a problem.

This was not to be. The equation for the period of a pendulum contains the number g, and that number is not exactly the same for every place on the earth's surface. A stone falls a trifle faster at the poles than at the equator. In an expedition of 1672, Richer measured the length of a seconds pendulum at Cayenne in French Guiana and found it to be about one-tenth of an inch shorter than a seconds pendulum in Paris. Halley also found that when he went from London to Saint Helena, it was necessary to shorten the pendulum so that it would continue to beat seconds. In 1673, Huygens published the theory of the pendulum clock in his *Horologium Oscillatorium* and explained the discrepancy in length of a seconds pendulum. Two causes are responsible for the lack of accuracy of a pendulum clock: the rotation of the earth and the flattening of the earth at the poles. The first cause has to do with centrifugal force. Imagine yourself on a merry-go-round traveling at high speed. A centrifugal force tends to throw you out, away from the center. As you move toward the center of rotation, however, the centrifugal force grows smaller. If you could confine your body to the precise center of rotation, there would be no outward force on you at all. The same sort of effect takes place on the earth, but to a much smaller degree, so small that only precise measurement can detect it. Referring to Figure 39, any point on the earth's surface makes one complete rotation on its north-south

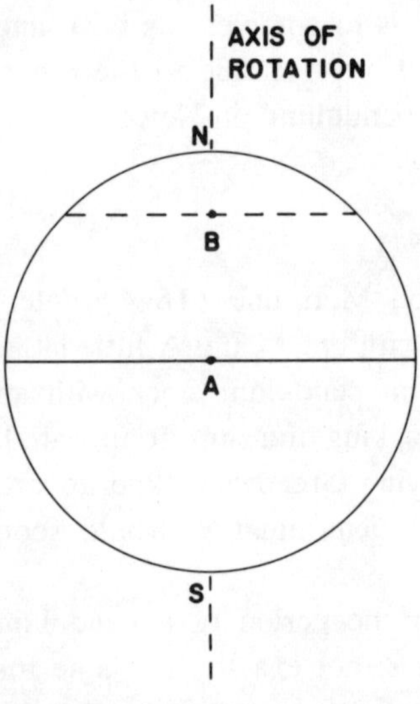

Fig. 39. Centrifugal force caused by the earth's rotation on its axis is greater on the equator (point *A*) than on a small circle (point *B*).

axis in 24 hours. A point such as *A*, on the equator, moves about 25,000 miles in that interval of time, so its speed is a little over 1000 miles per hour. A point such as *B* moves much more slowly, however. Because it is closer to the axis of rotation, it travels a much shorter distance in 24 hours. It is as though the point had moved closer to the center of rotation of the merry-go-round. Because point *B* moves more slowly than point *A*, the centrifugal force is lower at point *B*. At the north and south poles, the centrifugal force is zero.

At places such as *A* and *B*, the force of gravity must accomplish two things: It must pull a body toward the center of the earth; and it must overcome centrifugal force caused by the earth's rotation. The latter subtracts from its ability to do the former. And because centrifugal force is a variable quantity, depending upon latitude, the acceleration of free fall, *g*, also varies over the earth.

Huygens also proved that a second factor, the shape of the earth, influenced the numerical value of *g*. Using Newton's mechanics, he proved that the earth must be an oblate spheroid, a sphere that is slightly flattened at the poles. This would place a clock at the north pole slightly closer to the center of the earth than one at the equator, with a corresponding effect on *g*. The following table shows how *g* varies with latitude.

Latitude	Acceleration of Free Fall (feet/sec/sec)
0° (equator)	32.09
20°	32.11
40°	32.16
60°	32.22
90° (Pole)	32.26

Although the variation is slight, the accumulated error is too great to permit use of the pendulum clock in navigation without making suitable corrections. A pendulum clock runs about 100 seconds per day slower in the Caribbean than in Greenwich, England. Modern chronometers, by comparison, keep time with an accuracy of better than one second per day. Such a chronometer, based upon modern principles, was invented in 1765 by Pierre Le Roy of Paris. Le Roy's invention made use of new knowledge about the motion of springs.

The Law of Springs

Robert Hooke (1635–1703) was professor of mathematics and mechanics at Gresham College, a forerunner of the London University. We have already seen that he was convinced of the validity of the inverse square law of gravitation but was unable to connect it with the planetary motions. He felt, perhaps rightly, that he should have received more recognition from Newton for his part in the discoveries of the great man. This led to a forgivable desire to establish the priority of his discoveries. Soon after discovering his law of the spring, he issued it in the form of a short anagram, ceiiinosssttuv.

When the letters are unscrambled they read *ut tensio sic vis,* "as the extension, so the force." Translated into everyday language it means that the pull of a stretched spring is proportional to the distance it is stretched. It's easy to remember the law if you recall the appearance of an ordinary spring scale, with its uniform spacing between the weight markings. If a 1 pound weight stretches a spring ½ inch, a 2 pound weight will stretch it 1 inch, a 3 pound weight 1-½ inches, and so on. The stretch is just ½ inch per pound, unless we stretch the spring too far and destroy it.

The Oscillating Flywheel

If we attach a suitable weight to a spring, we can make it behave something like a pendulum, with the weight bobbing up and down about its rest position. Like the pendulum, however, the period of oscillation of a weight and spring depends on the number g. This makes the two methods equally unsuitable for use in clocks. But a spring and mass can be used to oscillate in a different manner that is independent of the pull of gravity. This can be done, for example, with a spirally wound spring and a small flywheel. One end of the spring is fixed and the other is attached to the flywheel. When the flywheel is rotated somewhat and released, it oscillates back and forth under the alternating force of the spring. This general principle is made use of in all mechanical clocks and watches made today. The flywheel and spring are independent of gravity because spring tension and the *inertia* of the flywheel, rather than its weight, determine the period of oscillation. The rotation of a perfectly balanced flywheel cannot be affected by the pull of gravity.

Like Huygens, Hooke tried to make a perfect marine clock. He failed because the expansion of metals by heat was not sufficiently understood at the time. To keep accurate time, the spring must exert precisely the same force on the flywheel over the temperature range normally encountered. Because it expands and contracts as the temperature changes, a spring tends to exert somewhat greater or smaller forces at different temperatures. This, in turn, causes the clock to run slow or fast depending upon the temperature. The problem was solved some seventy-five years after Hooke's discovery, by the compensating mechanisms illustrated in Figure 40. The flywheel consists of two circular segments fastened together as shown. These segments are made of two metals bonded together. The outer strip expands and contracts more than the inner strip under the influence of temperature changes. The circular bi-metallic strips curve in toward the center as the temperature increases. This reduces the inertia of the flywheel thereby compensating for the expansion of the spring. The

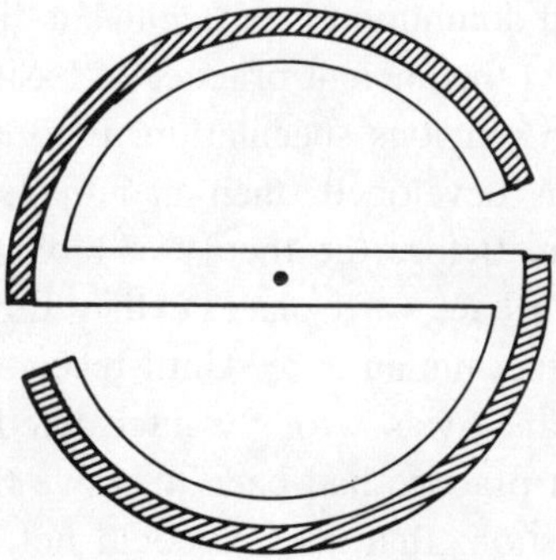

Fig. 40. Bimetallic compensation was the key to the accurate measurement of time. The flywheel is made of two metals, as shown above. As the temperature changes, the free ends move toward or away from the center, thereby compensating for other changes in the oscillating system. This produces a frequency of oscillation that is almost constant as the temperature varies.

combination is thereby able to maintain a constant period of oscillation over a wide temperature range.

The principles of Newtonian mechanics, through the application of inertia, had finally solved the problem of accurate timekeeping. Longitude determination became a simple matter even on long voyages. In the process of solving this problem, scientists had learned a great deal about the universe that is our home. The motions of the planets became known to an accuracy undreamed of by Copernicus or Ptolemy. Great improvements had been made to the telescope, permitting countless new celestial bodies to be observed. The laws of motion had been discovered and found to apply equally well throughout the universe. The *Principia* had established a mechanical model of the universe which is still valid today for the vast majority of scientific and everyday applications.

The full extent and revolutionary nature of the change that Galileo and Newton were working in men's minds was not apparent until the eighteenth century. The most dramatic element in that change was the conception of a physical or mechanical universe wholly independent of one's spiritual beliefs. Even Newton seems to have been unaware of that aspect of his work. He tells us that the regularities and stability of the solar system prove it "could only proceed from the counsel and dominion of an intelligent and powerful Being." Newton was repeating the ancient practice of basing all poorly understood phenomena on religious speculation. Before the theory of geological evolution was developed, men had the same attitude about the earth's formation. Before the theory of biological evolution (and even today in certain backward places), they thought the same way about the origin of the human race. Until recent times, almost every major scientific advance was wrongly interpreted as a blow against religion. The ancient practice had been to prove the existence of God by enumerating questions that science could not answer at the time. In retrospect, we can only conclude that the age-old practice of basing one's religious beliefs on an estimate of what science cannot do tomorrow was not only blasphemous, it was foolhardy.

Sound, Light and Waves

7

One of the great by-products of Newtonian mechanics was the invention of an accurate clock. Because man had learned to measure time with great accuracy, he had finally learned how to determine his location on the surface of the earth. The pendulum and the clock also brought with them the ability to measure time in small increments, and scientific inquiry spread to another phenomenon—sound. Side by side with sound, the nature of light aroused more and more interest. Several disturbing observations defied understanding on the basis of Newton's optics. It became clear that more was needed by way of explanation than could be accomplished by existing ideas.

Before we can understand sound and light, not to mention radio and the modern theories of the atom, we must come to grips with one of the most difficult concepts of classical science. Each of these phenomena depends upon the concept of *wave motion,* an idea that grew out of everyday life to explain the nature of sound and light.

Most of us are familiar with water waves. Drop a stone in a quiet pond, and circular disturbances go out from the center in ever widening circles. A cork floating on the pond bobs up and down as the crests and troughs move past it. Motion is certainly involved in the creation of water waves, but what kind of motion is it? Before we can answer that question, we must distinguish between two motions: the motion of the wavelike disturbances that move away from the stone, and the movements of individual water particles that give rise to the waves. Because no horizontal movements are imparted to the cork, the motion of the water can only be up and down. Water waves are called "transverse waves" because the water particles move at right angles, or transverse, to the direction of advance of the waves.

Water waves have always been familiar phenomena in man's everyday experiences. He was able to observe them carefully and discover their properties and peculiarities. In time he noticed that some of the unexplained properties of sound and light could be understood if the latter were assumed to move through space in the form of waves. But what have water waves to do with sound and light which show no visual indication of such motion? We can answer that question by reliving some of the thoughts and experiments that led to workable theories of sound.

Vibrating Strings

Sound was much more clearly understood in ancient times than light. Aristotle, whose views on physical phenomena were almost always wrong, correctly assumed that the sound from a vibrating string reaches our ears by movements of the intervening air. This was little more than conjecture, however, because exact measurements were rare in ancient times. Aside from Ptolemy's study of refraction, and Archimedes' work in mechanics, the laws of vibrating strings seem to be the only earthly subject that induced the Greeks to make scientific measurements.

The Pythagoreans are credited with having discovered the relation between the length of a vibrating string held at constant tension and the note emitted. When the length of the vibrating section is halved, the note produced is an octave higher. In music, two notes an octave apart bear a striking similarity to each other and are known by the same letter-name. Whether they are played simultaneously or sequentially, the effect is pleasant and harmonious. When the Pythagoreans reduced the vibrating portion to two-thirds of the whole string, it gave a note higher than the original by an interval that musicians call a "fifth." When the vibrating portion was reduced to three-quarters, a note called a "fourth" was produced, and so on. In the musical scale of *C*, the interval *CC*, which includes eight notes, is an octave. *CG* (five notes), is a fifth, and *CF* (four notes) is a fourth.

The Meaning of Pitch

The modern study of sound began with Galileo whose discoveries concerning the oscillations of the pendulum led him to investigate the vibrations of strings. His major contribution was the discovery that the pitch of a note depends upon the rate of vibration of the sound-producing object; that is, upon the number of vibrations occurring in a unit of time. He discovered that important relationship with the help of a simple but ingenious experiment that anyone can repeat. He moved a sharp or pointed piece of iron across a plate of brass so that a distinct note was produced. Examining the marks on the brass, he noticed a series of fine scratches on its surface rather than a single continuous scratch. The fine lines were an equal distance apart as though the sharp point of the instrument had come into contact with the brass at precisely regular intervals as the note was sounded. When he produced a high note, by a quick motion, he noticed that the lines were close together. When the note was lower in pitch, the lines were farther apart. Galileo concluded that each fine scratch corresponds to one vibration or oscillation of the iron instrument. These vibrations must be the cause of the sounds he heard. By counting the number of scratches made in a unit of time, he was able to associate the pitch of a note with the number of vibrations per unit of time necessary to produce it. With experiments of that kind, Galileo discovered the relation between the rate of vibration for a certain reference note (called the fundamental note) and the fourth, the fifth and the octave above it. He found these rates of vibration to be in the ratio of $1:\frac{4}{3}:\frac{3}{2}:2$. These ratios are listed in the table below. Also listed are the corresponding ratios discovered by the Pythagoreans for the relative *lengths* of vibrating strings.

For the moment, we will merely point out the surprising fact that the product of each pair of numbers on the same line is equal to a constant number—one. The table tells us that we must halve the length of a string to produce a note an octave higher than the fundamental, but we must double the number of vibrations per second to produce the same interval. It is clear that these two properties of vibrating systems are related to each other in an inverse manner.

	Relative length by Pythagoreans	Relative number of vibrations per unit time by Galileo
Fundamental	1	1
Fourth	$\frac{3}{4}$	$\frac{4}{3}$
Fifth	$\frac{2}{3}$	$\frac{3}{2}$
Octave	$\frac{1}{2}$	2

Sound Waves

To understand the relationship between string lengths and frequency, let's consider how a vibrating device, such as a violin string, can produce a sound that travels through the air. The motion of a vibrating string is smooth and continuous, but it will help our understanding to consider four different aspects of *phases* of its motion. To begin with, imagine that time has stopped at the precise instant when the string is about to begin its upward motion, the position it occupies at rest. As shown in Figure 41(a), the string then lies in a straight line between the supports *A* and *B*. The dots above the string represent a few of the evenly spaced particles of air. In Figure 41(b), the string has reached a phase in its oscillation corresponding to its maximum upward excursion. The upward motion has compressed the air slightly above the string. This is spoken of as a "compression" of the air. In part (*c*), the string has returned to its neutral phase and the high-pressure disturbance has begun to move away from the string. In part (*d*), the string is in the lowest phase of its motion and the region just above the string has acquired a partial vacuum, or *rarefaction,* which is produced as follows. As the string moves down in part (*d*), many particles of air are swept out of the region above

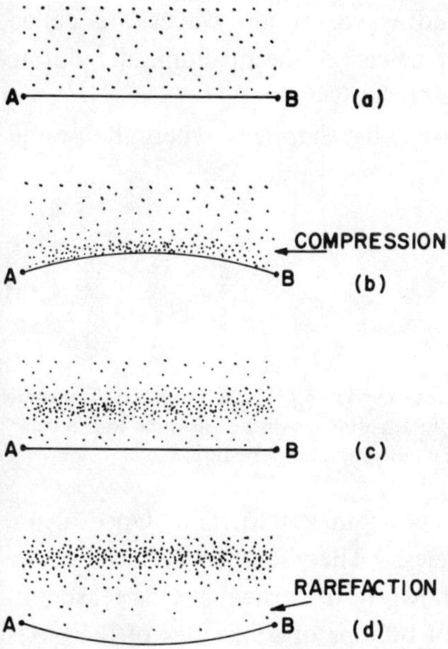

Fig. 41. A vibrating string produces pressure or density variation in the air: *(a)* the initial conditions; *(b)* the string moves up compressing the air above it; *(c)* the string returns to its neutral position but the compression moves upward at the speed of sound; *(d)* the string moves down leaving a rarefied region above it.

it and others cannot instantaneously move in to take their place. This effect produces a region having a relatively few particles of air. As the string moves through each succeeding phase of its vibration, it produces a continuous variation in pressure from a magnitude somewhat above normal atmospheric pressure to somewhat below normal. The same sequence of pressure variations occurs below the string, but these have been omitted for clarity.

Transverse and Longitudinal Waves

In the course of time, the pressure disturbances of a sound wave (compressions and rarefactions) are propagated from the string just as water waves move away from the stone that produced them. There is an important difference, however, between the two kinds of waves. Water waves are transverse waves; each particle of water moves alternately up and down, at right angles to the direction of wave motion. But if we could watch the particles of air as a sound wave passes by, we would see them move back and forth in the *same* direction (of motion) as that of the sound wave. Sound waves are called "*longitudinal*" waves because the particles of the medium, air, vibrate along the direction of propagation of the wave.

Figure 42 will help us visualize what happens when the sound

Fig. 42. A sound wave *(a)* consisting of four cycles of pressure variation. The same sound wave is shown in *(b)*, a short time later, after moving a distance to the right. Points marked C are compressions, regions of maximum pressure.

waves from a vibrating source move from left to right. Once again, the little dots stand for air particles. (There are really many more particles, of course, than can be shown in the drawing. One teaspoonful of air contains many billions of billions of molecules of the gases that constitute air.) What we see in Figure 42 are two "snapshots," the second (*b*) taken slightly after the first (*a*). The places marked *C* are the locations of maximum air pressure, the places where the air molecules are most closely packed together.

It is important to understand that each individual particle of air is restricted to a narrow range of back and forth movements. We should think of the wave motion as a rapid propagation of the pressure maxima of Figure 42 away from the source of sound. This is analogous to water waves which may travel thousands of miles although each individual water molecule never leaves its original vicinity.

Wavelength and Frequency

The wave motion illustrated in Figure 42 is similar to the motion of water waves, or to that of an oscillating flywheel in a clock, in that it is *periodic*. This is merely a way of saying that it repeats a given condition at regular intervals in the course of time. As the sound wave moves past a fixed point, that point experiences a regular sequence of pressure variations as successive compressions and rarefactions move by. Periodic repetition is a basic property of many kinds of wave motion.

Referring again to Figure 42, the distance in space between a pair of adjacent *C*'s (places of maximum compression) is called a "wavelength." It represents the portion of the wave that contains all of the pressure variations and corresponds to one pressure *cycle*. It is analogous to the distance betwen successive crests of a water wave. The points marked *C*, in fact, are also called "crests," just for that reason. The wavelength can also be measured between successive troughs or between any two corresponding places on successive cycles.

It is clear from Figure 42 that an ear fixed in space will "count" a certain number of crests in one second as the sound wave passes by. That number is called the "frequency" of the wave. It is this quantity, the number of oscillations or vibrations per second, that determines the subjective pitch of a tone. The higher the frequency, the higher the pitch.

With that background, we can understand how frequency and wavelength are related. Imagine a sound source vibrating at a frequency of 5 cycles (of pressure) per second. This frequency is much too low to be audible, but the same logic holds for sounds of any frequency at all. Figure 43 is a "snapshot" of such a sound showing 5 cycles or wavelengths of pressure variations. (The Greek letter λ, lambda, is almost always used to represent wavelength.) Because the frequency is 5 cps (cycles per second), precisely 5 cycles will move past any fixed place in one second. An ear at point *A*, for example, will "count" five wavelengths going by in one second. But a wavelength occupies a given length or distance in space; a distance that can be measured in feet, inches or any other unit of length. The length of one wavelength for a tone of 5 cps happens to be approximately 226 feet. The total length represented in Figure 43, therefore, is about 1,130 feet. The sound disturbance at point *B* must travel 1,130 feet in one second to reach point *A*. The velocity of the sound disturbances, therefore, is about 1130 feet per second. Experiment shows that all frequencies travel through air at that speed. In general, we

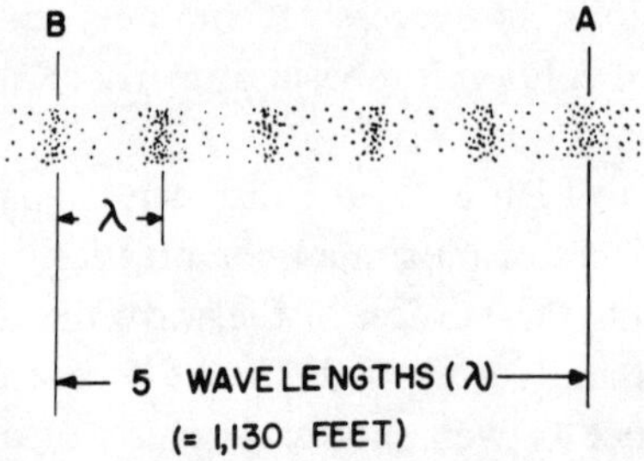

Fig. 43. Wavelength (λ) of a sound wave. One wavelength is the distance in space between two adjacent points of equal pressure. For the example above, five wavelengths occupy 1,130 feet.

can say that frequency, wavelength and the velocity of sound are related by the following simple law:

$$\text{wavelength} \times \text{frequency} = \text{velocity}$$
$$\lambda \times \mathrm{f} = \mathrm{c}$$

This explains why the product of the ratios on page 110 turned out to be equal to one. The wavelength produced by a vibrating string depends upon its length, if all other factors such as tension and string material are kept the same. Changing the length of the string changes the resulting wavelength (λ) in the air and produces a note of different frequency (f). But the velocity given in the equation above is the same for all frequencies (or wavelengths). So any change in λ must be compensated for by a change in f such that their product (c) does not change. λ and f are said to be in inverse proportion to one another. As one goes up, the other must go down in the same ratio. The products of the Pythagorean lengths and the Galilean frequencies were equal to one rather than c, because they were in the form of ratios rather than absolute frequencies.

Overtones

Galileo's student Mersenne discovered that a vibrating string often produces more than one note at the same time. The fundamental tone is, of course, the most obvious when the string is first plucked, but after it begins to die out it is easy to hear notes of higher pitch which continue a little longer than the fundamental note. René Descartes (1596–1650) suggested that these harmonics or overtones might be caused by various portions of the string that vibrate on their

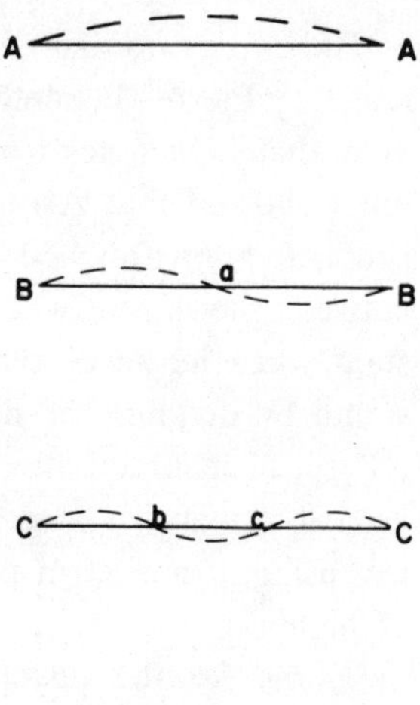

Fig. 44. Vibrating strings and harmonics. String *AA* is vibrating at its lowest or fundamental frequency. *BB* and *CC* are vibrating at twice and three times the fundamental frequency respectively. *DD* is vibrating at both the fundamental frequency and the third harmonic.

own account. In 1700, Joseph Sauveur (1653–1716) proved experimentally that a vibrating string can produce a number of related tones simultaneously.

Sauveur showed that a string may oscillate along its whole length, as in *AA* of Figure 44, or it may oscillate along parts of the string which move in opposite directions, as in *BB* and *CC*. In the last two examples, the moving portions of the string are separated by the *nodes a, b* and *c*—points where the string is at rest. In *AA*, the string produces its fundamental frequency; in *BB* and *CC*, it produces higher frequencies called overtones or harmonics. The second harmonic of the sound wave, *BB*, has twice the frequency and half of the wavelength of the fundamental, *AA*. The third harmonic, *CC*, has three times the frequency and one-third the wavelength. The frequency and wavelength of any succeeding harmonic are related in precisely that way. Sauveur also showed that two types of vibration can exist simultaneously, as in *DD*. In fact, that kind of oscillation is the usual sort except where special precautions are taken to prevent the occurrence of harmonics.

Measuring the Speed of Sound

The earliest measurements of the speed of sound through air were made by Pierre Gassendi (1592–1655). He wanted to test Aristotle's view that high notes travel faster than low notes. Gassendi's experiment showed that Aristotle was wrong again. He fired a cannon to produce a low-pitched sound and measured the time that elapsed between the instant when a distant observer saw the flash and the instant when he heard the explosion. He then calculated the speed of sound by dividing the distance between cannon and observer by the difference in time between the flash and explosion. Time was measured by means of a seconds pendulum. Gassendi then repeated the test using a musket in place of the cannon in order to obtain a note of higher pitch. His measured velocity, 1473 feet per second, was the same for the pistol and cannon shots. Mersenne repeated the experiment and measured 1380 feet per second. The modern figure is 1130 feet per second at room temperature. In 1740, Bianconi made measurements in summer and winter and found that sound travels faster through warm air than through cold air.

The Medium of Sound

Since the time of Aristotle, it had been generally accepted that sound travels through the air. Nevertheless, a few scientists of the seventeenth century believed that the presence of air might not be necessary to account for the propagation of sound waves. With his invention of the air pump, however, Otto von Guericke (1602–86) proved that sound does travel through the air. He suspended a bell by a thread in a glass container or "bell jar." The air was then pumped out of the enclosure and a clockwork mechanism within the container caused the bell to sound. He noticed that the sound of the bell became weaker and weaker as the air was pumped out. As air was readmitted into the container, the sound grew louder and louder. These experiments were the first proof that air is the medium through which sound propagates to the ear. Guericke also proved that sound can be transmitted through water and solid materials. Joseph Priestly (1733–1804) later showed that sound travels through gases other than air. The table below gives the approximate speed of sound in a few everyday substances.

Material	Velocity of sound (feet per second)
Air	1,130
Helium	3,230
Hydrogen	4,290
Kerosene	4,350
Water	4,920
Brass	15,400
Iron	19,600
Aluminum	21,100

The Most Simple Sound

We have described sound waves as periodically varying pressure disturbances that move through the air. But what sort of disturbances are they? Putting the question another way, do all sounds produce the same pattern of pressure changes? The answer is "no"; the exact pattern of pressure variation depends on the exact nature of the sound. A pistol shot produces a pressure impulse of short duration that moves away from the pistol at the speed of sound. This moving sound has regions of undisturbed air ahead of and behind it. As it turns out, this is one of the more complicated sounds to investigate. Equally complex are the sounds of a symphony orchestra or the sounds of speech. But all sounds—no matter how complex—have one thing in common: All are merely mixtures of much more simple sounds.

The building blocks of which all sounds are constructed are called "sinusoidal waves," or sine waves for short. "Sinusoidal" is an adjective meaning "sine-like," so once again the trigonometry of Hipparchus comes to the aid of a new science. As we shall see, the sine wave represents the simplest kind of periodic motion found in nature. It is of great fundamental importance to all of the exact sciences.

Before we can understand what a sine wave is or how it is related to sound, light, radio waves, or a fast-moving electron, it will be helpful to understand what a sine wave is *not*. First of all, a sine wave is not a physical thing. It is merely a mathematical or graphical *representation* of some changing physical quantity. It is what science likes to think of as a mathematical abstraction. Looking at a wiggly sinusoidal curve on a piece of paper cannot tell us what a sound wave looks like; it merely tells us, for example, how the pressure at a fixed place changes in the course of time as a sound wave passes by.

To illustrate, imagine a string vibrating at 100 cycles per second. All of the harmonics are suppressed in some way so that a pure tone of that frequency is produced. This is the kind of vibration shown in *AA* of Figure 44. Now let's suppose that a pressure-recording instrument is located a few feet from the string. The pattern of pressure changes measured by the instrument would look like the sine wave drawn in Figure 45. The straight horizontal line represents the normal atmospheric pressure. Whenever the sine wave is above that line the pressure is higher than normal; when it is below the line, the pressure is lower than normal. The distance above or below the straight line at any instant tells us the extent to which the pressure is different from normal atmospheric pressure. The distance, in other words, is proportional to the pressure variation caused by the wave.

The intersection of the two straight lines is called the "origin" of the graph and it represents the initial conditions at the instant when the sound first reaches the instrument. The passage of time is represented graphically by the distance to the right of the origin. Starting at the origin, the pressure rises to a maximum at *A* (a compression) and then falls back to the starting pressure at *B*. Then it drops to a minimum pressure at *C* (a rarefaction) followed by a rise once again to atmospheric pressure at *D*. These pressure variations constitute one cycle of the vibrating string as was discussed in connection with Figure 41. Because the frequency of the sound wave is 100 cps, it takes 1/100 second for one cycle of pressure variations to move past the instrument. The graph shows only two pressure

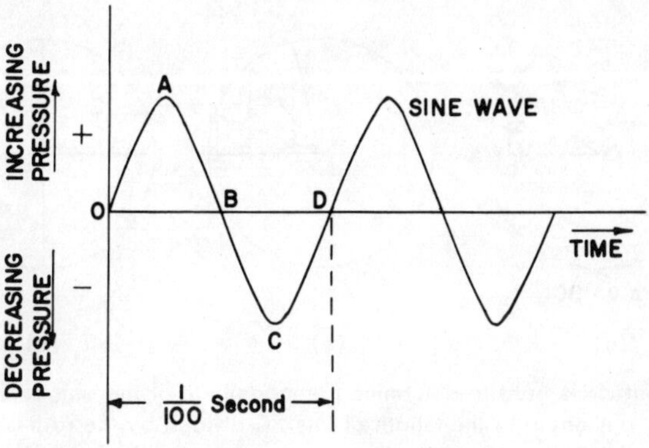

Fig. 45. Pressure variations of a 100 cycle-per-second sound wave. The pressure variations are recorded at a fixed point in the course of time as the wave passes by.

cycles, although other cycles will follow in time to the right on the graph as long as the string vibrates.

The shape of the sine wave does not change in the course of time, but the magnitude of its pressure swings depends upon the intensity

Fig. 46. Representation of a sound wave that increases in intensity at time T.

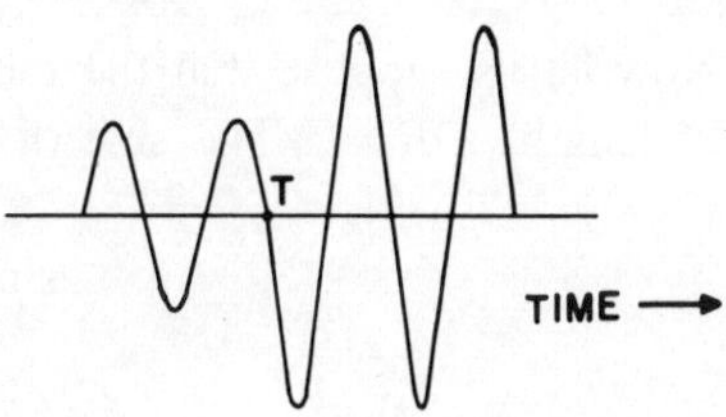

of the sound. Figure 46 shows how a sine wave might represent a sound whose intensity increases suddenly.

The Nature of Sine Waves

Thus far, we have discussed sine waves in general terms. Now we are ready to understand where the idea came from, and why they are called sine waves. Figure 47(a) shows a circle drawn about its center *O*. Lines *OA* and *OB* are equal to the radius of the circle, and form an angle *a* with each other. If we draw a line from *B* perpendicular to *OA*, we have a right triangle *OBC* which contains the angle

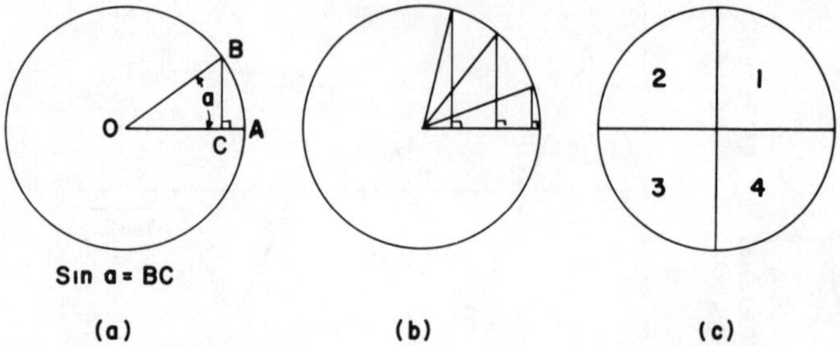

Fig. 47. A unit circle used to determine the magnitude of the sine of an angle. The sine of angle *a* is equal to the length of line *BC* divided by the radius of the circle, *OB*. Since the latter is unity, then sin *a* is equal to the length *BC*. That length depends upon the magnitude of the angle as shown in *(b)*. In *(c)*, the circle is divided in four equal parts or quadrants.

a. You will recall from Chapter 3 that the sine of angle *a* is equal to the side opposite the angle divided by the hypotenuse.

$$\sin a = \frac{BC}{OB}$$

Now let us suppose that the radius of the circle is equal to one unit of length: $OB = 1$. The sine of *a* then becomes

$$\sin a = \frac{BC}{1}$$

$$\text{or, } \sin a = BC$$

That is, the sine of *a* can be represented by the length of the line *BC* in a circle of unit radius. As shown in part (*b*), the corresponding length can be used as a convenient relative measure of sin *a* as the angle varies from 0° to 90°. To apply this method to angles larger than 90°, it is necessary to divide the circle into four portions or "quadrants" as shown in part (*c*). We will now examine the angles in the second, third and fourth quadrants in turn.

In Figure 48(*a*) the angle *b* has a magnitude somewhere between

90° (right angle) and 180° (a straight line). As shown in the diagram, it is the angle between *DO* and *OA*. By analogy with what was done in the first quadrant, we can define the sine of angle *b* as the perpendicular line *DE*. (Sin *b*, therefore, equals sin *c*.) We can now represent the sine of any angle between 0° and 180° as the length of an appropriate line in our unit circle. The sine of an angle as great as 180° has little meaning, of course, in connection with ordinary triangles. But we are redefining the concept of the sine of an angle in order to do a completely different job.

Figure 48(*b*) shows how angles between 180° and 270° are handled. The sine of angle *d* is equal to the sine of angle *e* which is

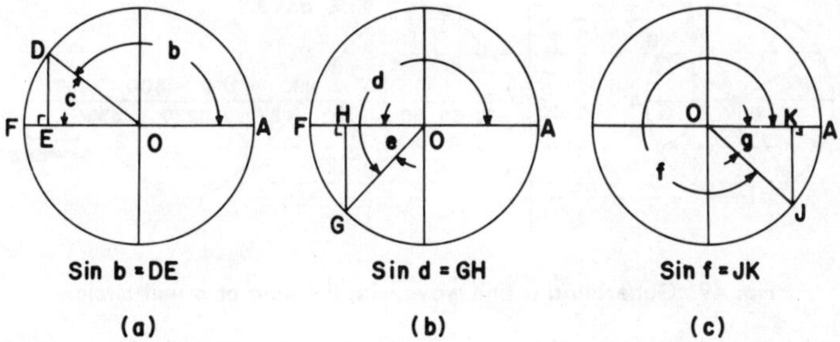

Fig. 48. Unit circles defining the sine of an angle in the second, third and fourth quadrants.

equal to the line *GH*. Similarly, the sine of angle *f* in Figure 48*c* is equal to the line *JK*. This makes it possible to cover all angles up to 360°. By considering additional revolutions of the radius around the unit circle, the sine of any angle—no matter how great—can be represented by the length of a line. The sine of 360° + a°, for example, is the same as the sine of a°. The sines of such angles will proceed through precisely the same variations during each subsequent rotation around the circle. This regular repetition of line lengths is analogous to all kinds of simple oscillatory motion found in nature.

Figure 49 shows how a sine wave can be constructed graphically with the help of a unit circle. The horizontal line on the graph is marked off evenly in increments of 30°. The distance of the sine wave above or below the horizontal line at a particular angle is equal to the length of the line representing the sine of that angle in the unit circle.

To illustrate how the sine wave is drawn, consider the angle *BOA* which is drawn equal to 30°. The sine of that angle is line *BC*. The length *BC* is then used to draw the point on the graph above the 30° mark. Angle *DOA* is equal to 60°, and the sine of 60° is equal to length *DE*. This length is used to draw the point on the graph above the 60° mark. Angle *FOA*, 90°, has a sine equal to line *FO*, and that length appears above the 90° mark on the graph. In that way, a series of points is drawn corresponding to the sines of various angles. When many angles are plotted on the graph, a smooth curve can be drawn through all of the points. We call that curve a sine wave.

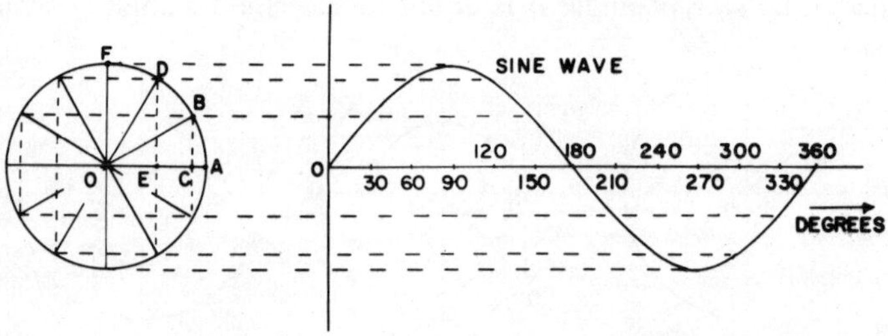

Fig. 49. Generating a sine wave with the help of a unit circle.

Sine waves are important because many periodic oscillations are sinusoidal in nature. Consider, for example, a weight hanging at the end of a spring. Let's define the zero position of the weight as the place it occupies when at rest. If we start the weight oscillating up and down, its displacement from the zero position in the course of time can be accurately described by a sine wave like that of Figure 49. Similarly, the pressure variations of a sound wave follow the same sine-wave pattern. In Figure 50, the upper part of the diagram gives us a satisfying pictorial representation of a sound, but the sinusoidal curve beneath tells us precisely how the pressure varies. In addition to sound waves and oscillating weights, sine waves can be used to describe radio waves, light waves, clock pendulums, oscillating flywheels, X-rays, gamma rays, alternating electric currents, and even moving electrons and other atomic particles. Wave motion is a fundamental concept that permeates all of science. It is quite impossible to understand science without a knowledge of waves.

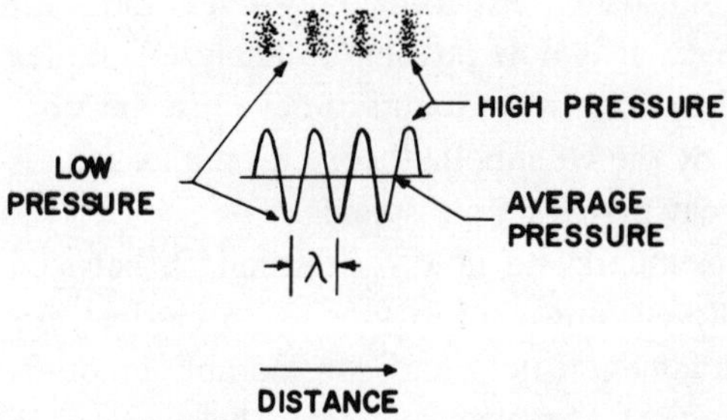

Fig. 50. Sine wave representation of a sound wave in space.

The Building Blocks of Sound

Scientists were shocked in the early part of the nineteenth century when the French mathematician Joseph Fourier (1758–1830) proved that any motion—including that of a vibrating musical instrument—can be represented as a sum of different sinusoidal motions. These sinusoidal motions have different frequencies, and different amounts of swing (amplitude), but each individual motion (called a "component"), can be represented as a sinusoidal curve. In the later nineteenth century, the German physicist Hermann von Helmholtz applied Fourier's ideas to sound. Sounds are produced by moving bodies, so the resulting sound waves ought to follow the same pattern of changes as the motions that caused them. To illustrate, imagine a machine gun firing periodically at a rate of five-hundred shots per second. Helmholtz tells us that the series of pressure impulses that reaches our ear is nothing more than a great number of simultaneous sine waves that add up to produce the complex pressure wave we hear. Each of these sine waves travels through the same air at the same time and at the same speed without affecting the others in any way. No matter how complicated the motion of a sound source may be, the resulting sound wave is merely a collection of many independent sinusoidal sound waves. Each behaves as if it were produced alone. Each travels through the air as though the others did not exist. And all add in the ear to produce the complex sound we hear.

This concept of the nature of complex sounds is more than a simple method of calculation or a convenient way of thinking about them.

Fourier's ideas have been illustrated experimentally over and over again in many fields of science. It is now possible to analyze complex sounds electronically, and they are always found to have just the sine-wave components predicted by the Helmholtz theory. Complex sounds are merely the sum of a great many simple sounds.

This theory enables us to understand how we distinguish between the same note played on different musical instruments. A clarinet, for examples, generates the fundamental note and considerable amounts of the odd-numbered harmonics. The even-numbered harmonics are much weaker, especially in the lower range of the instrument. A clarinet playing *G* below middle *C* has a fundamental frequency of 196 cps, as well as goodly amounts of $3 \times 196 = 588$ cps, $5 \times 196 = 980$ cps, and so on. When playing that note, the clarinet produces very little sound at frequencies of 2×196 cps or 4×196, etc. The flute, on the other hand, has a sound-wave recipe concentrated almost wholly in the fundamental and first two or three harmonics. Brass instruments such as the trombone and trumpet emit considerable amounts of sine waves at all frequencies up to the fifteen harmonic or higher. So it is the particular mixture of sine waves that gives each particular sound its distinctive auditory character.

Beats

When we listen to two simultaneous sounds that have nearly—but not exactly—the same frequency, we are aware of a sensation called "beats." The two sine waves blend into a single note having the frequency of the higher pitched sine wave. The amplitude or loudness of the combined note varies up and down in loudness at a slow rate. This rate is equal to the *difference* in frequency of the two sine waves. Suppose we are listening to two sounds having frequencies of 100 and 99 cps. We hear a single note whose sound level varies periodically at a rate of 100–99 or 1 cycle per second.

Joseph Sauveur explained beats as the periodic coincidence of the vibrations producing the two notes. To illustrate, consider the two upper sine waves in Figure 51. Each is produced independently by a different vibrating tuning fork, and the two sine waves are slightly different in frequency. Because they are different in frequency, the higher frequency note in (*b*) "catches up" with the lower frequency note in (*a*) periodically so that their maximum intensities coincide. Whenever that happens, the ear hears a relatively loud sound. At other times, a minimum of one note coincides with a minimum of the other and we hear a softer sound. As shown in (*c*), the two sounds add in the ear to produce the effect of a single note whose sound intensity fluctuates at a rate equal to the difference in frequency of the original sine waves. Those fluctuations in loudness are called "beats."

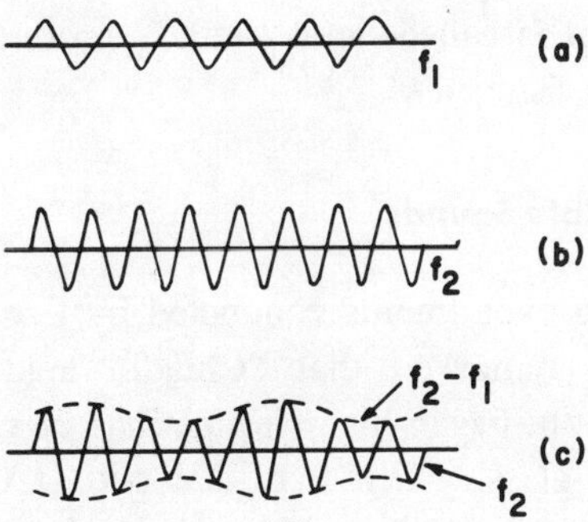

Fig. 51. The generation of audible beats when two sound waves of almost identical frequency enter the ear.

Most of the notes of the piano are produced by two or three strings, and each set of strings is tuned quite accurately to have the same frequency. This is done by adjusting the "beat frequency"—the difference in frequency of any two strings—to zero. Nevertheless, you can probably hear slow beats—about one or two per second—by striking a single key on the piano. The jangly sound that we associate with a barroom piano comes from beats that are too rapid for the ear to follow, or count.

Beats are most easily heard when the two sounds involved are relatively free of harmonics. It is easier, for example, to hear the beats produced by two flutes or softly blown pop bottles than those produced by a piano or two trumpets. But in any event, it is important to understand that beats are produced by the ear in the process of hearing. No portions of the vibrating systems that produce the sounds are vibrating at the beat frequency.

Combination-Tones

The beats studied by Sauveur are heard only when the notes are of nearly equal frequency. If the interval between the notes is increased gradually, the beat frequency soon becomes too great for the beats to be distinguished. If we go on increasing the interval, the difference in frequency between the two notes eventually reaches the range of human hearing and becomes audible. Like beats, this combination-tone, as it is called, is a "sound" produced by the hearing mechanism of the listener.

How to Make Visible Sounds

After witnessing experiments conducted by Ernst Chladni (1756–1827), Napoleon remarked that "Chladni made sounds visible." About 1785, Chladni began his investigations of vibrating flat plates, made of metal or glass, which were clamped at the center or edge. He made the plates vibrate by rubbing their edges with a violin bow. Then he hit upon the idea of sprinkling sand on the horizontally clamped plates. When the plates were bowed, he found that the sand shifted from the vibrating to the nonvibrating parts of the plates. The shifted sand formed patterns, or Chladni sound figures, which displayed the nodal lines of the plate. These lines are analogous to the nodal points of vibrating strings—points at which a string's motion is zero. By studying the sand patterns, Chladni was able to discover the way in which the various portions of a plate vibrate in order to produce a specific sound. A few of his sound figures are shown in the accompanying diagram. The first figure shows the sand pattern when the square plate is clamped at the center and bowed at a corner. As with the vibrating strings, the portions of the plate on opposite sides of a nodal line move in opposite directions at any particular instant. In the second figure, the bowing is done at the center of one of the sides instead of at a corner. The third figure shows the sand patterns when the plate is clamped at *A* and bowed at *B*. The portion of the plate within the circle moves in the opposite direction to the outer portion. Chladni produced and studied a great

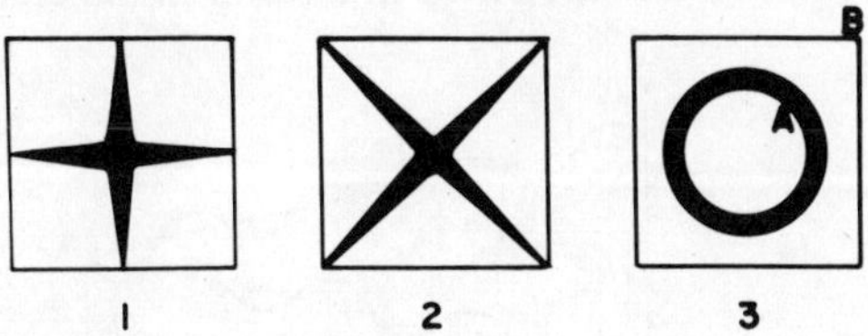

Fig. 52. Visible sound figures formed by placing sand on vibrating plates. Figure 1 is clamped at the center and bowed at a corner. In Figure 2 the bowing is at the center of a side. In Figure 3 the plate is clamped at A and bowed at *B*.

many of these figures and helped to clarify the different modes in which a plate or diaphragm can vibrate.

The Interference of Sound

Waves, whether of sound, light or water, are little affected by objects that are small compared to their wavelength. A water wave, for example, flows right by a small piling almost as though it weren't there. Ocean waves have a wavelength of many feet, which is considerably larger than the diameter of a pile sticking out of the water. Sound wavelengths are quite another matter. Some of them are large compared to everyday objects while others are small. The lower notes on the piano keyboard have wavelengths of thirty or forty feet while the high notes have wavelengths of only a few inches. While a low note passes around an ordinary object easily, a high note may cast a "shadow" behind the same object. Sound waves exhibit a mixed behavior as far as ordinary objects are concerned. Even a person's head may cast a shadow for high frequency notes. When he sings a low note, the sound level behind him is almost as great as in front of him, but if he whistles a high note the sound is considerably stronger in front. Similarly, when we walk around a corner from a hi-fi speaker, we continue to hear the low frequency notes but the high frequency notes are all but lost. Sound shadows of that kind make it difficult to study sounds in an ordinary room.

Another difficulty has to do with reflections. Sounds are reflected back and forth from walls and objects in a room, and the sound we hear is a complex mixture of sounds arriving from many paths. For these reasons, sound is often studied in a room made of special sound-

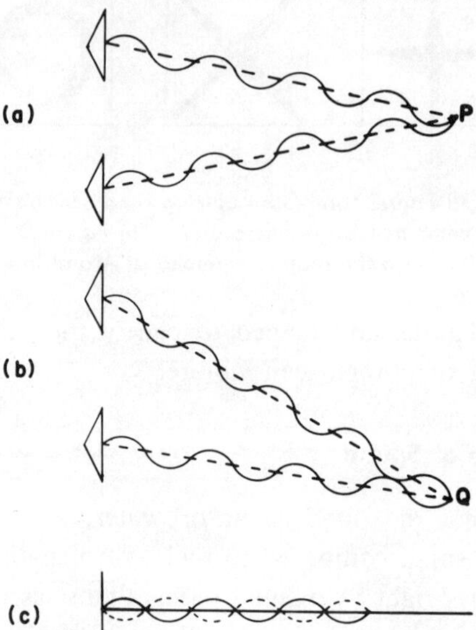

Fig. 53. The interference of sound. Two sound waves arrive at point *P* in phase or in step and reinforce each other to produce a sound that is louder than either one alone. In part *(b)* they arrive at point Q out of phase and cancel each other. Parts *(a)* and *(b)* are "snapshots" of the pressure variations in the space of interest. In part *(c)* the pressure variations are plotted as they would be received at point Q in the course of time. Because the two sound waves always produce equal but opposite pressure variations, their net effect is zero.

absorbing walls, a room free of echoes called an "anechoic" chamber. If we imagine ourselves in such a room we can perform a hypothetical experiment that illustrates the interference of sound.

Suppose our room contains two loudspeakers a few feet apart as shown in part (*a*) of Figure 53. Each loudspeaker produces sine waves of exactly the same frequency and both sine waves are precisely in step with each other. That is, the points of maximum pressure at a given instant occur at the same distance from the speaker for each sine wave. Such sine waves are said to be *in phase agreement.* A suitable instrument, such as a human ear, placed at point *P* would notice that the two sine waves arrive in phase (with one another). They started out in phase, and traveled the same distance to reach point *P*, so they arrive in phase. Because the ear receives two sine waves whose pressure variations are precisely in phase in the course of time, the effect is as though a single sound had arrived having twice the sound level. Turning either sound on or off merely changes the audible level of the sound.

Now let's consider what happens at a different point *Q* as in part (*b*) of the diagram. Point *Q* is a little nearer to one speaker than to the other. That difference in distance is equal to one-half wavelength, so a wave crest from one speaker will arrive at point *Q* just as a trough arrives from the other. In fact, the two sine waves will add up to zero at any instant. This cancellation is illustrated in part (c) which shows the two sine waves arriving at point *Q* in the course of time. Sine waves which oscillate out of step with each other are said to be "out of phase." The two sine waves interfere with each other at point *Q* such that no sound at all can be heard. At points intermediate between *P* and *Q*, the two sine waves are neither in phase agreement nor completely out of phase and the sound cancellation is only partial. As we move our ear from *P* toward *Q*, we hear a gradual reduction in sound until it disappears completely at *Q*.

If we continue beyond *Q*, the sound gets louder again as the two sine waves approach a second "in-phase" condition. At an appropriate point, the distances from our ear to each loudspeaker differ by precisely one wavelength. Under these conditions, the crests of the two sine waves arrive in phase once again and add to produce the loudest sound—just as they did at point *P*. If we move our ear still further in the same direction, we hear a successive variation in sound level as the two sine waves arrive alternately in phase and out of phase. This phenomenon is called the interference of sound. Interference is common to all kinds of wave motion and explaining it is a major success of wave theories in general.

The sounds we hear in an ordinary room are never completely cancelled by interference, but the direct and reflected sounds constitute an extremely complicated interference pattern. This pattern is different for each frequency because of the variation in wavelength. Some frequencies may reach the ear at a particular point and add to produce a loud sound, while others may cancel partially at the same point and produce a soft sound. We are accustomed to these effects in our everyday lives and they do not usually impair the intelligibility of speech or our enjoyment of music.

Resonance

There are many familiar examples of resonance in everyday things. A pendulum swinging at a constant rate is a resonant system. Other examples are a weight bobbing on a spring, a vibrating string, or the vibrating plates that produce Chladni's sound figures. These are all examples of mechanical resonance, a condition of a body when it is vibrating at a *natural frequency*.

We have hinted earlier that the frequency of a vibrating string depends upon such factors as tension and string material. Mersenne studied these factors quantitatively, and Brook Taylor (1685–1731) devised a formula that shows how they determine a vibrating string's natural frequency. In its modern form, Taylor's formula is as follows:

$$\text{frequency} = \frac{1}{2l}\sqrt{\frac{T}{M}}$$

where l is the length, T the tension or pull on the string, and M the mass-per-unit length of the string.

Taylor's formula gives the *fundamental* frequency of a vibrating string. As we have seen, a string can also vibrate at any multiple of that frequency—its overtones or harmonics. Similarly, any other mechanical system can sustain oscillations at a set of harmonically related frequencies determined by the system's dimensions and other physical properties.

Under suitable conditions, a volume of air can also be made to oscillate at its natural frequency. This is precisely what happens when we blow across the top of a milk bottle. We can compare this oscillation to the bobbing of a weight on a spring. Air has weight, and air has a certain "springiness" or compressibility. With these properties it can be made to oscillate in a manner analogous to the more familiar resonant systems.

Music versus Noise

Musical instruments are really complex resonators. Consider the wind instruments, for example. Depending upon their size and shape, they emphasize certain frequencies that are put into them by the mouthpiece. If we distinguish between the instrument and its mouthpiece, we can think of a kind of division of labor. The lips pressed over the mouthpiece produce many sounds, and the appropriate fundamental and harmonic tones are emphasized by the properties of the air column in the instrument. This is naturally a great simplification of the theory of musical instruments, but it helps us to understand how resonance fits into everyday things.

Because musical instruments depend upon resonance, the sounds they produce are related harmonically. At the opposite extreme is noise, a mixture of sine waves having random frequencies and amplitudes. As we all know, the world is full of all sorts of noises. When we hear a seemingly unrelated assortment of sounds, we call it noise. Indeed, the distinction between music and noise is sometimes hard to define. One man's noise may be another man's music. This confusion was illustrated recently by a British radio station. The announcer told his listeners that the piece to follow was a new composition by the noted young composer so-and-so. What followed was an ear-splitting conglomeration of noises in the tradition of a troop of Boy Scouts on a picnic. Various objects and pieces of furniture were banged or dropped during the performance to provide suitable emphasis. Strangely enough, two respected music critics liked the "composition." We can only conclude that music blends very slowly into noise as the frequency components become more random.

At the opposite end of the sound spectrum from pure sine waves, we have a "pure" confusion of sound called "gaussian noise" or "white noise." It is named after Karl Friedrich Gauss, who first studied the random nature of this noise. It is called "white" noise because it contains all frequencies of sound, just as white light contains all frequencies or colors of light. White noise has a sound like the *hiss* or *shh* we hear when a sensitive radio set is tuned between stations with its volume up full.

Decibels: Little Numbers for Big Things

The ear is capable of distinguishing sounds over an extremely great range of intensities. If we stand a few feet from a sound source having a power of about 100 watts, the sound level seems painfully high. Yet the average ear can still hear the sound if the power is reduced ten million-million times! Over this entire range of sound levels the ear can make useful comparisons concerning the intensities of different sounds.

The ear accomplishes this feat by sensing ratios of sound levels rather than differences in level. To illustrate, consider two experiments in which a sound level is changed. In the first, the sound power is increased from 1 watt to 2 watts, and in the second it is increased from 10 watts to 11 watts. In both examples the difference in power level is 1 watt, but the ear notices a much greater change in loudness for the 1 to 2 watt comparison. To get the same change in subjective loudness for the higher power, the level would have to go from 10 watts to about 20 watts. In other words, it is the *ratio* of two sound levels that determines the ear's subjective comparison of loudness.

This principle was discovered in 1834 by the German physicist E. H. Weber who extended it to include all of the senses. Weber's law tells us that a stimulus must be increased by a constant fraction of its intensity or magnitude to be barely noticeable as different. In 1860, Gustav Fechner gave us a more general rule relating sense perception to stimulus. He tells us that a sensation (such as apparent brightness, loudness, etc.) increases by a fixed amount whenever the stimulus (such as light or sound intensity) is increased by a constant factor. For instance, each time we double the sound intensity, we increase the subjective loudness by a fixed amount. Although neither law holds true exactly, they both emphasize the importance of *ratios* rather than *differences* of sound intensities or power levels.

We take advantage of this principle by using a special number system invented to describe ratios. If one sound is ten times as powerful as another, we say it is 10 decibels, or 10 db for short, more powerful than the first. If it is 100 times more powerful, it is said to be 20 db stronger than the first. Each time the sound is *multiplied* by ten, we *add* 10 db to thc decibel figure. A table of sound ratios and corresponding decibel numbers is given below:

Ratio of Intensities or Power Levels of Two Sounds	Decibels
1	0 db
10	10 db
100	20 db
1,000	30 db
10,000	40 db
100,000	50 db
1,000,000	60 db
1,000,000,000	90 db
1,000,000,000,000	120 db
0.1	−10 db
0.01	−20 db
0.001	−30 db
0.000,001	−60 db

The general rule for converting power or intensity ratios to decibels is as follows:

$$\text{number of decibels} = 10 \log_{10}\frac{P_1}{P_2}$$

where P_1 and P_2 are the two levels involved.

The faintest sound that we can hear has a sound intensity of about one ten-millionth of a billionth of a watt falling on an area of one square centimeter. (There are 2.54 centimeters in one inch.) The gentle rustle of leaves is a sound about 10 db greater. An average whisper has a level of about 20 db with respect to the faintest sound. The night noises of a city have a level of around 40 db. Our average department stores run about 60 db and ordinary speech varies between 60 and 70 db. Heavy traffic reaches a level of 85 db and a riveter across the street produces about 100 db. Hammering on a steel plate two feet away produces a level of 115 db.

This fantastic range of sound levels covers sounds varying in intensity by more than one-hundred billion times. The use of the decibel notation enables us to calculate and discuss such great ratios with numbers that are of convenient size. For the same reason, decibels have been carried over into the field of electricity and radio where similarly large ratios must be dealt with. Their use in connection with national budgets would not be without merit.

A Note about Waves

We have discussed periodic motion of fluids, solids and gases in this chapter and have noticed that wave theory offers an intuitively satisfying explanation of observed phenomena. We have seen that circular motion can be used to generate sinusoidal curves that are somehow related to the way that nature does things. In fact, sine waves seem to represent a simple kind of periodic motion that enables us to analyze and understand much more complex motions. In later chapters, we will find it necessary to apply the theory of waves to light, radio and matter in an attempt to grasp even more subtle methods of nature. If the concept of waves and their behavior is still new and illusive, a rereading of this chapter up to this point may be worthwhile. Much of science is concerned with the motion of energy and matter, and underlying both is the concept of waves.

The Importance of Size

The wave nature of sound was readily apparent in Newton's time. The vibratory motion of a sound source could be felt; there was a ready medium, air; the observation of interference fitted nicely with wave theory; the "bending of sound" into the shadow (called "diffraction") was analogous to the action of water waves; and the resonance of musical instruments could be explained and understood. With the exception of the medium of propagation, air, each of these effects depends upon the size of things. We have noticed that sound and water waves have lengths that are of about the same size as the things around us. This comparative equality was the key to an early understanding of sound. As long as the wavelength is comparable in size to the obstacles it meets, the effects of interference or diffraction are relatively large and easy to detect. It is not difficult to notice that sound bends around a corner.

An undersanding of light came with greater difficulty because, as we now know, the wavelengths of visible light are unbelievably small in comparison to ordinary things. A particular yellow light in the visible spectrum has a wavelength of about twenty-three millionths of an inch! This is the yellow light we see when table salt, sodium-chloride, is placed in a colorless gas flame. The frequency of this light is about five hundred and nine million million vibrations per second! Because the wavelength of light is so small, light casts a shadow which seems sharp on casual observation. It is not a simple matter to find evidence of the diffraction of light—the bending of light into the shadow. Similarly, one does not ordinarily connect everyday visual observations with the interference of light or with the effects of resonance. Nevertheless, Hooke gave a crude wave theory of light in 1664 and Newton also ascribed certain wave-like characteristics to light. Although Newton came close to developing the wave theory of light, he did not succeed in doing so—probably because it did not occur to him that the wavelength might be so exceedingly small.

The Diffraction of Light

Toward the middle of the seventeenth century, it was apparent to all scientists that light travels in straight lines or rays. The great successes of the new science of optics were ample evidence of the truth of the so-called *law of rectilinear propagation* of light. Human vision, the laws of perspective, lenses and mirrors all could be explained and understood on the basis of straight rays. Yet, there were disturbing questions left unanswered. Where do the colors come from in soap bubbles? Or why do we see colors in thin plates of mica or in a thin film of oil on quiet water? Even more disturbing was a brilliant discovery made by Francesco Grimaldi (1618–1663), a Jesuit professor of mathematics at Bologna.

In a book published after his death, in 1665, Grimaldi described a new property of light which he was unable to explain. He observed that when light passes through small holes or slits, or when it grazes narrow obstructions like needles and hairs, it produces unusual fringes of colored light where least expected. He also found that light *does* sometimes bend into the shadow. He called these effects *diffraction* to distinguish them from *refraction*.

Grimaldi experimented with a narrow beam of sunlight that entered a dark room through a small hole. He passed the beam through two small circular openings, or apertures, *AB* and *CD*, as shown in Figure 54. After passing through the apertures, the light illuminated

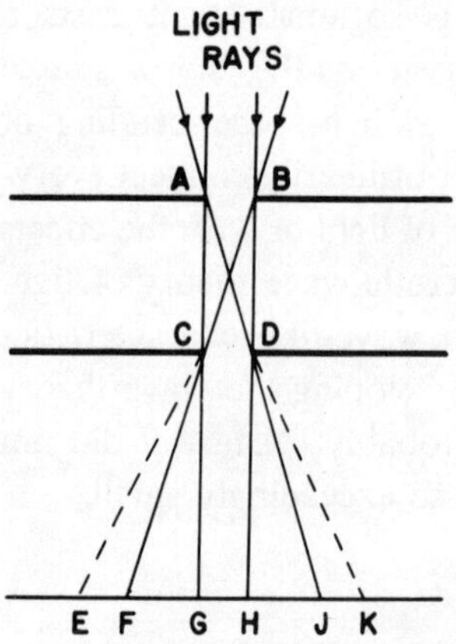

Fig. 54. The diffraction of light. The rays illuminate a greater region *(EK)* than can be explained by straight-line or rectilinear propagation of light *(FJ)*.

the screen at the bottom of the diagram. Grimaldi made careful measurements of the hole diameters and distances involved and concluded that the beam should illuminate an area from *F* to *J* in the diagram. A ray from *B* to *C* would reach as far to the left as point *F*, and a ray *AD* would reach point *J* on the right. If the law of rectilinear propagation were valid, all points beyond *F* and *J* should be in darkness. He found, however, that the beam of light produced the larger spot on the screen corresponding to the area between *E* and *K*. He concluded that light is diffracted or bent into the shadow, and this diffraction can be observed when careful measurements are made.

Grimaldi also experimented with tiny opaque bodies placed in the path of his beam. He found that the shadows cast by such bodies were bordered externally and internally by colored bands of light parallel to their edges. He accurately described these colored bands and pointed out that the color patterns were not the same as those produced by the refraction of light in lenses and prisms. He compared the colored bands to the waves caused by a stone dropping in the water, and supposed that light is a kind of fluid in a state of undulating or wave motion. He also suspected that light traveled at a great but finite speed.

In 1665, the same year Grimaldi's book was published, Hooke published his *Micrographia*. One of the more important subjects discussed was the iridescent colors of thin transparent films, such as flakes of mica, soap bubbles, blown glass, oil on water and mother-of-pearl. Hooke noticed that the rainbow colors in thin flakes of mica depend upon the thickness of its various portions. Variations in thickness lead to corresponding changes of color which recur in a regular order, much like a number of rainbows placed one above another. He was able to produce similar rainbow effects by pressing two flat plates of glass together with a thin air film between them. Although Hooke failed to find a relation between thickness and color, he came to the important conclusion that reflections from *both* surfaces of a film play a part in producing the colors. He also repeated Grimaldi's experiments and came to the conclusion that light is "nothing else but a peculiar motion of the parts of the luminous body" that produce motion in a fluid surrounding the body. Although diffraction was by now an experimental fact, there was still no adequate theory capable of explaining it.

Römer and the Speed of Light

Although most scientists of the time believed that light traveled at a finite speed, all early attempts to measure it met with failure. The first successful measurement of the speed of light was made in 1672 by the Danish scientist Olaus Römer (1644–1710) working at the Paris Observatory. Römer measured the period of revolution of one of Jupiter's moons very accurately and found the Jovian "month" to be about 42½ hours long. This was the time required for the innermost satellite to make one complete revolution around the planet. He also noticed that the length of the Jovian month was somewhat variable. He found that the satellite's period (of revolution) was somewhat longer when the earth was receding from Jupiter than when the two planets were approaching each other. We can understand why this should be so with the help of the accompanying diagram. Imagine Jupiter and the earth to be moving counterclockwise in their orbits as shown in Figure 55. Jupiter's moon is observed at a particular place in its orbit when the planets are in the places marked (1). The light that an observer sees at that instant must travel a certain distance from Jupiter's moon to earth. Now imagine that the moon

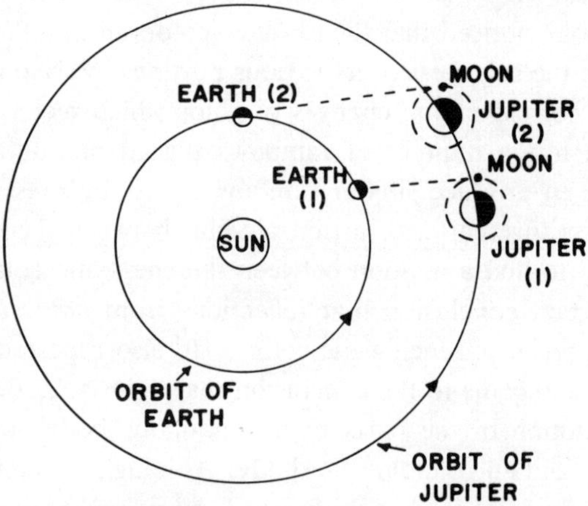

Fig. 55. Römer uses a moon of Jupiter to calculate the speed at which light propagates through space.

has completed one revolution around Jupiter. During this interval of time, the two planets have moved to new positions in their orbits marked (2). (The planetary movements are greatly exaggerated for clarity.) In order for light to pass from the moon to earth, it must now travel a somewhat longer distance than would have been necessary if the planets had not moved. The light that signals the completion of a Jovian month is delayed to a greater degree than the light that signals its beginning, because of the greater distance the former must travel. This makes the apparent period of the moon somewhat longer than the true figure. Similarly, the period is somewhat shorter than normal when the two planets approach each other.

On the basis of these observations, Römer calculated that the speed of light must be about 120,000 miles (or 193,120 kilometers) per second. The presently accepted figure is 299,792.5 kilometers per second. Römer also estimated that it would take eleven minutes for light to travel from the Sun to the earth. The correct figure is now put at about eight minutes and twenty seconds.

Wavelets of Light: A New Idea

Römer's discovery proved that light was not propagated instantaneously, but that it was some kind of process taking place at a finite rate. Attempts to explain refraction were usually based on the assumption that this process—whatever it might be—takes place at a different rate in different materials. Grimaldi's discovery of diffraction, and Hooke's work with iridescence seemed to imply that light involves a kind of wave motion. All of these ideas were molded into a brilliant theory by Huygens in his *Traité de la Lumière* ("A Treatise on Light") which was published in 1690. After almost three hundred years, Huygens' *Traité* is still highly recommended reading.

Huygens thought of a luminous body such as a candle flame as a vast number of minute particles (we would say molecules) which communicate impulses to the neighboring ether. Unlike sound, these impulses originate with each individual particle rather than with the luminous object as a whole. Because the motions of these particles are not related to each other, the many light impulses occur at random times. The "ether" through which they travel is not air—because light travels through a vacuum—but an all pervading medium made up of minute, hard, elastic particles. Each of these particles transmits any light impulses it receives to the particles it touches, but does not encounter any permanent displacement itself.

You can illustrate this idea by arranging a few similar coins in a straight line on a table. If you then slide another coin into one end of the line, the coin at the opposite end moves away from the pack. But all of the intermediate coins remain substantially in their original positions. From experiments of this kind, Huygens had learned that many such impulses could cross each other in more complicated

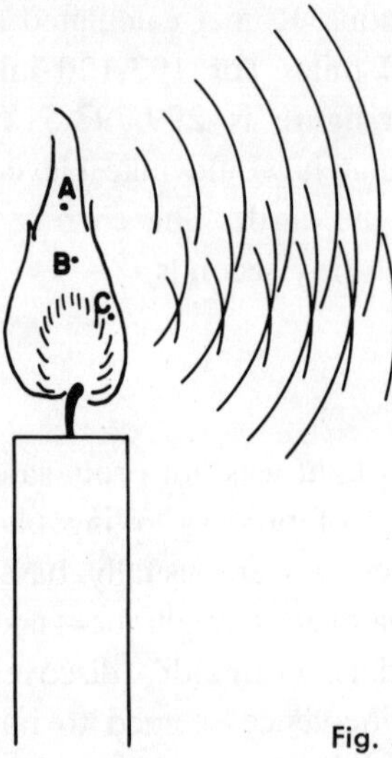

Fig. 56. Huygens' wavelets of light.

arrangements of particles without mutual interaction. Using these ideas, he developed a theory of light that explained most of its known properties.

As shown in Figure 56, Huygens imagined that each particle of a

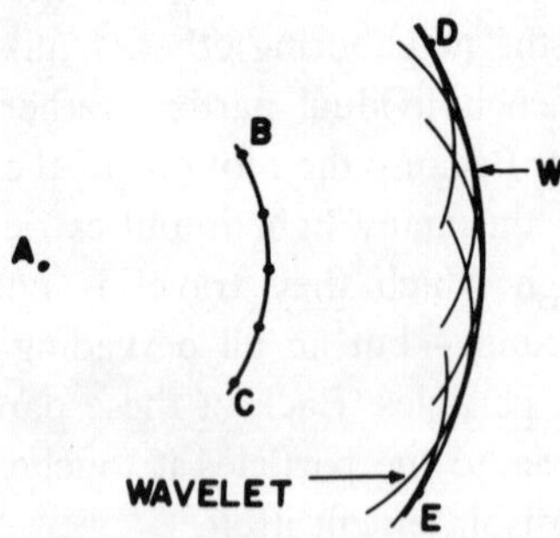

Fig. 57. The construction of wave fronts. According to Huygens, each point on the wavelet *BC*, gives rise to new wavelets tangent to a new wave front *DE*.

flame, such as *A*, *B*, or *C*, sends out a set of concentric, spherical impulses or wavelets. Each moves through the compact particles of the ether without disturbing the others, just as a complex sound moves through the air. Now let's consider a single point of light, *A*, in

Figure 57. The wavelet *BC* is one of many coming from the source of light. According to Huygens theory, each point on this wavelet can be thought of as a source of new spherical wavelets which are tangent to the curve *DE*. These secondary wavelets are too weak to produce any effect except in the region where they reinforce each other—on the surface *DE* which touches all of them. *DE* is then a new wave front of the wavelet.

Refraction

Using these ideas, Huygens was able to account for the rectilinear propagation of light and the existence of shadows. The wave theory also makes it possible to understand refraction, the bending of light

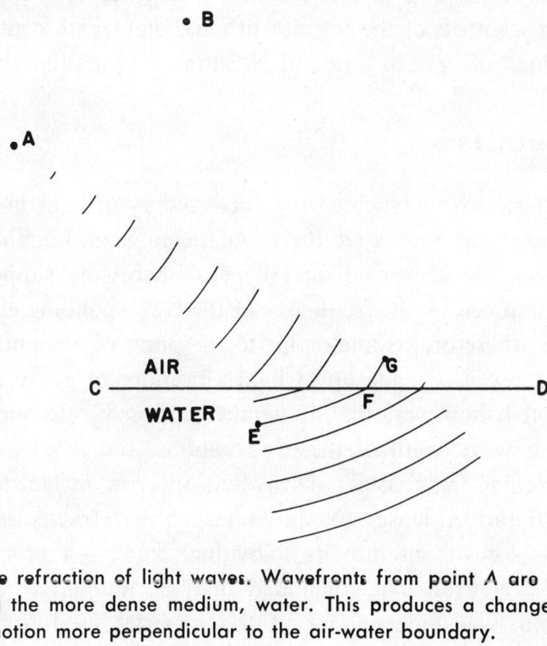

Fig. 58. The refraction of light waves. Wavefronts from point A are slowed down on entering the more dense medium, water. This produces a change in direction toward a motion more perpendicular to the air-water boundary.

that takes place when it passes obliquely from one medium to another. Figure 58 shows a series of concentric wave fronts moving through the air from a point-source of light *A*. (The direction of motion of each wave is perpendicular to its wave front at each point.) These eventually meet the surface of a second medium, water, at *CD*. As a portion of each light wave, such as *EF*, enters the water, it slows down relative to the portion of the wave in air, *FG*. This causes the light waves to change direction, all of them seeming to have originated at a point, such as *B*. The result is a series of waves moving in a direction more nearly perpendicular to the surface. Because light is always refracted toward the perpendicular when it enters a denser material, the wave theory implies a *reduction* in the speed of light in progressively more dense materials. The corpuscular theory, as we have noted, implies just the opposite relationship for the speed of light. This distinction would ultimately swing science toward the wave theory, when it became possible to measure the speed of light in different materials. But in the absence of any clearcut experimental proof, most scientists of the seventeenth and eighteenth centuries continued to think of light in terms of Newton's corpuscular theory.

Color-Corrected Lenses

One of those who advocated the wave theory of light in the eighteenth century was the great Swiss mathematician Leonhard Euler (1707–1783). He improved on Huygens' theory by supposing that color is determined by the frequency of the corresponding ether vibrations. Color, therefore, is analogous to the pitch of a sound. Sunlight is white because it is made up of light vibrations of every frequency. Euler doubted, however, that it would be possible to measure the frequency or wavelength of these virbrations.

On a practical front, Euler discovered an error in Newton's work that led to improved lenses. When white light is refracted in a prism, the beam is spread out into its individual colors—a process called "dispersion." Newton had concluded that the amount of spreading, or dispersion, was independent of the material used to make the prism. It would not be possible, therefore, to combine lenses of suitably chosen materials to eliminate the effects of dispersion. He believed, wrongly, that *chromatic aberration* (the colored fringes in lens systems), could be eliminated only by resorting to reflecting telescopes.

Euler believed, also wrongly, that the human eye was free of chromatic aberration. He suspected that this apparent lack of dispersion was caused by a suitable combination of several different transparent

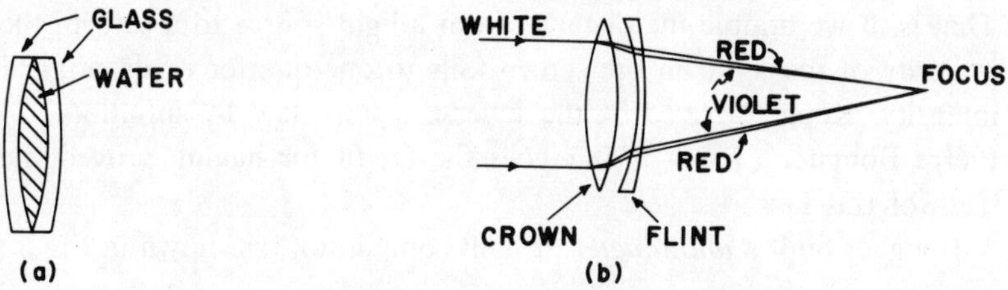

Fig. 59. Achromatic lens combinations. Color fringes are practically eliminated by bringing all wavelengths of light to the same focus.

media in the eye. Although his premise was wrong, his intuition was right, and he succeeded in making a compound lens free of chromatic aberration. As shown in Figure 59, he assembled two concavoconvex glass lenses which were cemented together at the edges. Water was placed between the two in order to effect the necessary compensation. When the various radii of the lenses were suitably ground, the red and violet rays came together at the same focus.

Modern lenses make use of Euler's principle. Two or more lenses are combined to form an "achromatic lens combination" so that the dispersion of one will be annulled by the other. The individual lenses are made of such materials as crown and flint glasses having different dispersive powers. A typical achromatic lens doublet employs a converging lens of crown glass and a diverging lens of flint glass. As shown in the diagram, white light is refracted by the crown glass lens, resulting in a certain amount of dispersion. Only the red and violet rays at the extremes of the spectrum are shown for clarity. The flint glass lens completely overcomes the dispersion of these two rays without overcoming all of the refraction. Both rays then come together at the focus. A practical lens combination of this kind was constructed in 1758 by John Dollond (1706–1761), a London optician. Dollond showed experimentally that it was possible to reduce chromatic aberration to negligible proportions, so the refracting telescope once again became reinstated in the good opinion of scientists.

The Inverse Square Law of Light

Like many other fundamental laws of nature, the intensity of light varies inversely as the square of the distance from the source. That is, if we double the distance from a light source to a screen, the intensity of the light on the screen falls to one-quarter of its original intensity. Kepler arrived at the inverse square law by intuition. To Pierre Bouguer (1698–1758) goes the credit for having proved the truth of this law.

Bouguer built a *photometer* or light comparator, as shown in Figure

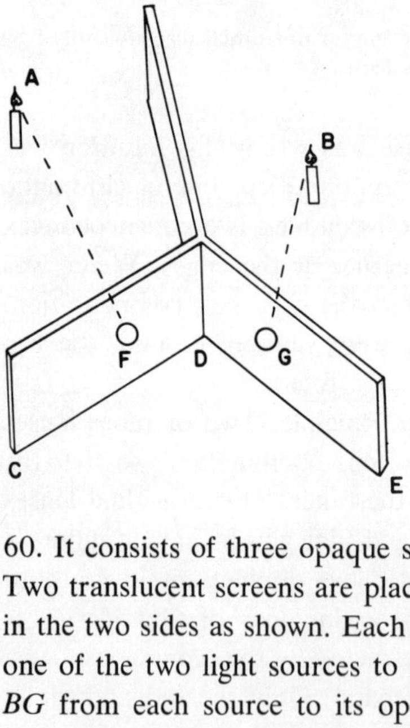

Fig. 60. Bouguer's *photometer*.

60. It consists of three opaque screens arranged in the shape of a *Y*. Two translucent screens are placed over the round openings *F* and *G* in the two sides as shown. Each of these openings was illuminated by one of the two light sources to be compared. The distances *AF* and *BG* from each source to its opening was adjusted until the screens at *F* and *G* appeared equally bright.

By suitable measurements, Bouguer was able to determine that the intensity of a light source falls off as the square of the distance. To illustrate, suppose we have five identical candles, each of which produces the same illumination at the same distance when tested in the photometer. If we place four candles at *A* so that the distance *AF* is 20 inches, and the other candle at *B*, the distance *BG* will be 10 inches when the screens seem equally bright. The illumination produced by a candle at *A* must be one-quarter as great as the illumination of a candle at *B*. But point *A* is only twice as far from the screen as point *B*. The illumination, therefore, depends inversely on the *square* of the distance. A doubling of the distance reduces the illumination to a quarter of its initial brightness. Similarly, tripling the distance results in one-ninth of the original brightness. In mathematical terms, the intensity, *I*, is inversely proportional to the square of the distance, r:

$$I \propto \frac{1}{r^2}.$$

Unexpected Lines in the Spectrum

Newton's experiments with dispersion, or the spreading of light into its component colors by a prism, indicated a continuous change of hues from red to violet. No one color predominated and each blended gradually into the next. A Scottish physicist, Thomas Melvill (1726–1753) produced an entirely different kind of spectrum by placing various salts in a flame. He used an opaque screen with a small hole in it to confine this light to a narrow beam, then passed the light beam through a prism to observe its spectrum. For ordinary table salt, he found that the spectrum had a bright yellow spot that predominated in a spectacular manner. This yellow spot was much brighter than the adjacent colors, and the transition at its edge was immediate rather than gradual. We now know that the preponderance of that particular shade of yellow is characteristic of sodium in table salt. Melvill found that various materials emitted other characteristic colors when placed in a hot flame. These observations were not understood at the time, but eventually gave birth to a major field of research, *spectroscopy*. The telling of that story, however, will have to wait until a later chapter.

The Interference of Light

The wave theory of light that gained universal acceptance in the nineteenth century was developed by the English physician Thomas

Young (1773–1829). One of the considerations that led him to the wave theory was the fact that light seems to travel at the same speed whether it is intense or weak. He believed that if light consisted of particles in motion, they would naturally move faster to produce a more brilliant effect. He knew that sound exists in the form of waves, and that soft and weak sounds travel at the same speed. This induced

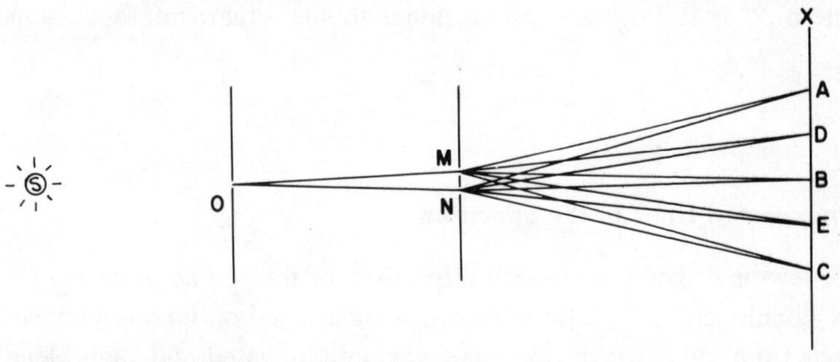

Fig. 61. Young's experiment to illustrate the interference of light. The two screens on the left select those light rays from the sun that emerge from holes *M* and *N* in phase agreement with each other. Reinforcement takes place at points *A, B* and *C,* where bright spots of light occur. Dark spots occur at points *D* and *E* for which the distances are just right to produce cancellation of light.

him to experiment with narrow beams of light along the lines originated by Grimaldi over one hundred years earlier.

Young set up an apparatus similar to that shown in Figure 61. The vertical dimensions of the diagram are highly magnified for clarity. To simplify our understanding of interference, let's imagine a source of light at *S* which emits light of only one color—a monochromatic light source. This might be a nonluminous flame into which some table salt is fed. Some of that light passes through a small hole *O* in the first screen. This opening serves as a point source of monochromatic light which illuminates two closely spaced pinholes *M* and *N* in the second screen. Waves sperad out in all directions from these apertures and overlap on the screen *X*. The straight lines from *M* and *N* to the screen *X* are rays drawn perpendicular to wave fronts which are not shown. The three apertures *M, N* and *O* limit the light striking the screen *X* to that originating at the same small portion of the light source. This means that the waves emerging from *M* and *N* came from the same oscillating atoms and must vibrate in synchronism. In scientific terms, the waves are said to be in phase agreement with each other. When the waves reach the screen *X*, they reinforce each other and produce brightness at certain places marked *A*, *B*, and *C*, and annul each other and produce darkness at other places marked *D* and *E*. Reinforcement occurs where the two waves arrive at the screen in phase agreement; that is, where their respective crests or troughs arrive at the same instant. To reach *B*, for example, both waves must traverse the same distance, so they arrive at the screen in synchronism and reinforce each other. They also reinforce at points *A* and *C* which are located so that one wave must travel precisely one wavelength further than the other. Once again the waves arrive in phase agreement and a bright spot occurs. Between the bright spots are points *D* and *E* at which the two waves cancel completely, because they arrive at the screen out of phase. The distances between the apertures and point *D*, for example, differ by exactly one-half wavelength. A crest of one wave arrives simultaneously with a trough of the other, and a dark spot is the result.

The alternate arrangement of light and dark spots or "fringes" on the screen is called an "interference pattern." Such patterns usually consist of many dark and light fringes as the phase relationship of the two waves passes alternately through places of reinforcement and annulment.

The distance between fringes is extremely small, and they are usually observed through a low-power telescope. If the pinholes are forty thousandths of an inch apart, the yellow fringes of sodium light are spaced about ten thousandths of an inch apart on a screen one yard away.

The Interference of White Light

When white light is used instead of a monochromatic light, each color produces its own interference fringes. Those for red light are spaced about twice as far apart as those for violet because of their difference in wavelength. The total pattern is the combined result of all of the colors. We can see similar brilliantly colored designs by looking at a distant light source through a piece of silk or fine-mesh screen, the meshes acting as closely spaced apertures.

Young's brilliant explanation of the interference of light in 1801 started a decline of the corpuscular theory. But this roused the corpuscular school to action, and Laplace succeeded in developing an explanation of interference on the basis of light corpuscles. This achievement would give heart to scientists one hundred years later when corpuscles reentered the theory of light. For the moment, however, corpuscles were approaching their eclipse. The phenomenon that overturned them completely was polarization.

Polarized Light

When light passes through certain materials, it is changed in a subtle but important way. This phenomenon, called "polarization," was first studied by Huygens. It can be observed when a beam of light passes through two slices of a crystal called tourmaline, cut parallel to the long axis of the crystal. As shown in Figure 62, we can see light through the crystals when their long axes are parallel. But if one is rotated slightly while the other is kept fixed, the light becomes dim. When the long axes are at right angles to each other, the light is completely stopped and darkness results. Rotation through another right angle restores the original brightness, and the light passes through once again.

At first, polarization was difficult to explain with either theory. Young, like the earlier scientists, supposed light to consist of longitudinal waves analogous to those of sound. These, as we shall see, cannot be polarized.

Water waves are transverse waves in which the water particles move at right angles to the direction of propagation of the wave. The motion of the particles is not random, however. Water waves are *polarized* in that the water particles can only move in one of the many directions at right angles to the wave motion. They are said to be "vertically polarized" waves because the water particles move up and down. Sound waves, which are longitudinal, cannot be polarized, because the motion of the air particles can only take place in one direction—the direction of propagation of the wave. If light waves were longitudinal, the "ether vibrations" could have no preferred direction. The tourmaline crystals could then have no directional properties with regard to light.

In 1817, Young suggested correctly that light waves might be transverse. August Jean Fresnel (1788–1827) then gave the wave

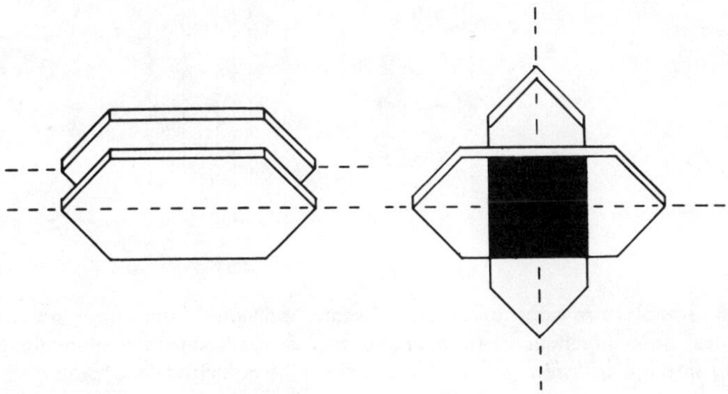

Fig. 62. The polarization of light. Tourmaline crystals allow light to pass through when their axes are in line. When one crystal is rotated 90 degrees, however, very little light gets through.

theory its modern form and provided a satisfying explanation of the experiment with tourmaline crystals. Although light waves are transverse, they are not necessarily polarized in the vertical direction as are water waves. Light can be thought of as a great number of waves moving through the same space simultaneously. Each is produced by the oscillation of an atom or molecule in an incandescent substance. Each point of a luminous substance vibrates at each instant only in one particular direction, but these directions are constantly changing for a particular point. The directions are also randomly different at the same instant for neighboring points. Thus, unpolarized light is really a collection of many polarized rays, in each of which the direction or plane of polarization is rapidly changing. This random polarization is illustrated in Figure 63(*a*). The observer should imagine a light ray being viewed end-on as it approaches him. The ray is advancing out of the paper at the central point of the diagram. According to Fresnel's theory, "ether vibrations" take place in all planes passing through the line of advance of the ray. These appear as straight lines 1–1, 2–2, etc., of which six are shown. The length of each line represents the amplitude or maximum excursion of a particle of the vibrating "ether." The light represented in part (*a*) of the diagram is

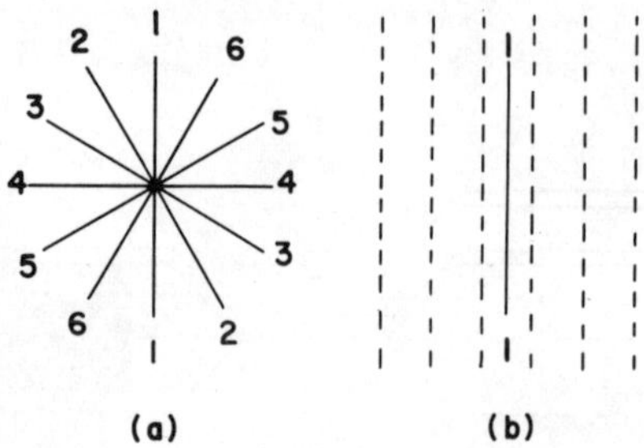

Fig. 63. Unpolarized light *(a)* consists of many individual light waves polarized in all conceivable directions at right angles to the light's direction of motion. After passing through a "grating," as in *(b)*, only light polarized in a preferred direction is evident.

called unpolarized light because there is no preferred direction of vibration of the transverse waves.

If unpolarized light passes through a material that acts as a grating, which is represented schematically by the dotted lines in part (*b*) of the diagram, only those vibrations taking place in one preferred direction will get through without difficulty. Others will be reduced in amplitude, or intensity, to an extent that depends upon their particular angle of polarization. Vibrations such as 4–4 at right angles to the grating are almost completely absorbed by the polarizing substance. Light that passes through such a grating is called "polarized light."

When light passes through the first tourmaline crystal in Figure 62, it becomes polarized in a direction perpendicular to the axis. If both crystals are oriented with their "gratings" in the same direction as in part (*a*), the polarized light passes through the second crystal with little further absorption. But if the two crystals are at right angles, as in (*b*), all angles of polarization are equally reduced in amplitude and almost no light can get through.

With the ability to account for polarization, the wave theory rapidly gained support. The final blow to the corpuscular theory occurred toward the middle of the nineteenth century when it became possible to measure the speed of light in different materials on earth.

Direct Measurement of the Speed of Light

In 1849, H. L. Fizeau (1819–1896) built the first apparatus for measuring the speed of light. The essence of his method is illustrated in Figure 64. Light from a source *S* falls on a semitransparent mirror *C*. Part of it passes straight through and part is reflected through a notch on the cog wheel and on to the mirror *A*. This light is reflected back through the notch and falls on the mirror *C*. Once again, part of the returning light passes through the mirror and is seen by the eye. The distance *AB* in the actual experiment is very much greater than *BC*.

In the description given above, the wheel is assumed to be stopped with a notch in the path of the light beams. If a tooth happens to be in the way, the light beam is stopped and the eye sees no reflected light. The cog wheel is then caused to rotate slowly, and the observer notices alternate periods of darkness and light as the notches and teeth move past the path of the light beam. As the speed increases further, the light that reaches the eye becomes dim and finally disappears when a certain speed is reached. To understand why this is so, imagine a light beam moving to the left and passing through a notch on the rotating wheel. This beam travels to the distant mirror *A*, is reflected, and returns to the vicinity of the wheel. During that interval of time, the wheel has rotated so that a tooth prevents the returning light from passing through and reaching the eye.

As the wheel is rotated faster, the light gets brighter again until at a second critical speed the brightness is once again at a maximum. Under these conditions, the wheel moves fast enough so that the returning beam is able to skip the obstructing tooth and return through the adjacent notch. A further increase in speed results in a dimming, and so on.

The speed of light measured in this way depends on the distance

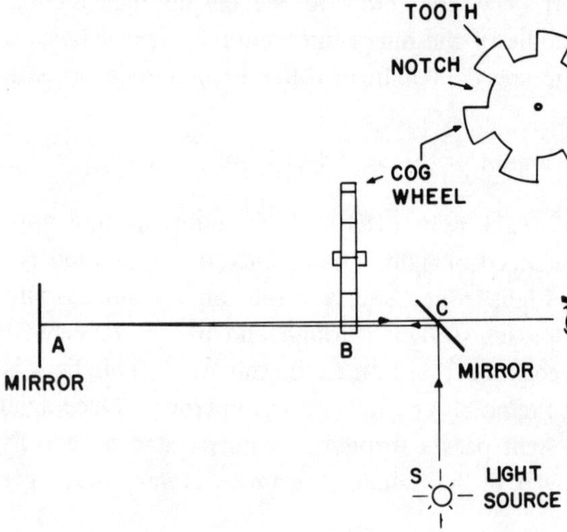

Fig. 64. Fizeau's apparatus for measuring the speed of light on earth.

D from the cogwheel to the mirror on the left, the speed of the cog-wheel *n* in revolutions per second, and the number of teeth *x* on the wheel:

$$\text{speed of light} = 4nxD$$

where *n* is the speed corresponding to the first dimming of the light. Using this method, Fizeau found the speed of light to be approximately 186,000 miles per second. Jean Leon Foucault (1819–1868) re-peated the measurement in 1850 using a system of rotating mirrors and obtained a speed agreeing well with Fizeau's. In addition, Foucault was able to show that the velocity of light in water was *less* than in air by the ratio of the index of refraction:

$$\text{Speed of light in water} = \frac{\text{speed of light in air}}{\text{index of refraction of water with respect to air}}$$

On page 53 we found than n = 1.33 for water, so the velocity of light in water is $186{,}000 \times \frac{1}{1.33}$ or about 140,000 miles per second.

The fact that light travels *slower* in more dense materials was the final evidence needed in favor of the wave theory of light. To Victorian scientists, this discovery disposed of the controversy between the corpuscular and wave theories, entirely in favor of the latter.

The Ether

The preponderance of experimental evidence discussed so far indicates that light *behaves as though* it were transverse mechanical waves in an elastic medium called the ether. This is quite a different matter from saying that light *is* that kind of wave motion. Even a few of the nineteenth century scientists felt instinctively the impossibility of transverse vibrations in an ether so rarefied that it can neither be felt nor detected with the most delicate instruments. Nevertheless, the question of mechanical possibility was quite forgotten because it was altogether insignificant compared with the ability of transverse waves to explain polarization, diffraction and refraction. It is one thing to ask what property of light is observed in polarization experiments, and quite another to ask how this property can be explained mechanically. The observed property is *true*, whether it can be explained or not. The ether could not be justified on any experimental basis, yet science felt the need of a medium for light, however unrealistic its properties might have to be. The wave theory accounted for the observed properties of light, but the ether accounted for nothing but the existence of waves.

We now believe that light is electrical in nature rather than mechanical, and modern science manages well without an ether. Nevertheless, the demise of the ether was a slow process that extended into the present century. In a later chapter we will discuss the factors that led to a new and more complete understanding of light and its relation to matter and energy.

Something in the Air

8

The previous chapters described the high points in man's successful assault on the problems of time and space measurement. With this success came the ability to navigate routinely over the entire globe. With the world his marketplace, civilized man now had easy access to the good things of life that were so unequally distributed over the earth. World trade had begun to alleviate the want and scarcity that had characterized man's entire history.

Despite his extraordinary success in navigation and timekeeping, man had not significantly increased the *number of kinds of things* that could be used to satisfy his wants or needs. Only a relatively few plants, animals, and natural substances are directly suitable for food or for the fabrication of useful products. His conquest over nature was one of transportation rather than of innovation of new materials.

At the close of the eighteenth century, man's knowledge of metals, glass making, pottery, dyeing, brewing, weaving and medicine scarcely exceeded that of the ancient peoples of the Mediterranean. He still used the same sail, plough, water wheel and grinding mill that were older than recorded history. The best silks still came from the Orient and the best tin from the British Isles. The only major inventions or new industries that modern man could claim were printing and the manufacture of gunpowder, optical instruments and accurate mechanical clocks.

Despite man's modest progress in developing new materials, the nineteenth century found him on the verge of a fantastic breakthrough. In the following years, the ordinary citizen would achieve a standard of living that was beyond the comprehension of the ancients. A major factor in that social revolution was the science we call chemistry. Our task now is to trace the steps that led to man's ability to create wonderful new products and materials out of the most unlikely substances of nature.

Chemistry of the Ancients

Primitive man collected useful materials from the earth and sea, from rocks, from strange smelling springs, from animals and plants, and he used them to make his metals, fabrics, paints, dyes, perfumes, pottery and medicines. Because the ability to make pure substances came much later, the quality of his products usually depended upon the peculiarities of the raw materials in his region.

Gold, silver and copper occur as metals in nature, and their discovery is lost in prehistory. Although all three were treasured by early man, it must soon have become apparent that copper was the most useful of the metals for making practical instruments. The invention of bronze—an alloy of copper and tin—did not come along until much later, perhaps 1500 B.C. Although ordinary copper is soft, it becomes much harder after cold-working—the hammering and bending involved in fashioning an instrument from the cold metal. The use of native copper marks the beginning of every metal culture in man's progress toward civilization.

Much later man learned to smelt the useful red metal from its ores. We can only speculate on the accidental discovery of the process. The green copper carbonate, called "malachite," was used in ancient Egypt as a cosmetic. If a bit of malachite happened to fall into the charcoal embers of a dying fire, droplets of copper would have formed. Malachite is changed into copper oxide by heat which, in turn, is reduced to pure copper when roasted with the carbon in charcoal. The removal of oxygen from a substance is called "reduction," possibly the first chemical process that man learned to use.

Although iron is not found in the native state, man became familiar with it in the form of meteoric fragments from outer space. For a long time, iron was more valuable than gold, because it was scarcer and much more useful. Beads of meteoric iron were used in Egypt for jewelry before 3400 B.C. Separating the metal from its ore seems to have been achieved by the Hindu's about 2000 B.C.

Once again, a number of pigments such as hematite are rich in iron in combination with oxygen. So the extraction of iron from the ore may have been accomplished by analogy with the smelting of copper. The tin used to make bronze can also be obtained from its common ores by removing oxygen. So early metallurgy began with the "reduction" of the ores now called oxides or carbonates by roasting them with charcoal in a closed chamber.

Many ores do not succumb to reduction alone, however. The ores we call "sulphides"—combinations of a metal with the element sulphur—require a prior step. Some of these ores can be reduced to the metal if they have been previously roasted in air. We call that process "oxidation," the addition of oxygen to a substance. It must soon have been observed that oxidation is facilitated if fresh air is continually supplied to the hot ore. The bellows then followed and with it came the first crude blast furnaces. It may also have been noticed that certain iron ores containing limestone more readily form a molten mass. These crafts were practiced thousands of years ago, but their true nature remained a mystery until the nineteenth century.

How to Materialize Invisible Spirits

In addition to reduction and oxidation, we also owe the process of *distillation* to the ancients. This process, which was known to the Alexandrians, is used to separate substances from each other on the basis of their individual characteristics. A traditional still is shown in Figure 65. The impure material to be separated is placed in the receptacle *A* and heated. Some of it then seemingly disappears or dematerializes. Though invisible, these ethereal substances are prevented from escaping by the *retort B*. They then materialize and collect in the *receiver C*. Today we know that various substances boil off as vapors at characteristic temperatures. By maintaining the mixture in *A* at an appropriate temperature, the constituent having the lowest boiling temperature can be separated out while others remain in the liquid state.

The nature of this process remained a mystery until recent times. When substances like alcohol or water were heated, they were thought to disappear and become "spirits"—a term that persists to this day. The task of the alchemist was to cause these "spirits" to materialize.

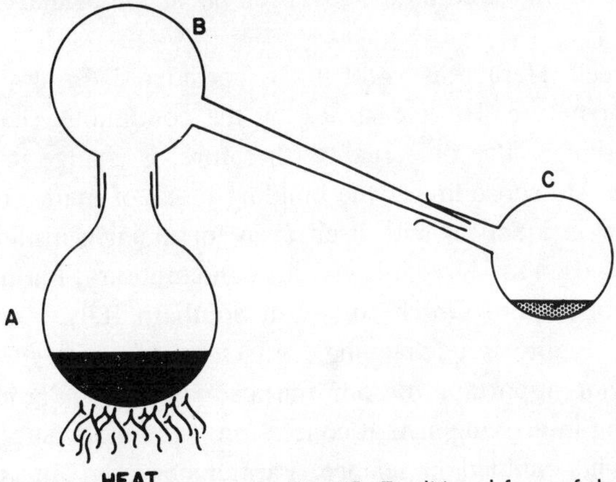

Fig. 65. Traditional form of the still.

In medieval times the substances we call gases and vapors were not recognized as the third state of matter. Only two states were acknowledged: the solid and the liquid. The ancient philosophers could not understand that the third state of matter is the connecting link between the smelting of metals and the distillation of alcohol. The physics of Aristotle taught that air has no weight, so neither spirits nor air could conceivably be a form of matter. As in the other exact sciences, an ancient Aristotelian hurdle would have to be cleared before order could emerge out of the chaos of alchemical recipes.

The Early Ideas of Matter

As we might expect, the science of chemistry has its historical roots in the minds of the early Greek philosophers. The first of these that is known to us is Thales of Miletus (about 600 B.C.), the great Greek geometer. In Thales we find the universal scientific desire to reduce the seemingly unrelated phenomena of nature to a certain unity. Thales believed that there must be *one primary matter* from which everything has developed by a gradual process of change or differentiation. He thought that *water* must be that primary matter. The line of thought that led Thales to that conclusion, unfortunately, has been lost to us. Neither do we know why his colleague, Anaximenes of Miletus (about 585–528 B.C.), advanced *air* as the primary matter of all things.

Another Greek, Heraclitus (480 B.C.), considered *fire* as the primary matter of nature. He was struck by the continuous change he noticed in all things. The only reality, therefore, is change itself. *To be* is *to change.* He chose fire as the building block of matter because it does not remain identical with itself even for a single instant.

Radically opposed to Heraclitus was his contemporary, Parmenides, a philosopher of Elea, a Greek colony in Southern Italy. According to Parmenides, reality is unchanging, and *change is nothing but an illusion.* It is not important for our purpose to follow Parmenides' line of reasoning in arriving at that conclusion. Like many subsequent philosophers who dabbled in science, Parmenides used unassailable logic to arrive at precisely the wrong conclusion. He denied *all* value to sense perception and argued that all truth must originate in the mind. We have already noticed the extent to which that contempt for the experimental method has hindered scientific development. Plato was its great exponent, but Parmenides was its author.

The Atoms of Democritus

The philosophy of Parmenides made a profound, if unfortunate, impression upon his contemporaries. Greek science had boldly set out to explain the changing phenomena in nature, and Parmenides had convinced his colleagues that these changes are impossible. The first task faced by Democritus (about 460 B.C.), therefore, was to explain how changes are possible at all. He believed that even if all change were an illusion, that illusion, at any rate, would require an explanation.

Democritus taught that the things which are unchanging and indivisible are *atoms*, the constituents of all matter. All else is a void or vacuum as we might say. By admitting the void, Democritus allowed his atoms to be separate and to move about. Change is nothing but the union or separation of unchangeable atoms. A substance can only change when its atoms unite with other atoms or become separated from each other. Whatever differences we perceive in a thing are based upon differences in size, shape or position of its atoms.

The atomic theory of Democritus was remarkably accurate, when we consider his complete lack of experimental evidence. More than two thousand years later, his ideas would develop into a physical theory that would be tested in the laboratory. But the Golden Age of Greece was not yet the time for such a development.

The Elements of Empedocles

According to Empedocles (about 450 B.C.), there are four basic constituents of everything—*air, fire, earth,* and *water*. Each of these elements is unchanging, and all substances contain them in varying proportions. A substance changes when the proportions of its constituent elements change.

The selection of earth, water, air and fire as the building blocks of all matter is really not as strange as it seems on first thought. They are obviously analogous to solids, liquids, gases and fire—a classification that comes naturally when we consider the material world. Neither should we imagine that the elements of Empedocles are the ordinary earth, water, air and fire that we know so well. In real life we find the elements only in combination. The liquid we know as water, according to the theory, contains a preponderance of elemental water, but also small amounts of the other three elements. The true elemental water is merely the *essence* of water—something we can never perceive.

The Seeds of Anaxagoras

Anaxagoras, who lived about the time of Empedocles, conceived of an unlimited number of different primitive elements which he called "seeds." His seeds were eternal and unchanging. A substance changes when its seeds become separated from each other, or when they unite with different seeds. The main difference between the theory of Anaxagoras and that of Democritus was the fact that every Anaxagorean substance contained all possible kinds of seeds. Their properties depended upon the type of seed that happened to predominate.

The Elements and Compounds of Aristotle

Although Aristotle was a poor physicist, he was an excellent observer. He firmly believed that any theory should be based on general experience. He attacked the atomic theory of Democritus at every opportunity, because it did not seem to fit with everyday observations. He felt far more comfortable with the fire, earth, water and air of Empedocles and accepted them into his own theory.

Where his predecessors had been content to talk about change in general terms, Aristotle became specific. In his books *On Generation and Corruption* and *On the Soul*, he gives a logical derivation of the four elements. He argues that touch is the primary sense—a notion that we have previously come across in connection with optics. The primary qualities of the primary elements, therefore, must be observable by the sense of touch. These primary qualities or properties are *cold* and *warm, wet* and *dry*. There are only six pairs of combinations of these four properties: warm-cold, warm-dry, warm-wet, cold-dry, cold-wet and dry-wet. The first and last of these combinations are trivial, so they can be neglected, leaving only four. As shown in Figure 66, the four elements are then supposed to be related to the four properties as follows:

warm-dry	fire
warm-wet	air
cold-dry	earth
cold-wet	water

Aristotle's derivation of the four elements of Empedocles was logical, and he was firmly convinced that it was based on experience. Whether we consider his arguments convincing or not, his ideas of change are certainly based on a more solid foundation than those of Democritus. Right or wrong, however, Aristotle's theory would be the basis of chemical thought for two thousand years.

Aristotle also differed from Democritus in his concept of a chemical compound. He argued that the formation of a common substance from the basic elements is more than a coming together of the elements. The elements themselves are changed when they combine with

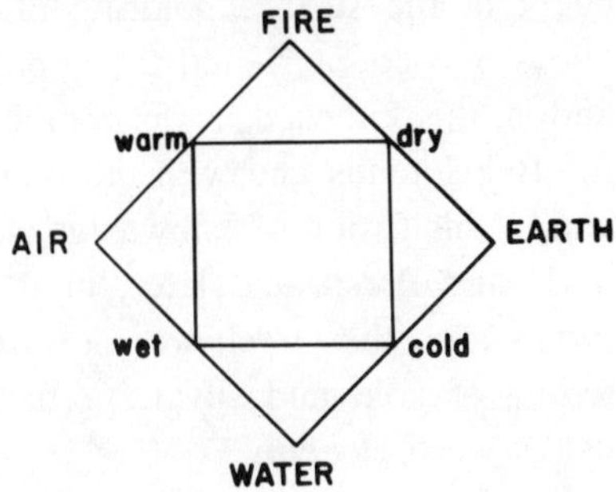

Fig. 66. The chemical elements of Empedocles and their properties as deduced by Aristotle.

each other. Chemical combination of elements is the "coming to be one of the changed reagents." This statement would later give rise to a great controversy concerning whether or not the elements continue to exist as such in chemical compounds.

Aristotle vs. Democritus

The Greeks were not content with a piecemeal explanation of nature. A theory had to grasp all phenomena in a single intellectual embrace. Although we did not mention it earlier, Democritus also proposed a kind of atom of which the human soul is made. The soul-atoms are fine and round, and can penetrate the entire body. They cause the body's motion and vital functions. Even thought could be reduced to a motion of these soul-atoms.

Democritus' system is classified as *materialism*, because everything in nature is explained on material principles. This helps us understand why Plato, Aristotle and most thinkers for two thousand years rejected his ideas. They could not accept a philosophy which reduced the noble achievements of the mind to a mere motion of material atoms. Not until the great scientific awakening of the seventeenth century would men begin to suspect that Aristotle had been wrong again.

Alchemy in Islam

Long before the Christian era, the Egyptians had acquired great skill in extracting and working with metals. They had also become adept in the surface coloring of metals and in the preparation of alloys that counterfeited the appearance of gold and silver. In Alexandria, this knowledge was combined with the astrology and magic of the Babylonians and with the philosophy of the Greeks. From Alexandria this fusion of knowledge, speculation and mysticism passed to Syria and Persia and later, in the seventh century, to Arabia. The Arabs took the Greek word *chemeia,* which referred to the counterfeiting of gold and silver, prefixed the Arabic article *al*, and gave us the word *alchemy*.

The fundamental premises of alchemy were as follows:

1. All matter is made up of a mixture of earth, air, water and fire in various mixtures.

2. Gold is the noblest and purest of the metals and silver is next.

3. Any metal can be changed into another, a process called *transmutation*, by changing the proportions of the four basic elements.

The alchemists believed that the transmutation of a base metal to gold could be achieved by using an elusive substance called the philosopher's stone. They also believed in an elixir of life and a panacea. The former would prolong life indefinitely and the latter would cure all of its ills. These objectives of alchemy were as real to the Arabian alchemists as is the synthesis of a new drug or fiber to the modern chemist. But we should not think of alchemy as merely a mass of fantastic superstitions. The alchemists have been adequately slandered by superficial evaluations of their accomplishments. If we bypass the inevitable mysticism and charlatanism, there is much of value to be found in Arabian alchemy. Its contribution—and it was an indispensable one—was the many centuries of trying all possible combinations of known substances; of rejecting those which do not interact, and slowly and methodically recording the recipes of those that do. Without this patient research—regardless of its motive—chemistry could not have become a science.

A typical alchemical recipe was that given for a pottery glaze by Jabir (c. 760–815), an Arabian alchemist who became known to Europeans as Geber. He writes in his *Book of Properties*:

Take a pound of litharge (a compound of lead and oxygen), powder it well and heat it gently with four pounds of wine vinegar until it is reduced to half its original volume. Then take a pound of soda and heat it with four pounds of water until the volume is halved. Filter the two solutions until they become clear and then gradually add the solution of soda to that of the litharge. A white substance is formed that settles to the bottom. Pour off the water and let the residue dry. It will become a salt as white as snow.

This material is white lead, used for centuries as a pottery glaze and as a paint.

The greatest of the Arabic alchemists was Rhazes (865–925) whose name means "man of Ray," a town in Persia. One of his books outlines the equipping of a laboratory. His suggestions for equipment include, among other things:

hearth	tongs	phials	stoves
bellows	shears	flasks	kilns
crucibles	pans	cauldrons	funnels
stills	scales	ovens	dishes
ladles	weights	filters	water-baths, etc.

A laboratory so equipped is hardly a mere wizard's den.

Rhazes was familiar with many chemicals, including possibly sulphuric and nitric acid. He was also the first to classify matter as either animal, vegetable or mineral. He classified the latter into six subclasses: (1) spirits, such as mercury or sulphur which either disappear or burn on heating; (2) seven metals; (3) six boraxes which include our borax; (4) eleven salts which include sea-salt, lime, potash and a few of our alkalis; (5) thirteen stones including malachite (copper oxide), hematite (iron oxide), gypsum (calcium suphate) and alum; and (6) six vitriols, certain compounds of a metal, sulphur and oxygen having a glassy appearance.

The Moorish alchemists that followed made further advances in the classification of chemicals. They recognized solutions of acids, salts and alkalis according to their effect on vegetable dyes that were used on fabrics. One of the dyes was called "litmus," now familiar to all students of chemistry. Litmus paper turns red when dipped in an acid, or blue when dipped in a solution of an alkali (our "bases" like lye and quicklime). The Moorish alchemists also gave recipes for the three important acids of modern industry: nitric, sulphuric and hydrochloric. These were made by distilling the vapors formed when various salts were heated.

European Alchemy

The sixteenth century saw the leadership of alchemy pass from Islam to Europe. Here it acquired a nobler aim under the influence of a Swiss physician, Aureolus Philippus Theophrastus Bombastus von Hohenheim, also known more concisely as Paracelcus (1493–1541). He was a man of violent, boastful and rebellious temper whose major contribution was a redirection of chemical effort into more fruitful channels. He turned his followers from the obsession of gold making, "the deceitful and mischievous art of alchemy," to a study of medicines. He taught that the true aim of alchemy should be the curing of human illness and disease through the study and devolpment of new drugs. His ideas inaugurated a new era in chemistry known as iatrochemistry or medical chemistry. This new field acted as a bridge between alchemy and the beginnings of an exact science of chemistry in the seventeenth century.

Chaos in Chemistry

The vital importance of air to chemical processes received its first recognition by Johann van Helmont who was born in Brussels in 1577. He showed experimentally that there are many kinds of gases and all are material in nature. He actually introduced the term "gas" which he derived from *chaos*. He had no way of collecting a gas and identifying it, so his descriptions were often incomplete.

Nevertheless, he became convinced that there are many kinds of gases all having different properties. This was quite an advance over the previous idea that air is the only substance of its kind. He enumerated many gases: windy gas (air), wild gas (carbon dioxide, the gas in soda pop), smoky gas, dry gas, etc. He described these gases in a rough kind of way based on their more obvious physical properties. He seems to have been the first chemist to suspect that the destructive power of gunpowder was caused by the gases it produces.

Jean Rey, who lived about 1630, was the first to recognize that air is somehow involved in the calcination (oxidation) of metals when they are heated in air. We now know that oxygen combines with the metal to form a new compound called an oxide, but this process was beyond the comprehension of Rey's predecessors. Rey tells us that a certain apothecary asked him why tin and lead increase in weight when calcined in air. In reply, Rey "devoted several hours to the question," during which he wrote twenty-eight essays on the subject which he let "slip from his hands." His major contribution to chemistry was the idea that all things, *including air*, have weight. Referring to the apothecary, he tells us that

> having placed two pounds six ounces of fine English tin in an iron vessel and heated it strongly on an open furnace for six hours . . . he recovered two pounds thirteen ounces of a white calx. Where did the extra seven ounces come from? Why . . . , from the air, which in the vessel has been rendered denser, heavier, and in some measure adhesive, by the vehement and continual heat of the furnace: which air mixes with the calx (frequent agitation aiding) and becomes attached to its most minute particles: not otherwise than water makes heavier sand which you throw into it and agitate, by moistening it and adhering to the smallest of its grains.

Despite the importance of Rey's idea, three things spoil it for us as they probably did for his contemporaries: (1) he gave no proof that the extra weight was caused by the air; (2) he described the air as combining with the white calx rather than the metal; and (3) he compares the process to a mixture of water and sand.

Although Rey missed the full significance of the union of air and a metal, he was convinced that air has weight. The followers of Aristotle argued that air is weightless because a bladder full of air weighs no more than a flat empty bladder. Rey replied that "only the veriest clown" would attach any significance to such a measurement. He pointed out that measurements of weight are normally carried out in the air or, infrequently, in water. "It is from this," he wrote, "that the error I have fought (that air is weightless) draws an argument that may dazzle feeble eyes, though not clear-seeing ones. For, weighing air in air itself, and finding no weight in it, they believed that it had none. But let them weigh water (which they believe to be heavy) in water itself, and they will find no weight in it either; the fact being that no element shows weight when weighed in itself." Rey had become the first scientist to clearly understand that Archimedes' principle of buoyancy applies to air as well as to a liquid. If we weigh an ordinary cooking pan in air, its weight is not increased by the amount of air it contains. It would weigh precisely the same amount if the pan were crumbled into a compact mass which contains little or no air. Notice, however, that Rey proved only that air *may* have weight; he did not prove that it *does* have weight. His reasoning merely demolished the Aristotelian arguments against the possibility that air might have weight.

Shortly after Rey's essays appeared, the air pump was invented. The weight of air might have been proved conclusively, if someone had merely weighed a container when it was full of air and when it was empty. But no one thought of doing it. The great discovery that air is weighty matter was made by Torricelli, a student of Galileo.

Why Nature Abhors a Vacuum

In Galileo's *Dialogues Concerning Two New Sciences*, he illustrates how an everyday observation—long overlooked by philosophers—might have led to a more accurate understanding of the air.

I once saw a cistern which had been provided with a pump . . . The stock of the pump carried its sucker (piston) and valve in the upper part so that the water was lifted by attraction and not by a push as is the case for pumps having the sucker placed lower down (the cistern). This pump worked perfectly so long as the water in the cistern stood above a certain level; but below that level, the pump failed to work. When I first noticed this phenomenon, I thought the pump was out of order; but the workman whom I called in to repair it told me the defect was not in the pump but in the water (level) which had fallen too low to be raised through such a height; and he added that it is not possible, either by a pump or by any other machine working on the principle of attraction, to lift water a hair's breadth above eighteen cubits; whether the pump be large or small, this is the extreme limit of the lift.

Galileo was referring to the common "suction" pump shown in Figure 67. Raising the piston as in part (*a*) increases the volume of the chamber, but the leather washer of the piston prevents air from entering the chamber from above. Water rises into the chamber be-

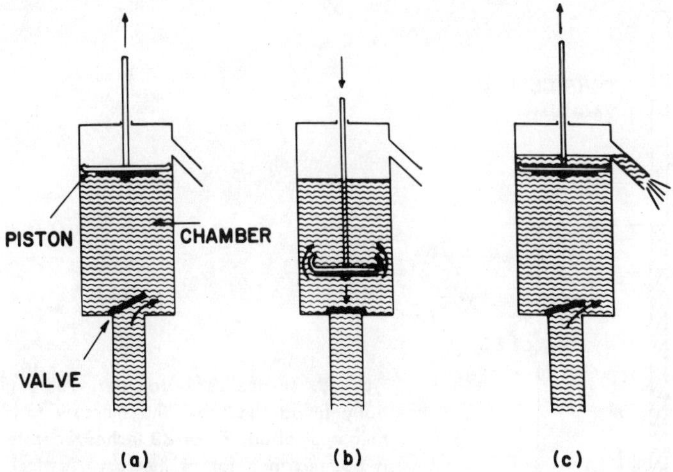

Fig. 67. Operation of the lift pump common in Galileo's time.

cause the atmospheric pressure on the surface of the water in the well is greater than the pressure in the chamber. On the downstroke, part (*b*), the valve is forced closed and the water pushes past the leather washer into the space above the piston. The next upstroke, in part (*c*), admits additional water into the chamber while discharging some of the water above the piston.

In Galileo's time, scientists believed that water is sucked into a pump because nature abhors a vacuum—the age-old *horror vacui.* Galileo was the first to wonder why this abhorrence stopped suddenly at a height of eighteen cubits, or about thirty-two feet as we would say. This was obviously an "effect" that cried out for a "cause."

This prompted Evangelista Torricelli (1608–1647), Galileo's pupil, to speculate on the height to which the *horror vacui* could raise a column of mercury. He knew that mercury is about fourteen times as heavy as water, so he suspected that the height would be one-fourteenth of the 18 cubits, or about 2½ feet. The experiment was performed by using a glass tube about 6 feet long which was sealed at one end. The tube was filled with mercury and the open end stopped with a finger. The tube was then inverted and placed with the stopped end dipping into an open pan of mercury (Figure 68).

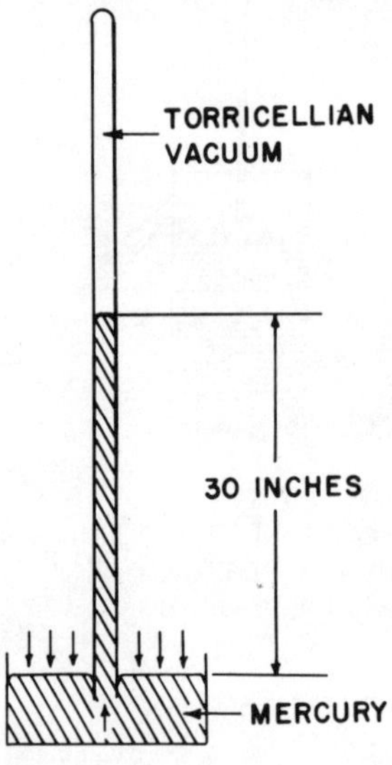

Fig. 68. Torricelli's barometer. Although initially filling the tube, the mercury falls to a height of about 29 or 30 inches when the end is opened in a jar of mercury. Torricelli concluded that the pressure of the atmosphere will only support a column of mercury that high.

When the finger was removed, the mercury in the tube sank to a height of about thirty inches above the surface of the mercury in the pan. No air could have entered the tube, so the clear space above the mercury column was subsequently named the "Torricellian Vacuum."

Torricelli suspected that the column of mercury was supported by the pressure of the atmosphere pushing on the free surface of the mercury. He attributed minor variations in the column's height from day to day to changes in atmospheric pressure. Later, Blaise Pascal (1623–1662) found that the height of the column was steadily reduced as the apparatus was carried up a mountain. Torricelli's instrument is now called a *barometer*.

The Principle of the Barometer

We can understand the principle of the barometer by supposing that air has weight. Imagine a column of air, as shown in Figure 69, having a cross section area of one square inch. The column extends in length

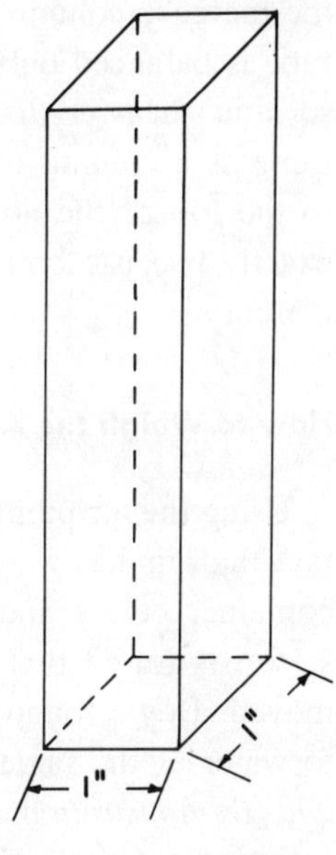

Fig. 69. A hypothetical column of air.

from the earth's surface to the top of the atmosphere. We now know that such a column of air weighs 14.7 pounds, if we neglect the small variations that occur from day to day. Every other column of air having the same cross section area weighs precisely the same amount. It is convenient, therefore, to speak of the pressure of the air—the weight, or force, acting on a unit of area. The pressure of the atmosphere is simply 14.7 pounds per square inch.

Notice that the pressure is independent of the cross section area we choose to consider. If we select an area of 10 square inches, the air column weighs ten times as much and the pressure is exactly the same.

We owe the next step in our understanding of the barometer to Pascal. In 1653 he showed that the pressure exerted on a fluid (a liquid or gas) is transmitted equally in all directions. He knew that the atmosphere pushes down on the open surface of the mercury with a pressure of 14.7 pounds per square inch (see Figure 68). According to Pascal's principle, this pressure is exerted everywhere in the mercury. It also acts in the mercury at the bottom of the tube. But there is no air in the top of the tube capable of pushing down on the mercury column. The atmospheric pressure at the bottom of the tube is balanced only by the column of mercury. Because mercury is so much heavier than air, a thirty-inch column weighs precisely the same as a column of air of the same cross section area that extends to the top of the atmosphere. The two columns balance each other exactly because both produce the same pressure on the reservoir of mercury.

How to Weigh the Air

Using the air pump, both von Guericke and Boyle later verified the fact that air has weight. This was done by weighing a metal or glass container before and after the air had been pumped out of it. Both scientists found that the container lost weight when the air was removed. Boyle found that "the proportion in gravity (weight) of air to water of the same volume will be as 1 to 938." In his *Discourses and Demonstrations on Two New Sciences*, Galileo measures the ratio to be 1:400. The true figure for air at room temperature and at sea level is about 830.

The Spring of the Air

Boyle published the results of his experiments with the air pump (his "new pneumatical engine") in 1660. These experiments suggested to him that air has a certain springiness. Pressure applied to a confined volume of air causes the air to become compressed into a smaller volume. When that pressure is removed, the air recovers its previous volume.

After reading Boyle's book, Richard Towneley suggested to the author an hypothesis "that supposes the pressures and expansions to be in reciprocal (inverse) proportion." He suspected, in other words, that doubling the pressure on air would reduce its volume to one-half. Boyle soon proved that Towneley's idea was correct, and it has become known as Boyle's law.

Boyle used a bent glass tube for his experiment as shown in part (*a*) of Figure 70. The short leg of the tube was hermetically sealed and the long end was open to the atmosphere. He then poured mercury in the tube and "by frequently inclining the tube," allowed some of

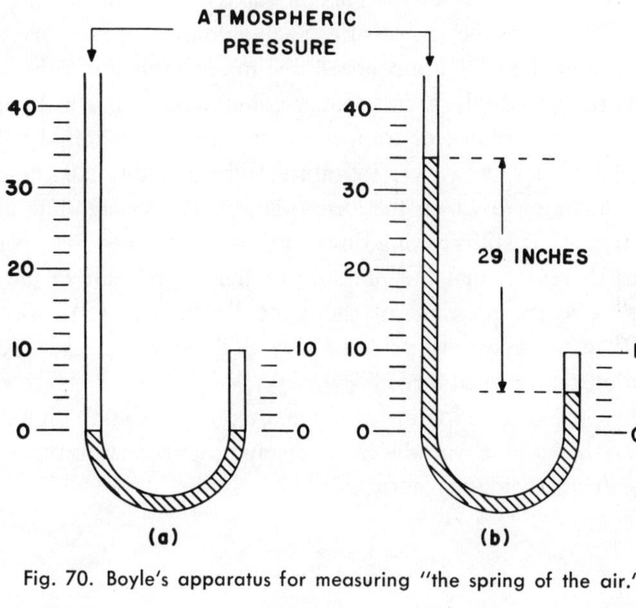

Fig. 70. Boyle's apparatus for measuring "the spring of the air."

the trapped air in the closed leg to escape. After much adjustment, the mercury rose to exactly the same height in each leg. The pressure of the air in the closed leg was then necessarily the same as the pressure of the atmosphere pressing on the mercury in the open leg. If the two pressures had been unequal, the mercury would not have risen so high in the leg having the higher pressure. Boyle concluded, therefore, that the air in the closed leg had a pressure equal to one atmosphere—14.7 pounds per square inch.

He then poured in additional mercury as in part (*b*) of the diagram until

> the air in the shorter leg was by condensation (compression) reduced to take up but half of the space it possessed . . . before; we cast our eyes upon the longer leg of the glass . . . and we observed not without delight and satisfaction that the quicksilver (mercury) in the longer tube was 29 inches higher than the other.

We can understand this result by recalling Torricelli's experiment. The air pressing on the mercury in the open leg of part (*a*) is equal in pressure to a column of mercury about 29 or 30 inches high, depending upon atmospheric pressure. Such a column of mercury would produce a pressure of 1 atmosphere, the initial pressure in the closed leg. In part (*b*), Boyle found that a column of mercury 29 inches high reduced the volume of trapped air to one half its initial volume. To this 29 inches, however, we must add an additional 29 or 30 inches of mercury to which the atmosphere is equivalent. The atmosphere, after all, is still pushing down on the mercury in the open leg. It's clear, therefore, that the pressure of the trapped air in part (*b*) is about twice the pressure of the same air in part (*a*). Also, the reduction in volume was produced by a doubling of the pressure acting upon the trapped air.

Boyle repeated the experiment for many pressures and volumes and found that the volume was always inversely proportional to the applied pressure. In mathematical terms,

$$\text{volume} \propto \frac{1}{\text{pressure}}$$

or,

$$\text{volume} = \frac{k}{\text{pressure}}$$

$$V = \frac{k}{p}$$

where k is a constant of proportionality depending upon the units in which volume and pressure are measured.

The Beginning of Scientific Chemistry

The alchemists, iatro-chemists and metallurgists had achieved important practical results by the middle of the seventeenth century, but they had done little to promote a scientific understanding of chemical processes. The development of the science of chemistry is a classic example of the interdependence of experiment and theory. It emphasizes the importance of cross fertilizing experimental facts with illuminating theoretical ideas that lead to understanding and unity. Although Aristotle's theories were sterile without experimental proof, the use of experimentation without theoretical guidance was blind. Before a science could grow out of the known mass of chemical facts, a theory would have to evolve that was in closer agreement with experience.

Three major stumbling blocks stood in the way of chemical evolution. The first was the still unrecognized fact that air—now a weighty substance—enters into many chemical processes as an unseen constituent or reagent. The second was the erroneous belief that fire is the universal element of "analysis" by which chemical compounds are separated into their elements. The third was the severe limitation placed on theorists by the small number of basic elements. The Aristotelians, or Peripatetics, still believed in the four ancient elements of Empedocles: earth, water, air and fire. The followers of Paracelcus favored three other "principles" or elements: salt, sulphur, and mercury. Depending upon personal preference, others selected one or two elements from each list.

The small number of basic elements led to an assignment of the same name to many different substances. Also, the lack of an agreed upon list of elements and compounds led to confusion and lack of mutual understanding among chemists. The fact that "occult qualities," "spirits" and "substantial forms" were invoked to explain chemical reactions confirmed the general view that chemistry was a kind of magic, or at best an art.

Galileo's research in mechanics had established a model of scientific inquiry in an allied science. It was now time for a critical reappraisal of the common chemical theories of the time. Also needed was a scientific method of conducting and interpreting chemical experiments. Boyle and Hooke stepped into this intellectual void and laid the foundations of the science of chemistry.

The "Sceptical Chymist"

Of the many books written by Boyle, the most important is *The Sceptical Chymist* which appeared in 1661. Boyle began by attacking the idea that there are only a small number of fundamental chemical elements. He compares such a preconception of the world to a man trying to decipher a book in which "he were acquainted with but three letters." "The Book of Nature," he observed, may need many more than three or four elements for its understanding. Boyle had finally swept away the Greek idea that chemistry, like geometry, could be developed logically from a few self-evident principles. He went on to divide all pure substances into two classes. One class contains those which had, so far, resisted all attempts to decompose them, or to break them up into substances that were simpler than themselves. These he called the true "elements." The second class of pure substances are those which, by one means or another, can be decomposed into simpler substances. These he called "compounds." The idea that theoretical preconceptions must give way to experimental facts had finally found a champion among the chemists.

Boyle also attacked the notion that fire is the universal method of separating all compounds into their elements. He pointed out that glass cannot be "analyzed" by fire, even though all chemists knew that it contains minerals (sand and lime) and the ashes of burned plants (soda or potash). He argued that even when fire does separate a substance into new chemicals, these are not necessarily elements. To illustrate, wood gives soot and ashes when burned, but oil, spirits, water and charcoal when distilled. All of these substances are surely not elements.

Boyle also noticed the importance of air in chemical reactions. If coal is burned in the open air it is calcined to ashes, but it is not calcined at all if it is heated red hot in a closed vessel. His interest in air and fire almost led Boyle to the discovery of oxygen. He realized that only a *part* of the air was necessary for combustion: "There may be dispensed through the rest of the atmosphere some odd substance . . . on account of which the air is so necessary to the subsistence of flame." With the help of his air pump, Boyle also showed that something in the air is necessary to animal life. Although he failed to identify these "parts" of the air, he did give birth to the idea of a vital substance contained in the air. One hundred years later, Lavoisier would show that oxygen is necessary both for life and for combustion.

Hooke and others extended Boyle's ideas in the fields of combustion and respiration. Hooke showed that the part of the air used up by a flame is the same part used up in breathing. He also demonstrated that a mouse would live longer in a jar of compressed air than in one at ordinary pressure. In 1663 Wren suggested that air contains "nitrous fumes" which sustain life, and Hooke extended the concept to include combustion. In 1665 Hooke showed experimentally that seeds need air for their growth.

Much progress had been made since the invention of the barometer and the air pump. Air had become known as a material substance that supports life and influences certain chemical reactions. It represented a third state of matter. Compressibility—as described by Boyle's Law—was a characteristic of matter in the airy state. The theory of matter had to take account of this fact. Democritus had taught that matter consists of discrete atoms separated by empty space, a concept that fitted well with the observation of compression. So the atomic view seemed more plausible than in earlier times. The time seemed right for a great theoretical development in chemistry, but it did not come until much later.

Heat, for example, was still a chemical element instead of a branch of physics. Unlike air, "spirits" were still not understood to be chemical substances in the gaseous state. The strokes of insight that did occur here and there were apt to be smothered in mere speculation and lost to other scientists. Communication was poor between chemists because of a tradition of secrecy and a lack of precise terminology with which to communicate. Improvements in these areas were just then taking place in mechanics under the influence of Newton and others. Nevertheless, chemists had no adequate sense of the advances made by seventeenth century physicists. It would take a century of groping before the precision of physics would find its way into the art of chemistry. The chemical revolution could not occur until chemists had learned enough physics to keep both sciences constantly before their minds.

The First Synthetics

9

Logs burning in the living room fireplace or in a campfire under the stars give a friendly, inviting warmth. But that is the extent of our dependence upon wood for heat. Coal, oil, gas and electricity have long since relegated wood to a minor role. It was not always so, however. Toward the end of the seventeenth century wood was still the basic source of energy for home and industry. And that source was rapidly being depleted. Great quantities of wood, or charcoal made from wood, were being consumed in the smelting of metals from their ores; in refining sugar; in making starch; in firing pottery, tiles, bricks and pipes; in drying malt and hops; in making soap, glass and gunpowder; and in countless other everyday activities.

As the need for timber increased, the land available for planting trees diminished. An anonymous authority tells us that in all countries near the sea, "most of the woods are consumed and the ground converted to corn and pasture." All industries that depended upon wood were forced into a search for *material substitutes.*

One of these was the smelting of iron. Unlike tin and lead, iron could not be separated from its ore with ordinary coal. But a substitute for charcoal was found in *coke,* a steel-gray, hard substance consisting almost entirely of the element *carbon.* An excellent reducing agent, coke is still indispensable in the smelting of iron. It is made by heating coal above 2000°F in a closed space for sixteen hours while coal gas and other substances are driven from the coal. The white-hot coke is then removed from the oven and quickly quenched in cold water to prevent it from burning up. Each ton of coal yields 1,400 pounds of coke and a large quantity of inflammable coal gas. Toward the end of the eighteenth century this gas came into general use for the illumination of buildings and streets.

Synthetic Pot-ashes

Other industries that found themselves searching for material substitutes were the manufacture of glass and soap. Substances called potash and soda (or soda ash) were widely used in these industries. Potash (now called potassium carbonate) was obtained from the ashes of burned land vegetation, and soda ash (called sodium carbonate) was derived from the ashes of plants that grow near the sea. These chemicals were formerly obtained by collecting water that had passed through the ashes, evaporating the water, and heating the residue in pots—hence the name "potashes."

Because of a shortage of these alkaline substances, the Academy of Paris offered a prize for a process of converting common salt, sodium chloride, to sodium carbonate, or soda. The prize was won by a French chemist, Nicolas Leblanc, who established a factory in 1791 for the manufacture of soda. The factory was confiscated by the revolutionary government in 1793, however, and in 1806, filled with despair, Leblanc committed suicide. That was the reward he received for giving the world cheap soap and glass.

Bleaching Water

Chemical research also received encouragement from other sources. The acid made from sulfur, oxygen and water (sulfuric acid) was useful in the bleaching of linen and the refining of precious metals. The middle of the eighteenth century found sulfuric acid being produced by new methods of mass production that greatly reduced its cost. Later in the century, it became the basic raw material of the glass and soap industries as a chemical of the Leblanc process.

In 1774 Scheele discovered a gas, chlorine, that was also derived from a chemical process involving sulfuric acid, and noticed its great bleaching power. Berthollet dissolved the gas in water in 1785 and found that the solution also had bleaching power. Berthollet's work founded the new bleaching industry.

White China

Another industry that created a demand for chemical knowledge was the manufacture of china and earthenware. Different colors in the finished product had long been obtained by using appropriate varieties of clay. But, then as now, the housewife insisted on pure white china. Difficulties often arose, however, because of metallic compounds in the earth. Certain clays could be rendered white on firing only by removing these impurities. It was well known, for example, that iron oxide imparts a yellow or red color to what would otherwise be white china. By the end of the eighteenth century, vendors of clay had learned to remove such impurities by chemical means.

The Search for Saltpeter

The manufacture of gunpowder shared in the fruits of chemical research. Of the three constituents of black gunpowder—charcoal, sulfur and saltpeter—the latter was always in short supply. Charcoal was derived from wood and sulfur from volcanic deposits. But saltpeter, or potassium nitrate as we would say, was of animal origin. Saltpeter, or nitrate, forms naturally in the decomposition of manure heaps.

There were few places in Europe where saltpeter could collect without being carried away by rain water. Consequently, saltpeter could be found only in cellars and stables where there was considerable animal waste-material undergoing decomposition. As early as 1540, French officers, called *salpêtriers commissionés*, were given the unenviable task of searching for and extracting saltpeter wherever it could be found. During the reign of Louis XIII (1610–1643) the annual saltpeter crop amounted to almost 2,000 tons. The need for the chemical increased further when in 1650 the German chemist Glauber showed that saltpeter can restore the fertility of exhausted soils.

It was not until 1775 that Lavoisier, the father of modern chemistry, was able to abolish the understandably unpopular practice of searching private houses for saltpeter. As comptroller of munitions, he had greatly increased the production of the chemical through improved methods of manufacture.

All of these accomplishments occurred before chemical theory found itself headed in the right direction. It was these accomplishments, in fact, that helped create a growing interest in chemical research. Both for this reason, and for the basic love of knowledge, some of the best minds of the time turned toward a study of chemistry. Because of their efforts, the end of the eighteenth century saw chemistry become an exact science on an equal footing with physics.

New Names for Old Things

Before we trace the great chemical discoveries of men like Priestley, Cavendish and Lavoisier, it will help to anticipate a part of the story—the part connected with nomenclature. The names used for various chemicals were ancient relics of the secretive and magical days of alchemy. Names such as "powder of algaroth," "salt of alembroth," "phagadenic water," or "colcothar" tell nothing about the *constituents* of a chemical, or the *relationship* between different chemicals. Terms like "oil of tartar by the bell," "oil of vitriol," "butter of antimony" or "flowers of zinc" are still more undesirable, because they lead to wrong ideas. Far from being related to oil, butter or flowers, such chemicals are, for the most part, violent poisons.

The reformed nomenclature that followed was the brain child of Guyton de Morveau, a French lawyer. Working with him were the great Lavoisier, Berthollet, who was Napoleon's chemistry instructor, and Fourcroy, a writer and leading figure of the Reign of Terror. Their new *Method of Chemical Nomenclature* appeared in 1787, and the reforms were first applied by Lavoisier in 1789 in his *Traité Elé-*

mentaire de Chemie ("Elementary Treatise on Chemistry").

The new nomenclature provided names that expressed the chemical nature or composition of a substance. As a first step, all chemicals were divided into two classes, the elements and compounds of Boyle. The elements included the most simple or fundamental substances "such as chemists to the present time have not been able to decompose." Compounds were substances made of two or more elements. Compounds should be named, therefore, "by the reunion of the names of those same elements." One of the elements given a new name was *oxygen,* from the Greek words for "acid" and "beget," because oxygen was wrongly believed to be a constituent of all acids. Others were *hydrogen, nitrogen* (in 1790 by Chaptal) and *carbon.* Sulfur retained its name and its acids were called sulfur*ic*, to designate the maximum possible oxygen content, and sulfur*ous*, for the acid containing less oxygen. According to the new method, the suffixes "ic" and "ous" indicate the greater or lesser oxygen content when a substance can combine with two different amounts of oxygen.

Oxygen, the most abundant element on earth, occupied a prominent place in the new system. Its compounds included first and foremost, the *oxides* formed when another element combines with oxygen. An important class of oxides were all *acidic,* that is, substances that dissolve in water to form acids which turn litmus paper red. These include the oxides of sulphur, carbon, nitrogen, phosphorus and other *nonmetals.* Opposed to the acidic oxides were the oxides of metals, which were called "basic," and which dissolved in water to form alkaline solutions. These included the oxides of potassium, sodium and other metals whose oxides in solution turn litmus paper blue. Some of these "basic oxides," like quicklime (calcium oxide), unite with water to form more complex compounds called "caustic alkalies."

When a basic oxide reacts with an acid, a substance called a "salt" is formed. These were named after the acid or acidic oxide from which they were derived. To illustrate, sulf*ate* is the general name for all salts formed by sulfuric acid, the suffix "ate" corresponding to the higher oxygen content. The term sulf*ite* refers to salts formed by sulfurous acid of lower oxygen content. Thus, the names "copper sulfate" and "copper sulfite" indicate the acid and metal from which they are derived. Similarly, *carbonic acid* is a solution of an oxide of carbon in water. The salts formed when carbonic acid reacts with basic oxides are called "carbonates." Thus, the basic oxide, calcium oxide, unites with carbonic acid to form calcium carbonate. Salts can also be formed by dissolving a metal, rather than its oxide, in a strong acid such as sulfuric acid. The same nomenclature applies to the resulting salts.

The new method of classifying chemicals is much more than a new set of names for things. It helps us understand how a chemical can be obtained or used, and what it has in common with related chemicals. Calling two chemicals litharge and galena tells us little or nothing about their chemical composition. But their new names, lead oxide and lead sulfide convey a wealth of information. First of all, the new names indicate that both contain *lead*. In addition, we know that many sulfides can be converted to oxides by heating them with oxygen in a closed space—a process called oxidation. So lead sulfide (galena) can be oxidized to yield lead oxide (litharge). The sulfur in lead sulfide combines with oxygen from the air and is given off as a gas, a compound of sulfur and oxygen. The free lead that remains is also oxidized to yield lead oxide. The lead oxide can then be converted to lead by heating it with carbon in a closed space—a process called reduction. Carbon extracts oxygen from the lead oxide and is given off as a gas, carbon dioxide. This leaves the pure metal, lead. Thus, a knowledge of ancient chemical processes along with the new names for chemicals gives valuable clues to the chemist. This reform was of greatest importance in stimulating the growth of chemical science.

New Symbols

The chemical reformation was completed in 1814 by the Swedish chemist, John Berzelius, who swept away the ancient alchemical symbols that had been used for chemical substances. Figure 71 illustrates some of the strange symbols and puzzling signs that were still in use in Lavoisier's time. Berzelius argued that "The chemical signs ought to be letters for the greater facility of writing . . . I shall, therefore, take for the chemical sign the initial letter of the Latin (or common) name of each chemical element." If the first letter is common to another element, he used both the initial letter and the first suc-

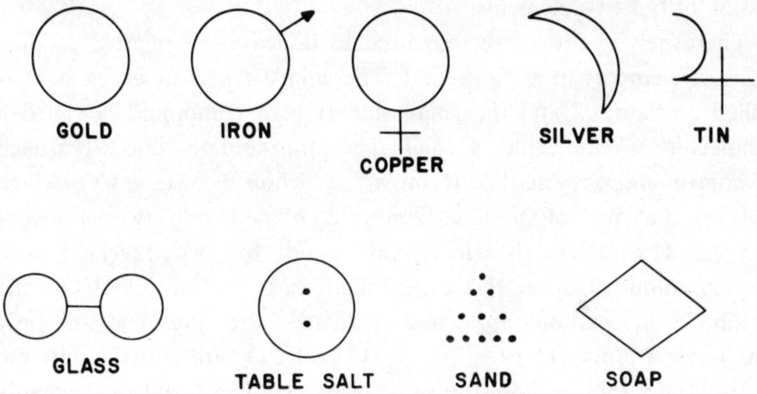

Fig. 71. A few alchemical symbols.

ceeding letter that they do not have in common. The new symbols are given below for a few common elements:

Element	*Symbol*	*Derivation*
Carbon	C	Common name
Copper	Cu	Cuprum (Latin)
Cobalt	Co	Cobaltum (Latin)
Calcium	Ca	Calx (Latin)
Chlorine	Cl	Common name
Chromium	Cr	Chrom (Greek)
Iron	Fe	Ferrum (Latin)
Potassium	K	Kalium (Latin)
Nitrogen	N	Common name
Oxygen	O	Common name
Phosphorus	P	Common name
Sulfur	S	Common name
Silicon	Si	Silex (Latin)
Tin	Sn	Stannum (Latin)
Zinc	Zn	Common name

Berzelius used these symbols to build up the symbols for compounds. Copper oxide, for example, becomes CuO—the symbol of the metal preceding that of the nonmetal. Similarly, zinc sulfide becomes ZnS. The ending "-ide" usually represents a compound consisting of but two elements.

These new symbols are also used to indicate the relative proportions of elements in a compound. The smallest part of an element is called an "atom," and the smallest part of a compound is called a "molecule." A molecule of CuO, therefore, contains one atom each of copper and oxygen. The formula for carbon dioxide is CO_2 which tells us that one atom of carbon is combined with two atoms of oxygen. The prefix "di-" in *di*oxide stands for two oxygen atoms. Carbon monoxide, on the other hand, has the formula CO. The "mono"- in carbon *monoxide* indicates the presence of only one oxygen atom. The two gases CO and CO_2 are entirely different even though they contain the same elements. The former is a deadly poison and the latter gives soda pop and champagne their pleasant sparkle.

The reformed chemical names and symbols were not used extensively until well into the nineteenth century. They will be helpful to us, however, in understanding the great discoveries of the eighteenth century.

The Weight of Heat

Of the four elements of Empedocles, earth, air, water and fire, the latter gave way with the greatest reluctance. The last concerted attempt to preserve the material nature of heat was the "phlogiston theory." According to the theory, all combustible materials—such as wood or charcoal—contain an inflammable element that is given off during burning, calcination (oxidation) and respiration. This inflammable element, phlogiston, was thought to be absorbed by the surrounding air.

The phlogiston theory, which persisted through most of the eighteenth century, was founded by the German chemists Becher and Stahl. Charcoal was thought to be practically pure phlogiston. When an ore is converted into metal by heating it with charcoal, the metal absorbs some of the phlogiston. Similarly, when a metal is roasted in air, it loses its phlogiston to the air and turns into a calx (its oxide as we would say).

$$\text{metal} = \text{phlogiston} + \text{calx}$$

The chemical equation given above presented a paradox. It was well known that a metal weighs *less* than its calx. How can a metal which has lost its phlogiston encounter an increase in weight? Most chemists didn't try to answer that question. A few of those that did attributed the lighter weight of the metal to the levity (or negative weight) of phlogiston! In spite of Newton's law of universal gravitation, chemists could believe that a chemical element—phlogiston—could have negative weight. This view was not universally held, however. Most chemists preferred to plead ignorance on such questions while believing, nevertheless, in the general idea of phlogiston. They used the concept when it was helpful and ignored it when it failed. It would be the task of Lavoisier to eliminate phlogiston from the table of chemical elements. But that is getting ahead of our story.

The Weight of "Fixed Air"

In 1756, the great Scottish scientist Joseph Black (1728–1799) published a brief but novel paper concerning *Experiments upon Magnesia alba, Quicklime, and some other Alcaline Substances.* His work was of extreme importance as the first quantitative study of a series of chemical reactions. His great triumph was in showing that the several chemical changes under study could be detected and understood by weighing the various chemicals involved.

During his experiments, Black discovered a gas different from air that could exist either as an ordinary gas, or in the solid state in combination with other substances. In this *combined* state, it could be transferred from one solid substance to another without escaping as a gas. The extension and development of Black's ideas was to be the main task of chemistry for the remainder of the century.

We can understand Black's achievement by following his experiments with chalk, our calcium carbonate, $CaCO_3$. He knew that chalk loses almost half of its weight when it is heated. The residue, calcium oxide, is called quicklime. He showed that the difference in weight was accounted for by an amount of "air" that is given off by the reaction. He used the apparatus illustrated in Figure 72 to collect the gas over water. The receiver is filled with water and inverted in a jar of water. As the gas evolves from the heated chalk it moves through the bent tube and rises in the receiver. This apparatus for

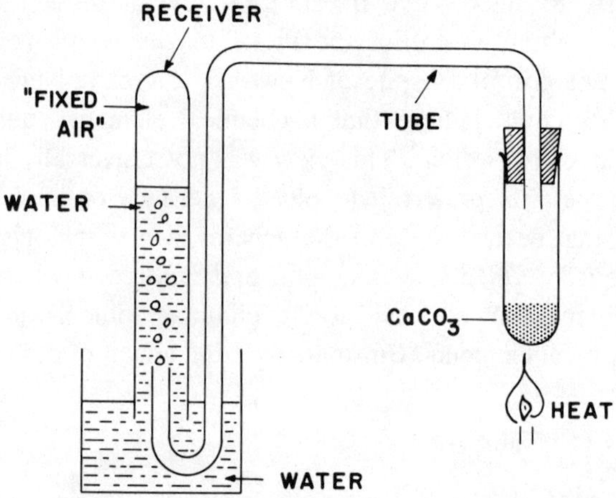

Fig. 72. A pneumatic trough used to collect "fixed air," our carbon dioxide and other gases.

collecting gases is called a "pneumatic trough," and was developed by Stephen Hales in 1727.

Black showed that the "air" evolved from chalk is different from ordinary air. Along with Hales, he called it "fixed air" and maintained that it resembled the "air" caused by combustion and breathing. We call it carbon dioxide, CO_2.

He then repeated his experiment quantitatively. He accurately measured 1 ounce of chalk and heated it to drive off all of the "fixed air." The resulting quicklime weighed about 0.56 ounce. He surmised that the missing 0.44 ounce was given off as "fixed air." We can express the reaction as a chemical equation.

$CaCO_3$	=	CaO	+	CO_2
chalk	=	quicklime	+	fixed air
calcium carbonate	=	calcium oxide	+	carbon dioxide
1 ounce	=	0.56 ounce	+	0.44 ounce

Notice that the same weight appears on both sides of the equation. Black was a pioneer in believing that matter can neither be created nor destroyed by a chemical reaction. This so-called "law of conservation of matter" has become a cornerstone of modern chemistry. Notice also that each particular kind of atom on the left of the equality sign is balanced by an identical atom on the right. There are one calcium, one carbon and three oxygen atoms on each side of the equation. This is also characteristic of chemical reactions.

Black now went a step further. He knew that chalk and quicklime are both basic or alkaline substances. Mixing either one with an acid produces a salt. So he mixed an acid with 1 ounce of chalk, increasing the amount of acid slowly until the combination was *neutral*; that is, until it would neither turn litmus paper red nor blue. Let's call that amount of acid *a* ounces. (He noticed that a gas, fixed air, was given off during this process.) Then he found the weight of quicklime (CaO) that would just neutralize the same amount (*a* ounces) of acid. That turned out to be about 0.56 ounces. (No gas was given off during this reaction.) One ounce of chalk and 0.56 ounces of quicklime neutralize the same amount of acid. But it was shown earlier that 0.56 ounce of quicklime is derived from 1 ounce of chalk by the action of heat. Hence, the carbon dioxide that is combined with quicklime in chalk, and liberated as bubbles when acid is poured on chalk, does not affect its power to neutralize an acid. A definite amount of chalk, whether heated into quicklime or not, neutralizes the same amount of acid.

Black then repeated a well known chemical reaction, the addition of water to quicklime to produce slaked lime, or calcium hydroxide.

CaO	+	H_2O	=	$Ca(OH)_2$
quicklime	+	water	=	slaked lime
calcium oxide	+	water	=	calcium hydroxide

It was well known that slaked lime could be changed back to the original material, chalk, by treating it with a mild alkali such as carbonate of soda. We would write the reaction as follows:

$Ca(OH)_2$	+	Na_2CO_3	=	$CaCO_3$	+	$2NaOH$
slaked lime	+	soda ash	=	chalk	+	caustic soda
calcium hydroxide	+	sodium carbonate	=	calcium carbonate	+	sodium hydroxide

By weighing each compound in each chemical reaction Black was able to show that his "fixed air" (CO_2) could be transferred from one substance (soda) to another (chalk) without escaping to the air. He also proved that the weights of the various substances could be accounted for by assuming any "lost" weight to have left a reaction in the form of "fixed air."

Black's research was important because of the light it cast on the nature of alkaline substances which had become indispensable to industry. The word "alkali" had come to mean any substance that neutralizes an acid. Quicklime (CaO), potash (KCO_3), and caustic potash (KOH), (obtained by boiling water, potash and lime), were all "alkalis." Black added the important knowledge that certain alkalis fall into closely related groups. He discovered the "family relationship" between potashes (potassium carbonate, KCO_3) and chalk (calcium carbonate, $CaCO_3$). Both liberate fixed air when acted upon by acids. Quicklime (CaO), on the other hand, does not. Today we say that the latter is an oxide, while the two former substances are *carbonates.*

The Invention of Soda Water

Soda water was invented by Joseph Priestley (1733–1804), one of the great amateur chemists. Priestley was born near Leeds and studied for the ministry. He somehow managed to support his family by working either as a schoolmaster, as a minister or as a private tutor. Because of his liberal religious and political views, he lived a troubled life and eventually saw his home and laboratory burned by a politically inspired mob. He migrated to the United States in 1794 and died in Northumberland, Pennsylvania.

Priestley's first chemical research dealt with a colorless gas that he collected from the bubbling vats of a nearby brewery. He noticed that the gas would extinguish burning pieces of wood and he suspected that it was the same "fixed air" that Black had studied fifteen years earlier. He found that the gas (carbon dioxide) was somewhat soluble in water, making "a glass of exceedingly pleasant sparkling water which could hardly be distinguished from Seltzer water." He later described his discovery to the members of the Royal Society and received their interest and praise. We now call it soda water, a weak acid solution of carbon dioxide gas in water.

Muriatic Acid

Soon afterward, Priestley made a great contribution to chemistry. He tried heating common salt with vitriolic (sulfuric) acid and collected "marine acid air" over *mercury* instead of water. His apparatus was similar to that of Black (Figure 72) except for the substitution of mercury for water. The colorless gas that he obtained (hydrogen chloride) has a pungent, irritating odor. He found that water sucked up this gas greedily—easily absorbing hundreds of volumes of it. No wonder no one had ever collected it before. The water in the conventional pneumatic trough had dissolved the gas almost as fast as it was produced. When dissolved in water, it is called muriatic or hydrochloric acid. A few years later it gave birth to the modern bleaching industry.

Using his new methods, Priestly discovered other important kinds of "air." In 1773 he discovered "alkaline air" or ammonia as we would say. Then he brought his new gases together, ammonia and hydrogen chloride, and saw an amazing reaction. The gases suddenly disappeared and were replaced by a white cloud that settled out as a fine white powder. Two pungent gases had combined to form an odorless white powder—ammonium chloride—which is now used in making batteries.

The Breathing of Animals and Plants

Priestley showed experimentally that plants continually purify the atmosphere, keeping it fit to sustain animal life. He found that sprigs of mint grew exceptionally well in air that had been contaminated by the breathing of animals. He also proved that such air is once again rendered fit for breathing by the action of the growing mint. He concluded that "the injury which is continually done to the atmosphere by the respiration of such a number of animals . . . is, in part at least, repaired by the vegetable creation."

Priestley also collected nitric oxide gas, nitrous oxide (or laughing gas, N_2O), and nitrogen peroxide gas, NO_2. In 1785 he prepared carbon monoxide gas (CO) by heating iron oxide with charcoal. But because the gas burned, to form CO_2, he mistook it for "inflammable air," or hydrogen in our terminology.

Dephlogisticated Air

The most important of Priestley's discoveries was that of "dephlogisticated air," or oxygen. He obtained it by heating the red oxide of mercury. (The discovery was made independently by Scheele at about the same time.) He found that this gas "is five or six times better than common air for the purpose of respiration, inflammation, and, I believe every other use of common atmospherical air." Also, "a candle burned in this air with an amazing strength of flame; and a bit of red-hot wood crackled and burned with a prodigious rapidity. . . ."

Priestley had no idea of the real nature of the gas he had discovered, oxygen. In order to support combustion, ordinary air was thought to pick up phlogiston given off by a burning object. Since his new gas did this so much better than ordinary air he believed that it was completely "dephlogisticated," hence the name.

At about that time, Daniel Rutherford, a student of Black at the University of Edinburgh, discovered the existence of two gases in the air. Many chemists, including Priestley, had allowed small animals to breathe the air in a confined volume until it would no longer support life. The volume of air decreased only slightly in such experiments because the animal's breathing replaced the used up oxygen by carbon dioxide gas. Rutherford absorbed this CO_2 with the help of quicklime and water (lime water) which has great affinity for the gas. After absorbing the carbon dioxide, the spent air had a volume only four-fifths of the original amount. The "air" that was left—nitrogen—was quite inert, and did not support combustion. Priestley demonstrated that the missing one-fifth was precisely the same gas as the "dephlogisticated air" (oxygen) that he had obtained from the oxide of mercury.

Priestley's discovery of oxygen marked a turning point in chemistry. In the fertile mind of Lavoisier, it would put an end to the phlogiston theory and direct chemistry along the correct scientific path.

Priestley's interest in electricity led to his discovery of a new method of studying gases. He mixed inflammable gases with oxygen and exploded the mixture by means of an electric spark. When he "sparked" carbon monoxide (CO) and oxygen, the residue was "fixed air." In modern terminology,

$$\text{carbon monoxide} + \text{oxygen} = \text{carbon dioxide}$$

For the first time, fixed air had been shown to be a compound of simpler substances.

The Discovery of Hydrogen

Henry Cavendish (1731–1810) a member of a British noble family, devoted himself completely to scientific research. He obtained "inflammable air" (hydrogen) by dissolving zinc, iron or tin in dilute sulfuric acid or hydrochloric acid. Using each of the two acids with each of the three metals, he collected six samples of the gas and watched each sample burn with the same pale blue flame. Since the same "inflammable air" was made in so many different ways, Priestley concluded that Cavendish had found the elusive phlogiston. The fact that hydrogen was lighter than air gave some credence to the idea. Phlogiston, after all, ought to have a negative weight. Cavendish did not realize that he had discovered a new element, hydrogen.

Exploding a Myth

Cavendish produced some of his "inflammable air" and, after exploding it with common air, he noticed that a dew was deposited on the walls of the container. Further test showed that about 423 volumes of "inflammable air" combined with 1,000 volumes of common air leaving about four-fifths of the original common air untouched. He concluded that "almost all the inflammable air, and about one-fifth part of the common air, lose their elasticity, and are condensed into the dew which lines the glass."

But what was the nature of the dew? Could it be water? After repeating the experiment on a large scale, he obtained 135 grains of a liquid which turned out to be water. Water, the element of the Greeks, was really a combination of two gases, "dephlogisticated air" (oxygen) and "inflammable air" (hydrogen)! Water was not an element, after all.

The idea of earth as an element had not been taken seriously for years. Air was also known to contain at least three simpler substances —nitrogen, oxygen and carbon dioxide—so, like earth, it could not be an element. Water now joined the list of compounds. The only remaining element of Empedocles and Aristotle was fire, disguised in its cloak of phlogiston. With the work of Lavoisier, heat began its slow transition from the list of chemical elements to its proper place as a branch of physics.

Chemistry Becomes a Science

Great chemists from Boyle to Cavendish had achieved considerable progress in their experimental researches on combustion and respiration. But each discovery in turn had been molded to fit an erroneous theory. Most of these theoretical gymnastics came to an end with Antoine Laurent Lavoisier (1743–1794), the chemist whose interpretations first suggested the real significance of burning and breathing. This new understanding created the science of chemistry.

Lavoisier was the first scientist to state explicitly the law of conservation of matter: "One may take it for granted that in every reaction there is an equal quantity of matter before and after the operation." Lavoisier specifically included the gases in his definition of matter. Because of his confidence in this important law, he used his considerable wealth to obtain the best weighing balances in Europe. With these balances he proceeded to lay phlogiston to rest.

He put an accurately weighed quantity of tin in a flask and sealed it so no additional air could enter during the experiment. Then he heated the flask until the tin was calcined. After that step he weighed the sealed flask again and found that the weight had not changed. This disproved the idea advanced by Boyle and others that weighty fire particles enter the calx through the flask. When he subsequently opened the flask, he noticed that air rushed into it. The flask now weighed *more* than when it was sealed. He then weighed the tin calx and found that it had gained in weight by precisely the same amount as the flask had gained after the inrush of air. Obviously, the tin combined with something in the air during calcination and gained weight accordingly.

As soon as Lavoisier learned of Priestley's discovery of "dephlogisticated air," the true explanation of his experiment became clear. Calcination, as well as burning and respiration was merely the union of a substance with a part of the air. He called that part "vital air" and, later, oxygen. The new compound formed by calcination weighed more than the original metal by the weight of oxygen that had combined with it. This was simple enough. No levity, no mysterious phlogiston, not even weighty fire particles—and the finest balances in the world to support his conclusion. Today, we describe the calcination or oxidation of tin with the following equation:

$$\begin{aligned} \mathrm{Sn} + \mathrm{O_2} &= \mathrm{SnO_2} \\ \text{tin} + \text{oxygen} &= \text{tin oxide} \\ &\quad \text{(stannic oxide)} \end{aligned}$$

Lavoisier also showed experimentally that "fixed air" is a compound of oxygen and carbon. He heated mercury calx (the red oxide of mercury) together with charcoal (carbon) and obtained "fixed air." He knew that oxygen would have been obtained if the carbon had not been included, so "fixed air" must be a compound of oxygen and carbon. He had previously burned a diamond in a closed glass vessel and had also obtained "fixed air." So diamond was much like carbon. Both combine with oxygen to form "fixed air."

$$C \quad + O_2 \quad = CO_2$$

diamond or carbon + oxygen = carbon dioxide

With the help of his balances, Lavoisier determined that carbon dioxide contains 72.1 per cent oxygen—a good approximation of the correct figure, 72.7 per cent.

Lavoisier also showed that phosphorus and sulfur combine with oxygen to produce acids when dissolved in water. This gave oxygen its name—"acid maker." The view of oxygen as a generator of acids had to be revised later when hydrochloric and hydrocyanic acids were found to contain no oxygen.

In 1783 Lavoisier confirmed Cavendish's discovery of the nature of water. He burned hydrogen and oxygen coming from separate tubes and produced a quantity of pure water. In the same year, he performed the experiment in reverse and obtained hydrogen from water. He passed steam over hot iron chips and collected hydrogen produced by the reaction. The iron removed oxygen from the steam (H_2O) releasing a quantity of free hydrogen.

iron + steam = iron oxide + hydrogen

By about 1785, the phlogiston theory was on the decline and the new chemistry was gaining acceptance under Lavoisier's leadership. It received its greatest expression in his classical text book *Traité Elémentaire de Chemie* which appeared in 1789. In it he explained existing chemical knowledge on the basis of his new ideas. Unfortunately, Lavoisier did not live to see the final triumph of the new movement. The publication of his book coincided with the French Revolution. Although he assisted the new government in its early days, the subsequent Reign of Terror "had no need for savants." He was tried on a trumped-up charge and beheaded on May 8, 1794—just one month after Priestley had fled from England. In losing Lavoisier science had lost the greatest chemist of the time.

Rebirth of the Atom

10

Just as modern chemistry began with Lavoisier, modern atomic theory —which touches all of physical science—began with Dalton. John Dalton (1766–1844) was a Quaker teacher of Manchester, England. His outstanding achievement was the explanation of known chemical facts on the basis of atoms and molecules.

The concept of atoms and molecules as the smallest particles of elements and compounds was far from new. But Dalton's atoms differed from those of earlier times. As a true scientist, he gave them only those specific properties demanded by chemical facts. They were not given vague properties that could be bent to suit any desired explanation. A few of Dalton's atomic symbols are shown in Figure 73.

While Dalton was speculating about atoms around the turn of the century, a great controversy raged concerning the nature of compounds. Berthollet, the French chemist, believed that the amount of each element in a compound was not necessarily constant. The proportions of hydrogen and oxygen in water, for example, might vary slightly depending upon the amount of each element that happened to be available when the water was formed. Berthollet's arguments were eventually refuted by another French chemist, Joseph Louis Proust (1755–1826). Using improved methods of preparing pure chemicals, he showed in 1807 that a compound always contains the same elements combined in exactly the same proportions by weight. This principle has become known as the "law of constant proportions."

Like Lavoisier's law (the conservation of weight in a reaction), Proust's law marks a milestone in chemical knowledge. Both laws tended to support the atomic theory that Dalton had first described in 1803. He taught that, "the ultimate particles (atoms) of all homo-

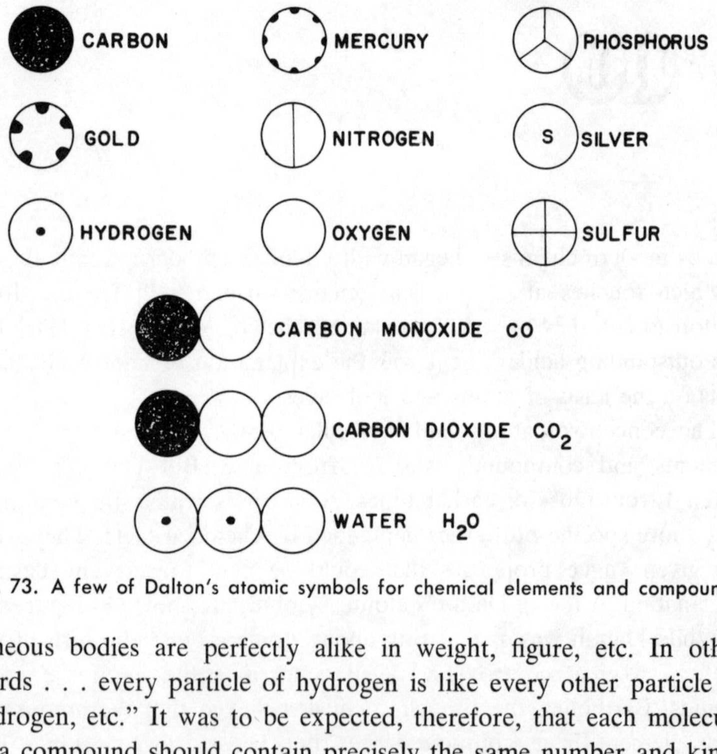

Fig. 73. A few of Dalton's atomic symbols for chemical elements and compounds.

geneous bodies are perfectly alike in weight, figure, etc. In other words . . . every particle of hydrogen is like every other particle of hydrogen, etc." It was to be expected, therefore, that each molecule of a compound should contain precisely the same number and kinds of atoms as its neighbors. And because atoms of the same element have the same weight, all molecules that contain them must be identical for the same compound. Similarly, because atoms are unchangeable and indestructible, there can be no loss of weight in a chemical reaction.

Weighing the Atom

Dalton went on to explain that "it is one great object of this work to show the importance and advantage of ascertaining the *relative weights* of the ultimate particles, both of simple and compound bodies (elements and compounds). . . ." Dalton knew that there was no hope of weighing individual atoms of different elements, but he did see a way of determining their *relative* weights. Although he made many mistakes in the determination of these relative atomic weights, his method was basically correct. To illustrate, Dalton believed incorrectly that one unit of weight of hydrogen combines with seven units of weight of oxygen to produce water. Assuming that the water molecule contains one atom of each element, an atom of oxygen would weigh seven times as much as an atom of hydrogen. If the atomic weight of hydrogen is called unity, the relative weight of oxygen ought to be seven. His conclusion was wrong, of course, for two reasons: (1) water contains *two* atoms of hydrogen for each oxygen atom; and (2) the proportions by weight of the two elements in water are in the ratio 8:1. If we take the weight of hydrogen to be one, then oxygen must have a relative weight of sixteen.

Dalton discovered another important law that greatly supported his atomic theory. Proust had shown previously that the same pair of elements sometimes combine in different proportions to form entirely different compounds. He found, for example, that tin and oxygen combine in either of the following proportions by weight:

Tin	Oxygen	
88.1%	11.9%	stannous oxide
78.7%	21.3%	stannic oxide

Proust failed to discover any relation between the two compositions, because he expressed the relative weights of the two compounds as percentages of the total weight. Dalton expressed the composition of each compound in a different way—he wrote the two weights of oxygen that combine with the *same weight* of tin. For example:

Tin	Oxygen	
100 grams	13.5 grams	stannous oxide
100 grams	27.0 grams	stannic oxide

Written that way, it is clear that the two weights of oxygen are in the ratio of 1:2. Dalton's law of multiple proportions tells us that when the same weight of an element combines with different weights of another element to form more than one compound, the different weights are in the ratio of small whole numbers, two, three, four, etc.

Dalton naturally concluded that stannous oxide (SnO) contains one atom of each element, while stannic oxide (SnO_2) contains one atom of tin and two of oxygen. The law fitted in beautifully both with his atomic theory and with Proust's *law of constant proportions.*

Law of Combination by Volume

At the turn of the century, scientists knew that two volumes of hydrogen and one volume of oxygen can be "sparked" or exploded to produce water with no gas left over (Figure 74). This was a

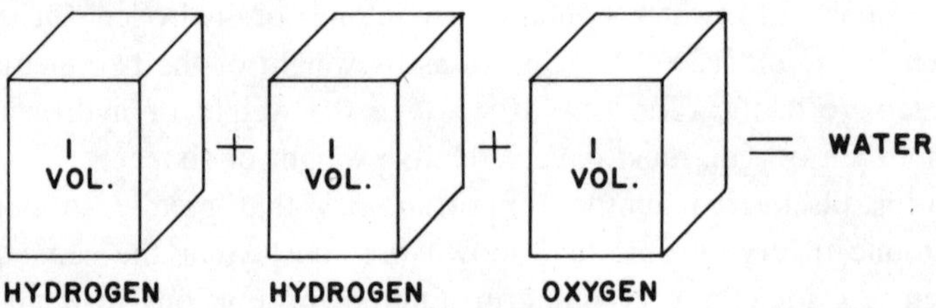

Fig. 74. The law of combination by volume applied to hydrogen and oxygen.

typical example of a general law that began to take shape. Gay-Lussac's chemical law of volumes states that the volumes of gases (assuming the same pressure and temperature) that combine in a chemical reaction are related to each other in the ratio of small whole numbers. Gay-Lussac tested the law with the most extensive and exact experiments on gas reactions yet made. By 1808, he had verified the law for ammonia (NH_3), hydrochloric acid (HCl), nitrous oxide (NO), nitric oxide (NO_2), and sulfur dioxide (SO_2).

He found that when gases combine, their relative volumes always bear a simple whole-number relation to each other *and to the volume of the resulting compound if it is a gas*. For example, one volume of oxygen combines with two volumes of hydrogen to form two volumes of water vapor; one volume of nitrogen combines with three volumes of hydrogen to form two volumes of ammonia gas, and so on.

Gay-Lussac then went a step too far and concluded that equal volumes of gases contain the same number of atoms. Dalton showed that this could not be true. Take nitrous oxide (NO), for example—a compound made of nitrogen and oxygen. Gay-Lussac would have written the formula for the reaction as follows:

$$N + O = NO$$

(1 atom of nitrogen + 1 atom of oxygen = 1 molecule of nitrous oxide.)

It will help to understand the implications of this equation if we consider the nature of a gas. A gas is a substance that consists of extremely small particles of matter separated by great empty spaces. The distances of separation are vastly greater than the dimensions of the particles. So the volume of a gas depends on only two factors: the *number* of particles and their *distance apart*. The volume does not depend upon the *size* of the particles since they are merely points for all practical purposes. Gay-Lussac was aware of these ideas. He concluded that a liter of nitrous oxide should contain the same number of particles as a liter of nitrogen or a liter of oxygen—even though a molecule of nitrous oxide is larger than an atom of either oxygen or nitrogen. He believed, therefore, that one volume (say a liter) of nitrogen should combine with one volume of oxygen to produce *one* volume of nitrous oxide. When the experiment was performed, however, he obtained *two* volumes of nitrous oxide; just twice as much as he could account for. It was as though one particle of nitrogen and one particle of oxygen had combined to produce two particles of nitrous oxide. If equal volumes of elemental gases contain equal numbers of *atoms,* one volume each of nitrogen and oxygen should combine to make only *one* volume of nitrous oxide instead of two.

This difficulty was explained in 1811 by the Italian, Amedeo Avogadro (1776–1856). He suggested that the atoms of some, but not all, of the gases go around in pairs. In other words, there might be molecules of gaseous elements as well as of compounds. The reaction for nitrous oxide could then be written:

$$N_2 + O_2 = 2NO$$

(1 molecule of nitrogen + 1 molecule of oxygen = 2 molecules of nitrous oxide or, 1 volume of nitrogen + 1 volume of oxygen = 2 volumes of nitrous oxide.)

The subscript 2 for nitrogen and oxygen indicates that each molecule of each element contains two atoms. Avogadro's law tells us that equal volumes of gases (at the same pressure and temperature) contain the same number of *molecules*. Despite the truth and importance of Avogadro's law, it was bitterly controversial at the time and was not accepted generally until revived some fifty years later.

Relative Atomic Weight

During the second decade of the nineteenth century Jons Berzelius (1779–1848) performed the work that made him the recognized leader of chemistry. An ardent believer in the atomic theory, he decided to determine the relative atomic weights of all known elements. To insure accuracy, he purified every chemical he used, not once or twice, but dozens of times. The difficulties under which this great Swedish chemist worked were tremendous. His "laboratory" consisted of two rooms in his house equipped with neither running water nor gas. He had neither furnace nor hood, and his sink had a pot waiting underneath to catch the flow when the plug was removed. Whenever he needed a platinum crucible, Berzelius had to borrow the only one in Sweden. He fabricated most of the other equipment he needed with his own hands. He became an expert glass blower, for example, just to construct his own apparatus.

Despite these handicaps, Berzelius purified and analyzed compound after compound until he had determined the composition of over two thousand chemical substances. After years of patient toil, he published the atomic weights of the forty different chemical elements known to the scientific world of his time. His table of the atomic weights stands to this day as a monument to his accuracy, skill and perseverance. Later, his work assumed even greater importance when a great Russian chemist, Mendeléev, noticed certain similarities among the elements.

Electricity from Metals

In the latter part of the eighteenth century, there was considerable interest in the shock-producing fishes such as the electric eel or *Gymnotus*. Accounts of them were given by John Hunter (1773), Cavendish (1776) and others. It was generally believed that their shocks were electrical in nature. This led Luigi Galvani (1737–1798)

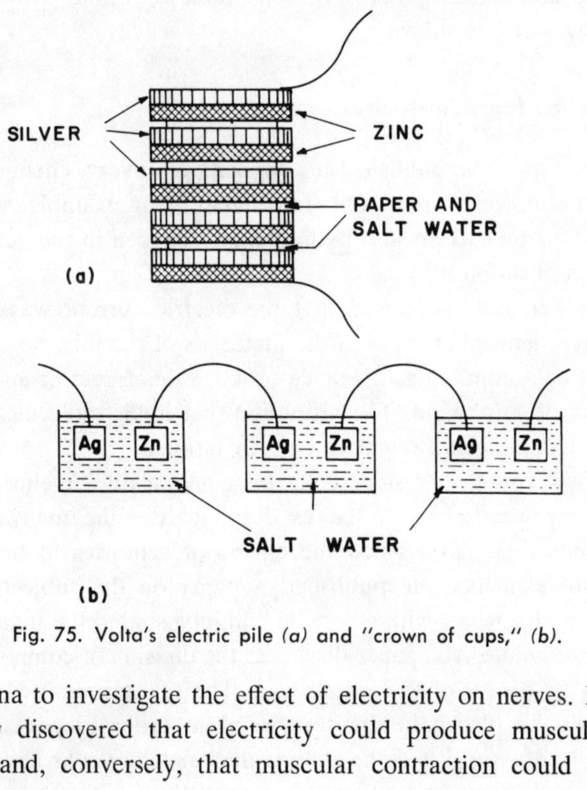

Fig. 75. Volta's electric pile *(a)* and "crown of cups," *(b)*.

of Bologna to investigate the effect of electricity on nerves. He subsequently discovered that electricity could produce muscular contraction and, conversely, that muscular contraction could produce electricity.

In following up Galvani's work, Alessandro Volta (1745–1827) of Pavia, made an important discovery. In 1800, he found that a pair of dissimilar metal coins in contact can produce an electric current. He immediately constructed his famous voltaic pile, the first electric battery. Figure 75 shows its construction. The pile is a series of alternate disks of silver and zinc, although other metals can be used. Each pair is sandwiched between paper strips moistened with salt water. The current is taken off the top and bottom disks by metal wires. These are connected to any external object through which it is desired to pass the current. The same results were obtained with Volta's famous *couronne de tasses* (1800), or "crown of cups," as shown in the lower part of the diagram. The pairs of plates of silver and zinc are placed in cups of salt water or dilute acid and connected by wires as shown.

New Metals from Electricity

As soon as Volta published his amazing discovery, chemists rushed to use this powerful new tool. In England, for example, water was decomposed by current into hydrogen and oxygen in the very year of Volta's publication.

It was generally supposed that the electric current was somehow related to chemical changes in the materials of the pile. So the attraction between atoms in a substance took on the aspect of an electrical attraction of some kind. The attraction that holds molecules together had previously been thought to be gravitational.

The new voltaic pile stirred the imagination of Berzelius. Perhaps this mighty weapon might be used to analyze the many stubborn compounds that had resisted the efforts of centuries to break them down into elements. He published a paper on the subject with his friend von Hisinger—with whom he had long shared the indispensable platinum crucible. The paper discussed the division of compounds into elements by means of the voltaic pile. The two terminal wires of the pile would be placed in the compound in order to pass a current through it. He suggested the theory that metals would always move to the negative terminal and nonmetals to the positive terminal.

A few years later, a young English chemist, Humphry Davy (1778–1829), sent the new electric current through various compounds and electrified the imagination of the scientific world. Using the research of Berzelius and Volta's new electric pile, he isolated many strange new elements having undreamed of properties. Potash and soda had been used for countless centuries, yet no scientist had the slightest inkling of the metallic elements hidden within them. Davy built a powerful voltaic battery of one hundred and fifty cells and sent the current flowing through some molten potash. He soon saw the magnificent decomposition he had hoped for. Just as Berzelius had predicted, globules of a silvery metal collected on the negative terminal and immediately caught fire giving off an intense light. Being an excellent experimenter, he was soon able to carry out the decomposition so that the new metal did not catch fire when it came in contact with oxygen. When he placed the globules of metal in water, they skimmed about like water bugs and burned with a lavender light. Because the new metal came from potash, it was called "potassium." The electrical process of obtaining it is called "electrolysis."

Between the years 1807 and 1808, Davy passed his electric currents through six substances that had never been analyzed and broke them down into elements. The elements were oxygen, on the one hand, and six new metals. The compound substances were the alkalis, potash and soda, and the so-called alkaline earths, baryta, strontia, lime, and magnesia. He called the corresponding metallic elements potassium, sodium, barium, strontium, calcium and magnesium, deriving the names from the old names for the corresponding substances. A new method of analysis, electrolysis, had entered the practice of chemistry. It was a magnificent tool that would give humanity a wonderful new metal—aluminum.

The Law of Mass Action

Berthollet had been wrong in his controversy with Proust about the Law of Fixed (or Constant) Proportions. With Proust's victory in that battle assured, chemistry rejected another idea of Berthollet that contained the germ of an important discovery. Berthollet's really important point is that chemical reactions do not always proceed until all of the original chemicals are "used up." This is true even if the quantities of the original substances are in precisely the correct theoretical amounts to achieve a complete reaction. To understand Berthollet's point, let's consider a reaction that chemists now call "reversible." Black had shown that when chalk is heated, it changes to lime by giving off carbon dioxide to the atmosphere. The reaction of the lime kiln can be written as follows, where the arrow means *yields:*

chalk	$\rightarrow$lime	+fixed air
$CaCO_3$	$\rightarrow$$CaO$	+CO_2
calcium carbonate	$\rightarrow$calcium oxide	+carbon dioxide

But the same three chemicals can be involved in just the opposite reaction. If lime is dissolved in water, it produces a solution called "lime-water." When carbon dioxide is bubbled through lime-water, it unites with the lime, and chalk powder settles to the bottom of the beaker. Because chalk is not very soluble in water, it cannot remain in solution. It is said to form a "precipitate" or to "precipitate out" of the solution. The chemical reaction can be written as follows:

chalk	$\leftarrow$lime	+fixed air
$CaCO_3$	$\leftarrow$$CaO$	+CO_2
calcium carbonate	$\leftarrow$calcium oxide	+carbon dioxide

Why does the arrow point to the right in the lime kiln reaction, and to the left for the beaker of lime-water? (The water does not enter into the reaction. The same reaction would take place if a flow of CO_2 gas were passed through powdered CaO, though it would take longer.) There is certainly nothing absolute about the way these substances react with each other.

In the lime kiln reaction, the chalk changes completely to lime because the CO_2 is *volatile*—it is a gas that *immediately leaves the region of the chemical reaction.* Because it does so, it cannot combine with the newly formed lime to change some of it back into chalk. The lime water reaction goes to completion, because the chalk leaves the solution as a precipitate. Additional CO_2 passes through the lime solution until all of the lime combines with CO_2 and settles out as chalk. Another way of putting it is to say there is an excess of carbon dioxide in the reaction.

Another example will help to make the point clear. When steam (water vapor, or oxide of hydrogen) is passed over red hot iron, iron oxide and hydrogen are produced:

$$\text{Iron} + \text{steam} \rightarrow \text{iron oxide} + \text{hydrogen}$$
$$3Fe + 4H_2O \rightarrow Fe_3O_4 + 4H_2$$

Three atoms of iron (Fe) plus four molecules of water vapor yield one molecule of iron oxide and four molecules of hydrogen gas (H_2). Conversely, when hydrogen gas is passed over heated iron oxide, we get steam and metallic iron.

$$\text{Hydrogen} + \text{iron oxide} \rightarrow \text{steam} + \text{iron}$$
$$4H_2 + Fe_3O_4 \rightarrow 4H_2O + 3Fe$$

It is clear that this too is a reversible reaction. Once again, what determines the direction of the arrow? In the former reaction, there is a great *excess* of steam and the hydrogen that is formed is swept away by the flowing steam and cannot react with the iron oxide. In the latter reaction, the hydrogen is present in abundance, and the water vapor is carried away by the flowing hydrogen. It is prevented, therefore, from reacting with the metallic iron. By changing the relative concentrations of the steam and hydrogen, a chemist can make the reaction take place in either direction.

Suppose we heat all four substances in a closed vessel. Which reaction takes place? The answer is *both*. Steam will react with iron, and hydrogen will react with iron oxide. Eventually, a condition of balance or equilibrium will come about. The two reactions will then take place simultaneously so as to nullify one another. Neither reaction will go on to completion. The amounts of iron and iron oxide produced depend on the concentrations of hydrogen and steam. The greater the concentration of hydrogen, the greater the amount of iron. The greater the concentration of steam, the greater the amount of iron oxide. The Law of Mass Action (or of Equilibrium) helps us understand why. The law states that the *rate* or speed of either reaction depends upon the concentrations of the reacting substances. If there is an excess of steam, the steam+iron reaction speeds up. If there is an excess of hydrogen, the hydrogen + iron oxide reaction speeds up. Depending upon the relative concentrations, one of the reversible reactions speeds up and the other slows down until a condition of equilibrium is reached. There is then a definite relationship between the concentration of the steam and the concentration of the hydrogen.

Many reactions, like that of the lime kiln, go on to completion because one of the products of the reaction is volatile—that is, a gas. When table salt and sulfuric acid combine, one of the products is volatile HCl which escapes.

$$\begin{array}{llll} \text{Salt} + & \text{Sulfuric acid} & \rightarrow \text{Sodium sulfate} & + \text{Hydrogen chloride} \\ 2NaCl + & H_2SO_4 & \rightarrow Na_2SO_4 & + 2HCl \end{array}$$

Equilibrium cannot be established because of the escaping HCl and the reaction goes to completion. Other reactions go on to completion because one of the substances is insoluble:

$$\begin{array}{llll} \text{Barium chloride} & + \text{sodium sulfate} & \rightarrow \text{Barium sulfate} & + \text{salt} \\ BaCl_2 & + Na_2SO_4 & \rightarrow BaSO_4 & + 2NaCl \end{array}$$

Barium sulfate precipitates because it is insoluble. So additional barium sulfate is continually formed to take its place until one of the initial chemicals is used up.

Even if none of the chemicals is precipitated or volatile at ordinary temperatures, it is still possible for some reactions to go to completion. This can happen if the desired end product has a much lower boiling temperature than the other liquids involved. For example, the following is an early method of preparing nitric acid:

$$\begin{array}{llll} \text{Sodium nitrate} & + \text{Sulfuric acid} & \rightarrow \text{sodium bisulfate} & + \text{Nitric acid} \\ NaNO_3 & + H_2SO_4 & \rightarrow NaHSO_4 & + HNO_3 \end{array}$$

Nitric acid boils at 86°C (157°F), a temperature much below the boiling point of sulfuric acid (330°C or 646°F). This makes it possible to distill HNO_3 from a mixture of $NaNO_3$ and H_2SO_4 in an apparatus like that shown in Figure 76. Because of its lower boiling point, HNO_3 distills from the solution in preference to H_2SO_4.

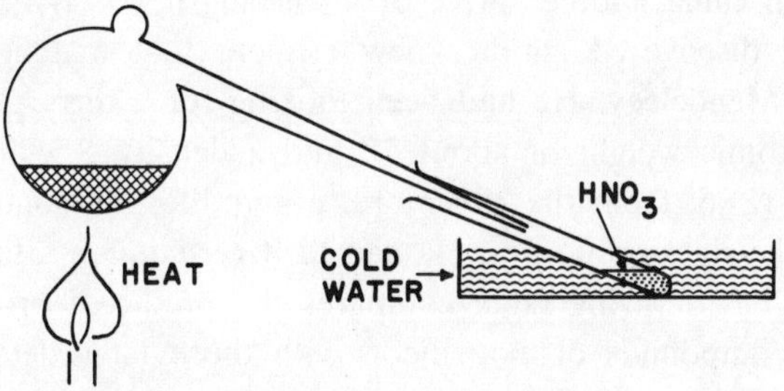

Fig. 76. An apparatus for distilling nitric acid (HNO_3) from a mixture of sodium nitrate and sulfuric acid in the heated container..

The Siberian Prophet

In the year 1869, a professor of chemistry at the University of St. Petersburg made a startling prediction. To the list of sixty-three known chemical elements, at least three more would be added. He called them eka-aluminum, eka-boron and eka-silicon. Dmitri Ivanovitch Mendeléev (1834–1907) had performed no related experiments, and had never seen the new elements whose discovery he boldly predicted. He knew of no substances that contained them and had not the slightest idea of where they might be found.

Nevertheless, this young man from Tobolsk, Siberia, was brash enough to predict their chemical and physical properties. Eka-aluminum would be found to have properties similar to the metal aluminum. Its atomic weight would be sixty-eight times that of the hydrogen atom. It would be easily fusible, it would form chemicals called alums, and its chloride would be volatile. Then, in 1875, this new element was discovered. From a zinc ore, Lecoq de Boisbaudran isolated the hidden eka-aluminum. Repeated experiments proved that there had been no mistake. This new metal was, indeed, the element whose properties had been so accurately predicted by Mendeléev. The discoverer called it *gallium* after the ancient name of his native land, France.

There were many who were skeptical. Perhaps it was just a lucky guess. Had not the great Lavoisier said that it was impossible to predict the nature and number of the elements?

But then came another piece of disturbing news. Winkler in Germany had discovered another new element that matched the eka-silicon of Mendeléev. He had been looking for a dirty gray element with an atomic weight of about 72 and a density 5.5 times that of water. In 1886, from the silver ore, argyrodite, he isolated a metal with just the right properties. He named it *germanium* after his native country. The following table compares the predicted properties and chemical compounds of eka-silicon with those later determined for germanium.

	Predicted (eka-silicon)	Measured (Germanium)
Atomic weight	72	72.3
Specific gravity	5.5	5.47
Formula of oxide	XO_2	GeO_2
Specific gravity of oxide	4.7	4.7
Formula of chloride	XCl_4	$GeCl_4$
Boiling point of chloride	below 100°C	86°C
Formula of fluoride	XF_4	GeF_4
Formula of ethyl compound	$X(C_2H_5)_4$	$Ge(C_2H_5)_4$
Boiling point of ethyl compound	160°C	160°C
Specific gravity of ethyl compound	.96	.96

(*X* stands for eka-silicon.)

The accuracy of the predictions left Mendeléev's critics dumbfounded. Perhaps the young Russian was not a charlatan after all.

Two years later the world of science was completely convinced. From a laboratory in Scandinavia came news that Nilson had isolated eka-boron. He had found indications of the element in the ore euxenite and had kept working until the new element lay before him. Like its two predecessors, the new metal scandium exhibited every property that had been predicted for it. The data were now conclusive, and chemists sought to learn the new methods of Mendeléev.

The Periodic Law

Mendeléev had achieved his phenomenal chemical insight by a careful inspection of all known properties of the elements. He gathered together all of the data on the sixty-three elements of his time. He even included fluorine—known to be an element, although not yet isolated. This information was recorded on sixty-three cards, one for each element. Then he carefully examined all of the information.

Here were all the chemical elements, each one made up of a different kind of Daltonian atom. Their atomic weights ranged from 1 for hydrogen to 238 for uranium, the heaviest element. Some, like hydrogen, oxygen, nitrogen and chlorine are gases at room temperature. Others, like mercury and bromine, are liquids. Still others are solids. Some are soft metals, like sodium and potassium, and others are hard like platinum and iridium. Some are heavy like uranium and osmium, and others float on water like the metal lithium. Iron corrodes quickly in the atmosphere, but gold never tarnishes. Copper is red, iodine is gray, and phosphorus is white. Some elements unite with one atom of oxygen, others with two, and others still with three or four.

In this maze of diverse and seemingly unrelated data, Mendeléev found a beautiful kind of order. He noticed that all the elements seem to fall naturally into a number of groups on the basis of their properties. As he laid these groups of cards out before him, a striking relationship became clear. Except for hydrogen, the properties of the elements seemed to be related to their atomic weights in a way that repeats itself periodically. To illustrate, if we place hydrogen in a class by itself, we can list the lightest elements in each of Mendeléev's columns as follows:

Lithium	Beryllium	Boron	Carbon	Nitrogen	Oxygen	Fluorine
7	9.4	11	12	14	16	19

The numbers under the elements are the atomic weights as known to Mendeléev. These were the lightest known elements (except for hydrogen) and each seemed to be the head of a class of elements having similar properties. The next heavier element, for example, is sodium (23). Sodium resembles lithium quite closely in both its chemical and physical properties. Thus, Mendeléev placed it below lithium in the table of elements. Then, in normal sequence, came magnesium (24), aluminum (27.3), silicon (28), phosphorus (31), and sulfur (32), each falling miraculously beneath the element it resembles. The next row began with potassium (39) and, once again, each element seemed to belong in its place. In this way, he arranged all the elements into a pattern that told a story of their properties.

When the table was completed, he remarked at how well the elements fitted into their places. The extremely active nonmetals—fluorine, chlorine, bromine and iodine—all showed up in the same group, number seven. The active metals lithium, sodium and potassium were all in the first group. Mendeléev had discovered that the properties of the elements are repeated periodically, when they are arranged in order of atomic weight. He had discovered a profound new regularity among the elements.

But there was more to be learned from Mendeléev's *Periodic Table of the Elements*. Arranged in that way, the elements showed an astonishing regularity in the way they unite with oxygen. The elements in Group I unite with oxygen two atoms to one. The chemical formulas are Li_2O, Na_2O, K_2O and so on. All the atoms in the second group unite with oxygen, atom for atom: BeO, MgO, CaO, ZnO, and so on. The elements of the third group join on the ratio of two atoms to three: B_2O_3, Al_2O_3, In_2O_3, and so on. Similarly, regular ratios exist for the other groups of elements. How wonderfully simple. Knowing the properties of one element in a group is to know, in a general way, the properties of the other elements in that group.

In order to arrange the elements in their correct groups, Mendeléev found that gold (199) and platinum (198) had to be placed in reverse order in the table. With characteristic boldness, he proclaimed that the currently accepted weights of gold and platinum were wrong. Gold must be heavier than platinum. He was later vindicated when the atomic weights were determined more accurately: gold (197) and platinum (195). Gold is heavier than platinum just as Mendeléev had staunchly maintained.

The same problem arose for tellurium and iodine—and once again Mendeléev placed the elements in reverse order, the heavier tellurium appearing before the lighter iodine in the table. The critics harped on this inconsistency because repeated determinations of the atomic weights showed that tellurium (127.6) *was indeed* heavier than iodine (126.9). And yet the properties of the two elements made it clear that they belong in reverse order. Why was this so? Perhaps the future would explain this paradox. The difference in atomic weight, after all, is quite small—only about ½ of 1 per cent. The explanation became clear in later years when more was learned about the nature of the atom. The atomic weights were only *nearly* perfect as a method of ordering the elements. The principle of classification had to be sought not in the atomic weight itself, but in some other characteristic closely connected with it. As we shall see, it was later learned that the chemical properties of the elements depend upon the number of electrical constituents (electrons) of the atom. This number is closely related to the atomic weight, but other factors may affect an atom's weight without having an effect upon the number of electrons it contains.

The current practice is to assign each element an *atomic number,* beginning with 1 for hydrogen and continuing successively up to 92 for uranium, the heaviest naturally occurring element. These atomic numbers are used in the accompanying periodic table which has been brought up to date by including the elements discovered since Mendeléev's time.

The group of elements from atomic number 58 to 71 are listed below the table because they are not related to the elements below which they would ordinarily fall. Together with Lanthanum, 57, they have long been called the rare earth metals. Some of them are actually not rare. Cerium, 58, occurs more abundantly than lead, tin or mercury. The problem with these metals has long been the difficulty of extracting them from natural substances. They occur as compounds mixed together in monazite sand. Because they are so chemically similar, their separation is much too expensive for most commercial purposes.

Valence and Atomic Marriage

Mendeléev's great accomplishment was the Periodic Table which classified the elements logically into clearly recognizable families. One of the family characteristics made obvious by the table is called "valence."

THE PERIODIC TABLE OF

Period	Group 0	Group I	Group II	Group III
1		Hydrogen 1		
2	Helium 2	Lithium 3	Beryllium 4	Boron 5
3	Neon 10	Sodium 11	Magnesium 12	Aluminum 13
4	Argon 18	Potassium 19 Copper 29	Calcium 20 Zinc 30	Scandium 21 Gallium 31
5	Krypton 36	Rubidium 37 Silver 47	Strontium 38 Cadmium 48	Yttrium 39 Indium 49
6	Xenon 54	Cesium 55 Gold 79	Barium 56 Mercury 80	Lanthanum* 57 Thallium 81
7	Radon 86	Francium 87	Radium 88	Actinium 89

* RARE EARTHS: Cerium 58; Praseodymium 59; Neodymium 60; Promethium 61; Samarium 62; Europium 63; Gadolinium 64; Terbium 65; Dysprosium 66; Holmium 67; Erbium 68; Thulium 69; Ytterbium 70; Lutetium 71.

It was well known in Mendeléev's time that atoms of different elements combine in various ratios to form compounds. But these ratios are most specific, and depend upon the particular elements entering into combination. Hydrogen and chlorine, for example, can combine only atom for atom; an atom of hydrogen never combines with less than nor more than one atom of chlorine. So we write the chemical formula for hydrogen chloride as HCl. An atom of the metal calcium, on the other hand, can combine with two atoms of chlorine to form calcium chloride ($CaCl_2$). The atoms of calcium have twice the combining capacity of hydrogen atoms. An atom of aluminum can combine with three atoms of chlorine to give $AlCl_3$, showing that the combining capacity of aluminum is three times that of hydrogen. An atom of carbon can combine with four atoms of chlorine to form CCl_4, carbon tetrachloride. Thus, carbon has four times the combining capacity of hydrogen. The atoms of each element have a certain definite combining capacity—a characteristic that is called its "valence." Valence comes from the Latin noun *valentia,*

THE CHEMICAL ELEMENTS (1962)

Group IV	Group V	Group VI	Group VII	Group VIII
Carbon 6	Nitrogen 7	Oxygen 8	Fluorine 9	
Silicon 14	Phosphorus 15	Sulfur 16	Chlorine 17	
Titanium 22 Germanium 32	Vanadium 23 Arsenic 33	Chromium 24 Selenium 34	Manganese 25 Bromine 35	Iron 26 Cobalt 27 Nickel 28
Zirconium 40 Tin 50	Niobium 41 Antimony 51	Molybdenum 42 Tellurium 52	Technetium 43 Iodine 53	Ruthenium 44 Rhodium 45 Palladium 46
Hafnium 72 Lead 82	Tantalum 73 Bismuth 83	Tungsten 74 Polonium 84	Rhenium 75 Astatine 85	Osmium 76 Iridium 77 Platinum 78
Thorium 90	Protactinium 91	Uranium** 92		

** The man-made elements follow uranium.

meaning strength. By implication, a carbon atom has the "strength" of four hydrogen atoms because its combining capacity, or valence, is four times as great.

The valence of hydrogen is taken to be one, because a hydrogen atom never combines with more than one atom of any other element. It is a convenient standard, therefore, by which the valence of the other elements can be measured. Hydrogen combines with oxygen two atoms to one, for example, and produces water, H_2O. The valence of oxygen must then be two, because one atom of oxygen combines with two of hydrogen. Similarly, the valence of chlorine is one because the formula of hydrogen chloride is HCl. Chlorine is said to be *univalent* and oxygen is *bivalent*. Nitrogen, which can combine with hydrogen three atoms to one as in ammonia (NH_3) is said to be *trivalent*. Carbon, an atom of which can combine with four atoms of hydrogen (CH_4) is said to be *quadrivalent,* and so on.

The valence of an element can be thought of as the number of atoms of hydrogen that can be replaced by one atom of the element. Calcium oxide has the formula CaO, and hydrogen oxide, or water, has the formula H_2O. Calcium has a valence of two, therefore, because one calcium atom takes the place of two hydrogen atoms when their oxides are compared. Valence can also be thought of as the number of atoms of oxygen that combine with *two* atoms of the element. This rule is based on the fact that oxygen has a valence of two. The metals sodium and potassium, whose oxides are Na_2O and K_2O, have a valence of one because one oxygen atom combines with two atoms of either element. The oxides of calcium, barium and strontium (CaO, BaO, SrO) show these metals to have a valence of two.

Certain elements can have more than one valence. The oxides of nitrogen and phosphorus (N_2O_5, P_2O_5), which form nitric and phosphoric acids in water, tell us that nitrogen and phosphorus have a valence of five in these compounds. The same elements combine with hydrogen (NH_3, PH_3) to form ammonia and phosphine, but with a valence of only three instead of five. When an element such as nitrogen exhibits two different valences, the two numbers usually add up to eight, as in the examples above.

The idea of valence was first clearly stated by Sir Edward Frankland in 1852. It was based, in those days, on the way in which elements seemed to behave. Only during the present century was a theoretical basis found that could lead to an understanding of valence. We now know that it is connected with the sharing of electrons by atoms of different elements.

The Chemist's Helper

Ever since the discovery of carbon monoxide (CO), it has been known that the gas can be made to unite with oxygen to form carbon dioxide gas (CO_2). This reaction causes the blue flame we see over a burning bed of coals. It was learned near the end of the eighteenth century, however, that a trace of water must be present for the reaction to take place. The amount of water needed is minute, but its presence greatly accelerates the reaction. Measurements prove that the water does not enter into the reaction. There is just as much water present after the reaction as there was before, but its presence somehow facilitates the reaction.

In the early decades of the nineteenth century, many similar reactions were noticed, all having one common characteristic—the rate of chemical change was greatly increased by the presence of certain other substances in extremely minute amounts. Berzelius suggested, in 1838, that such substances speed up a chemical reaction by "arousing the slumbering affinities of the substances," thereby inducing them to combine. He introduced the term "catalysis" for this class of reaction. The substance that causes catalysis is called a "catalyst" or a "catalytic agent." Just how these catalysts function is still far from being completely understood. Yet, they have become an indispensable part of modern chemical industry. Hundreds are used to hasten or slow down important reactions or to increase the yield of processes that would otherwise be commercially impracticable. They have been called the chemical parsons that unite the elements in marriage, but remain uninvolved themselves.

The amount of a catalyst necessary for some reactions is unbelievably small. When sodium sulfite (Na_2SO_3) is dissolved in water, the oxygen of the air slowly oxidizes it to sodium sulfate (Na_2SO_4). But if a minute amount of copper is added, the process of oxidation is greatly speeded up. The effect of copper on the process can be noticed even if the water used for the sodium sulfite solution has been in a copper pan for a few moments. In fact, the catalyst is so efficient that a concentration of one ounce of copper sulfate in 600 million gallons of water contains enough copper to speed up the reaction! Catalysis is a supreme example of the infinite importance of the microscopically small.

Catalysts have been indispensable ever since the beginning of chemical industry. One of their more important applications has been in the manufacture of sulfuric acid (H_2SO_4). Sulfuric acid is a workhorse of our modern civilization. It is used in making fertilizer, in refining petroleum, in cleaning metals before plating, in electrical storage batteries, in the manufacture of rayon, dyes and explosives, and in countless other products. The amount of sulfuric acid used by a nation is always a good barometer of its standard of living.

Sulfuric acid manufacture begins with the gas sulfur dioxide (SO_2) which can be obtained in many ways: by burning sulfur in air, by heating iron sulfide (FeS_2), or as a by-product in the smelting of various ores. Sulfur dioxide (SO_2) must then be oxidized to form sulfur trioxide (SO_3), which combines with water to give sulfuric acid (H_2SO_4).

$$SO_3 + H_2O \rightarrow H_2SO_4$$
$$\text{sulfur trioxide} + \text{water} \rightarrow \text{sulfuric acid}$$

The major difficulty encountered in making sulfuric acid has always been the oxidation of sulfur dioxide to form sulfur trioxide. The reaction between sulfur dioxide and oxygen takes place very slowly at ordinary temperatures. When the temperature of the gases is raised, the reaction is indeed speeded up, but another problem arises. The increased temperature causes much of the sulfur trioxide to change

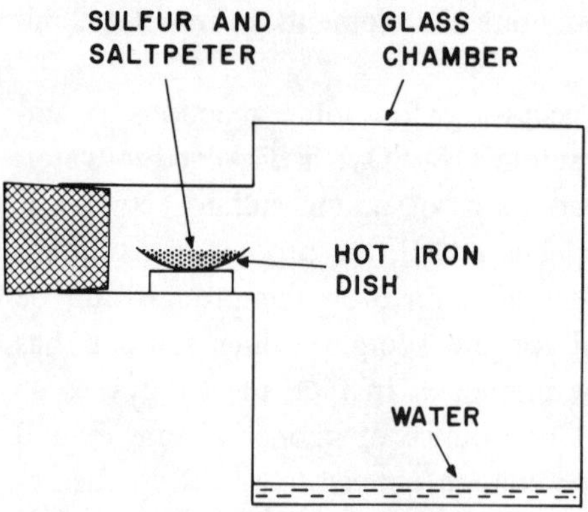

Fig. 77. The chamber process for preparing sulfuric acid.

back into sulfur dioxide and oxygen. The reaction is reversible, and high temperature favors the decomposition of sulfur trioxide, the desired substance. Unfortunately, a temperature that produces sulfur trioxide at a satisfactory rate results in an amount sulfur trioxide that is too small for commercial success of the process.

The difficulty was surmounted, undoubtedly by accident, in 1666 by Nicolas le Fèvre and Nicolas Lémery. They prepared the acid by burning sulfur in the presence of a little saltpeter, potassium nitrate (KNO_3). The process was first used on a large scale by Ward, in 1740, in a plant at Richmond near London. As shown in Figure 77, Ward used large glass chambers up to 66 gallons capacity with horizontally projecting necks. The bottom contained a small amount of water. In the neck was a piece of earthenware upon which was placed a red hot iron dish containing sulfur and potassium nitrate. The neck of the chamber was then closed with a wooden plug. When the sulfur was completely burned, the operation was repeated until the acid at the bottom of the chamber had the desired strength.

We know today that the "chamber process," as it is called, makes use of catalysts provided by the potassium nitrate. The heat of the red hot dish decomposes the potassium nitrate to produce the catalysts, nitric oxide (NO) and nitrogen dioxide (NO_2). In additon to these gases, the chamber also contains sulfur dioxide (SO_2), produced by the burned sulfur. The sulfur dioxide reacts with nitrogen dioxide to produce sulfur trioxide and nitric oxide.

$$SO_2 + NO_2 \rightarrow SO_3 + NO$$

$$\text{sulfur dioxide} + \text{nitrogen dioxide} \rightarrow \text{sulfur trioxide} + \text{nitric oxide}$$

The nitric oxide (NO) then combines with oxygen in the chamber to form nitrogen dioxide once again. The sulfur trioxide (SO_3) combines with the water in the chamber to produce sulfuric acid.

The catalysts NO and NO_2 are not used up in the process. They merely provide a series of intermediate reactions that occur easily, in place of a single direct reaction that occurs only with difficulty.

A modern method of converting sulfur dioxide to the trioxide uses a different catalyst, either platinum or certain compounds of the metal vanadium. At a temperature of about 450°C (about 840°F), the oxidation of sulfur dioxide to the trioxide takes place rapidly in the presence of the catalyst, producing almost 100 per cent sulfuric acid when combined with water.

A knowledge of catalysis is also responsible for the enormous use of butter substitutes. Animal and vegetable oils and fats are, essentially, compounds of glycerine and acids such as stearic and oleic acids. Some of these compounds are solid fats while others are liquid oils. The glycerine compound of stearic acid, for example, is a main constituent of beef suet, while that of oleic acid is a liquid found in olive oil. The only difference between the two acids is the amount of hydrogen they contain. Oleic acid (which produces oil) differs from stearic acid (which produces hard fat) only in that it contains less hydrogen. The former is called an "unsaturated" compound, while the latter is referred to as "saturated."

Catalysis accommodates our preference for hard fats by facilitating the chemical reaction that adds hydrogen to oleic acid, thereby producing saturated stearic acid. The catalyst that performs the task is finely divided nickel. In the presence of this metal, various liquid oils combine with gaseous hydrogen and are converted into solid fats—a process called "hydrogenation."

There is a simple catalytic reaction that can be performed by anyone in his own home. First, try to set fire to a lump of sugar with the flame of a match. The sugar will char, but will not catch fire. Now rub the sugar in cigarette ash and try again. The ash acts as a catalyst, and the sugar catches fire quite easily and burns rather briskly. In hundreds of similar ways, catalysts assist our chemists in the manufacture of countless products so essential to our modern way of life.

Death of the Vital Spirit

Carbon which is familiar to us in its three physical forms—charcoal, graphite and diamond—is the most prolific of the elements. Its compounds are so numerous—counted in the hundreds of thousands—that their study has developed into a separate branch of chemistry, *organic* chemistry. The name survives from an earlier and less scientific time when chemists drew a sharp distinction between compounds derived from living organisms and those from the lifeless mineral world. They believed that the compounds of carbon that come from plants and animals could only be synthesized with the help of a special form of energy inherent in the living cell—the so called "vital force."

In 1828, a young German scientist, Friedrich Wöhler (1800–1882), synthesized the first organic compound outside a living organism. The substance, urea ($NH_2 \cdot CO \cdot NH_2$), had been discovered in urine in 1773 by Rouelle. As soon as he was sure of his ground, Wöhler wrote to his teacher, Berzelius, "I must tell you that I can prepare urea without requiring a kidney of an animal, either man or dog." The chemical world was astonished at the news. Perhaps this magnificent new achievement would lead to the synthesis of the sugars, the proteins, or to the substance of life itself.

But the voices of a few still sounded a feeble protest. Wöhler had begun, after all, with ammonia (NH_3) which originally was of animal origin. It may have carried along enough of the "vital force" to give rise to urea. Although the argument went on, the protest was soon lost in the great excitement of the discovery. Today we know that ammonia can be produced synthetically from inanimate nitrogen and hydrogen. Wöhler's great achievement had driven the last vestiges of magic from the theory of chemistry. The synthesis of urea finally banished the alchemical spirits from science. The ancient fallacy of mysterious forces persists today only in astrology and other kinds of superstitious nonsense.

Isomers: Chemistry's Unidentical Twins

One of the first new compounds prepared by Wöhler while studying under Berzelius was silver cyanate. At the same time, another young German chemist, Justus von Liebig (1803–1873), was at work in the laboratory of Gay-Lussac in Paris. Liebig had always been interested in fireworks, and he was busy with the study of the explosive chemicals known as fulminates. During these researches, Liebig prepared a strange new compound having the same atomic composition as Wöhler's silver cyanate, yet of vastly different physical and chemical properties. Here was a most puzzling development. How could two compounds made of the same kinds of atoms in precisely the same proportions have different properties? Apparently the old rule, "one composition, one compound," could no longer be accepted as an article of faith.

The two young men communicated with each other and became lifelong friends. When Berzelius learned of the discovery, he recognized its tremendous importance and coined the term "isomers" to designate chemical compounds having the same composition but different properties. A great interest was aroused in isomerism and soon many substances were found having dozens of isomers. More than one hundred compounds can be produced, for example, by combining nine carbon atoms, ten hydrogen atoms, and three oxygen atoms. The isomerism of carbon compounds helps to explain the enormous number of compounds in organic chemistry—a number that is increasing at a rate of several thousand each year.

Chemical Architecture

Dalton had conceived of a molecule as a number of atoms linked together to form a definite molecular pattern. The discovery of isomerism by Liebig led to the recognition that the properties of a compound depend upon the particular arrangement of the atoms in a molecule and not merely upon its composition. Just as bricks can be arranged in various patterns, so also can atoms be linked in different ways within the molecule. Each different structure produces a compound having different properties.

Our knowledge of an isomer is complete only when we know the internal structure or *constitution* of the molecule. Only then can a reasonable method be devised to *synthesize* a compound, or build it up from simpler materials. A great advance in this direction was provided in 1858 by Archibald Couper and Friedrich Kekulé. They showed that isomerism could be understood by drawing *graphic* formulas for isomers. Their idea was to represent the union of two atoms as a line (or dot), according to the valence of the atom. We can illustrate the concept by drawing the graphic formula for the hydrocarbon compound methane, CH_4. Methane contains one carbon atom and four hydrogen atoms.

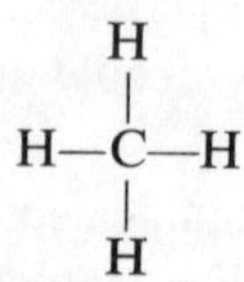

Carbon has a valence of four and hydrogen a valence of one. We can think of a carbon atom as having four individual places that have a tendency to link with other atoms. Hydrogen, with a valence of one, has only one such place. Each line, or bond, therefore, represents a valence of one. The carbon atom has four valence bonds associated with it and each hydrogen atom has one such bond.

Another hydrocarbon in the same series as methane is propane C_3H_8 which can be represented graphically as follows:

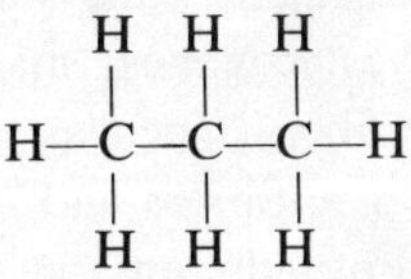

This formula can also be written in the simpler form, $CH_3 \cdot CH_2 \cdot CH_3$, known as a "structural" or "constitutional" formula. Each dot in the formula represents a bond between adjacent carbon atoms in the molecule. The arrangement given above is the only one possible for the hydrocarbon propane. But two atomic arrangements are possible for butane, C_4H_{10}: normal butane and isobutane.

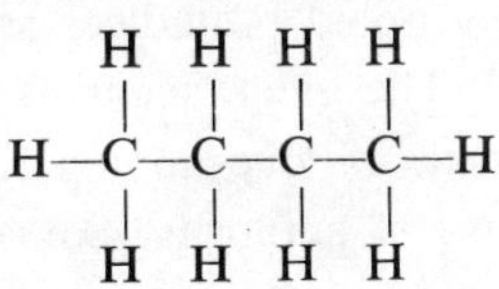

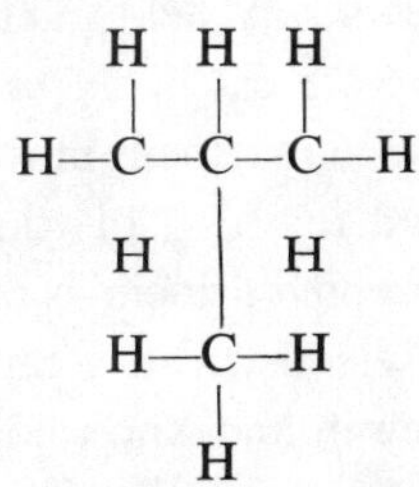

$CH_3 \cdot CH_2 \cdot CH_2 \cdot CH_3$
(normal butane)

$CH_3 \cdot CH \cdot CH_3$
$\cdot$
CH_3
(isobutane)

Butane has a *straight chain*, and isobutane has a *branched chain*.

The number of possible isomers increases rapidly as the number of carbon atoms increases. There can be three pentanes (C_5H_{12}), five hexanes (C_6H_{14}), nine heptanes (C_7H_{16}), and eighteen octanes (C_8H_{18}).

The compounds discussed above are called *saturated* hydrocarbons, because the carbon atoms are linked together by only one valence bond. Saturated compounds can accept no additional hydrogen. In *unsaturated* compounds, a pair of carbon atoms can be linked by two or three bonds: $CH_2{:}CH{\cdot}CH_3$ (propylene) and $CH\vdots CH$ (acetylene).

Organic chemistry uses these structural concepts to synthesize the countless new products that have made our lives so enjoyable. It has given us antibiotics, anesthetics, and many other wonderful drugs; thousands of synthetic dyes; countless man-made plastics; new and better fuels; the "miracle fibers" such as nylon, Orlon and Dacron; synthetic rubber; improved fertilizers and insecticides for the farmer; new detergents; and a host of indispensable products that find application in every phase of human activity.

All of this was made possible by the stubborn persistence of one man to fight the established scientific dogma of his time—the universal belief that a "vital force" forever placed the organic compounds beyond the reach of man. The history of science is a recurring story of great men who had the insight and courage to question beliefs that were universally held. Yet the most deep-rooted prejudices seem to be the ones most likely to be overturned. The great scientists of the past have done their most significant work by questioning the obvious, and rejecting the dogmatic teachings of authority. Somewhere in the accepted principles of modern science is a fallacy, and some day a gifted scientist with a rare talent will search it out. And once again man's new found knowledge of nature will explode into a sparkling shower of gifts to all of mankind.

Atoms of Electricity

The idea of atoms of electricity, or "electrons" as we would say, developed rather slowly. The first step in the discovery of the electron was taken in 1833 by the great Michael Faraday (1791–1867) who began his scientific career as a bottle washer in the laboratory of Sir Humphrey Davy. It was in this same laboratory that Davy had used an electric current from the voltaic pile to isolate so many new metals.

One of Faraday's important contributions was the discovery that electricity seems to exist in multiples of small definite amounts. The amount of electricity in these electrons, to use the modern term, is always fixed—never more nor less than a specific amount. Faraday made this deduction from careful experiments in which an electric current was passed through various chemicals dissolved in water—the electrolysis of aqueous (water) solutions.

Faraday began with an apparatus similar to that in Figure 78, and found that some solutions provide an easy path for the electric current while others do not. Substances whose solutions conduct electricity are called "electrolytes" while those that do not are called "nonelectrolytes." Faraday tested many electrolytic solutions and found them to be decomposed in one way or another by the electric current. When copper sulfate was tried, for example, metallic copper was

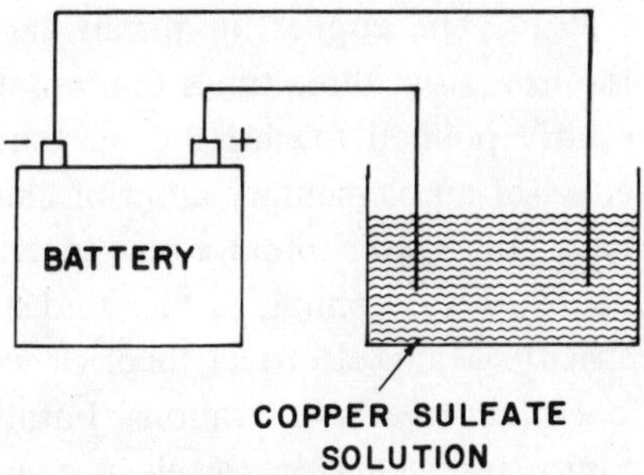

Fig. 78. Faraday's apparatus for studying the electrolysis of chemicals in water solutions.

deposited on the negative electrode in the solution. In all of these experiments, he found that the weight of the deposited metal is proportional to the amount of electricity passed through the solution.

Faraday then went a step further. He passed precisely the same current through several different solutions and found a relation between the number of atoms liberated by the current and the valence of those atoms. To illustrate, imagine that the same amount of electricity is passed through three solutions: hydrochloric acid (HCl), copper sulfate ($CuSO_4$) and iron chloride ($FeCl_3$). If we limit our attention to the negative electrode in each solution, three elements are liberated: hydrogen, copper and iron. Hydrogen, being a gas, bubbles up through the solution to the surface where it can be collected and weighed. The metals, copper and iron, collect at the negative terminals and they, too, can be weighed. If we divide these three weights by the relative atomic weights of hydrogen, copper and iron, we obtain the relative number of atoms of each element liberated by the same amount of electricity. Suppose that x atoms of hydrogen are liberated. Then, we find that $\frac{1}{2}x$ atoms of copper and $\frac{1}{3}x$ atoms of iron have been produced. The relative numbers of atoms of these three elements are in the ratio $1:\frac{1}{2}:\frac{1}{3}$.

We now know that these numbers are related to the valence of hydrogen (1), copper (2), and iron (3) in the original compounds HCl, $CuSO_4$ and $FeCl_3$. These results suggest that each kind of atom is able to carry only a definite amount of electricity, a multiple of the charge on an electron. The smallest unit of electricity is associated with the hydrogen atom. The copper atom can carry exactly twice that amount, and the iron atom three times that amount.

Faraday's law clearly pointed toward the existence of a minimal quantity of electricity—of an elementary atom of charge. It also suggested that the valence bonds which hold atoms together are electrical in nature. Thus, chemistry was coming further under the influence of the science of electricity. Many electrical theories were advanced in succeeding years to explain these observations. Faraday conceived of compounds broken into "fragments" by the electric current. He named them *ions* (or "wanderers" from the Greek) because they "walked" with an electric charge to one electrode or the other in the solution. Others devised different theories, but none received universal support. The modern explanation was advanced in 1887 by a young Swedish student of chemistry named Svante Arrhenius (1859–1927). Far from being accepted immediately, Arrhenius' *ionic theory* set off a violent controversy that divided the world of chemistry. Much later, Arrhenius received the Nobel Prize for Chemistry in 1903 for his research. This is the greatest honor that a chemist can achieve.

Ions versus Atoms

On May 17, 1883, Svante was putting the finishing touches on his dissertation for the doctor's degree at Upsala University, the university of Berzelius. He had performed hundreds of experiments on electrolytes and was carefully sifting the results and writing his report. Then, suddenly, a new idea flashed through his mind, and he could not sleep that night until he had worked through the whole problem. The new doctrine that resulted kindled a flame that set fire to the entire chemical world.

His startling new idea was simply this. When a solid salt, like ordinary table salt (sodium chloride), is dissolved in water, a fundamental change takes place. The change—although invisible—greatly alters the properties of the substance. Both water and sodium chloride are nonconductors of electricity. But a solution of sodium chloride in water is a good conductor of electricity. Arrhenius said that the molecules of sodium chloride split up, or "dissociate," into the particles that Faraday had called "ions." He further stated that these ions were not caused by the electric current, but were present in the solution before the current was sent through. Only these ions take part in chemical reactions.

This unorthodox view almost cost Arrhenius his degree. Except for a few close friends, all of the scientific world rebelled at his revolutionary ideas. He argued that the two ions of the sodium chloride molecule were completely separate—free to swim about independently in all directions in the solution. No NaCl molecules were present; only sodium ions and chlorine ions existed in the solution.

If the molecules dissociate, the skeptics asked, then why is the solution clear and white? Where is the greenish yellow color of the poisonous chlorine atom? Arrhenius replied that the chlorine *ion* is different from the chlorine *atom* because the ion contains an electric charge. The chlorine atom has no electric charge and is electrically neutral. Dissociation had split the molecule of NaCl into ions rather than atoms. Today, we designate the chlorine ion as follows, Cl^-, with a single superscript minus sign to show that it carries a unit negative charge. Similarly, the sodium ion is written, Na^+, because it carries a positive charge. The number of units of electric charge on a particular ion is equal to its valence in the compound from which it came.

Chemical Radicals

Most positive ions consist of only one metallic element: Li^+ (lithium), Ca^{++} (calcium), Al^{+++} (aluminum). Once again, the charge on each ion corresponds to the valence of the corresponding element.

Sometimes an element has more than one valence. Iron, for example, can exist as the ferric ion (Fe^{+++}) or the ferrous ion (Fe^{++}). The ending "ic" stands for the higher valence.

Negative ions often consist of a nonmetal, either alone as for Cl^- (chlorine), O^{--} (oxygen), or S^{--} (sulfur), or in combination with other elements, NO_3^- (nitrate), SO_4^{--} (sulfate), CO_3^{--} (carbonate), ClO_3^- (chlorate), or OH^- (hydroxide). Although few positive ions consist of more than one element, the ammonium ion, NH_4^+, is quite common. Groups of elements, such as these, which can be transferred from one compound to another as a unit are usually called "radicals."

Through the ionic theory, the fields of electricity and chemistry became closely connected. The forces that hold atoms together seemed more than ever to be electrical in nature. When an electrolyte is dissolved in water, it dissociates into ions of equal but opposite electrical charge. These particles carry electricity through the solution during electrolysis.

The theory also explains why hydrogen and chlorine are liberated when a current is passed through a solution of HCl. The H^+ ions are attracted to the negative terminal where their charge is neutralized. This changes the H^+ ion into a hydrogen atom. Two hydrogen atoms join together to form an H_2 molecule of hydrogen gas which escapes from the solution. Analogous changes take place for the Cl^- ions at the positive terminal. If one of the ions in a solution happens to be a metal, such as copper, or silver, it is deposited on the negative electrode. Similar mighty transformations take place each time an inorganic acid, alkali or salt is subjected to electrolysis.

Electrolysis is used extensively, today, to plate expensive metals on objects made of cheaper metals. It is also used to plate corrosion-resistant metals such as chromium, cadmium and nickel on objects made of strong metals such as steel which would otherwise rust excessively.

Electrolyis is also responsible for the great aluminum industry. As far back as 1825, Wöhler had isolated a few pieces of aluminum in the laboratory, but the process was much too costly for commercial application. The common aluminum ore, bauxite, has the formula Al_2O_3 which is similar to iron oxide Fe_2O_3, or hematite. For centuries, chemists had known that carbon is able to remove oxygen from hematite to produce metallic iron. Yet carbon cannot tear aluminum from its chemical partner, oxygen. Aluminum has a much stronger attraction for oxygen than does carbon. Put another way, aluminum is a more powerful reducing agent than carbon.

If aluminum oxide were soluble in water, it might have been possible to separate the metal by passing a current through the solution. Unfortunately, bauxite does not dissolve in water. But even this difficulty can be overcome if the compound can be melted at a reasonably low temperature. Like water solutions, molten compounds can conduct electricity. Unfortunately bauxite melts at a very high temperature—a temperature much too high for commercial practicability.

The impasse was overcome in 1886 by the young chemist, Charles Martin Hall, who hit upon a unique idea. Perhaps he could find a mineral in nature that dissolves aluminum oxide at a reasonable temperature. Working in his father's woodshed, he tried one mineral after another. Then he came across a white glassy rock from Greenland called cryolite. It melted easily, and when bauxite was dropped into the melt it quickly dissolved into the molten cryolite. With great excitement, Hall passed an electric current through the hot solution and saw bright, metallic aluminum deposited on the negative electrode!

Using his new process, Hall organized a company for the manufacture of aluminum that was eventually called the Aluminum Company of America. It is still the largest producer of aluminum in the world. Using Hall's process, the world's producers make millions of tons of aluminum each year, and sell it for pennies a pound.

Hidden in the periodic table are many metals whose extraction from the ore is still costly and inefficient. Perhaps a flash of genius in the mind of some young chemist will soon release one or more of them to the service of man. The common rocks around us hold untold wealth for a scientist with the right idea.

Strange Lines in the Spectrum of Light

Along with electricity, the study of light eventually found its way into chemistry. In 1802, W. H. Wollaston (1766–1828) noticed dark streaks in the spectrum of sunlight—light that had passed through a prism. He assumed that these streaks were boundaries of the natural colors. In 1814, a Bavarian instrument-maker, Joseph Fraunhofer (1787–1826) built a powerful new optical instrument—called a spectroscope—and took a closer look. As shown in Figure 79, a beam of light is formed by means of a narrow slit and a lens. This narrow slit of light is passed through a prism and viewed through a small telescope. The light is refracted by the prism, each color being bent by a different amount. The observer sees as many lines (slit images) as

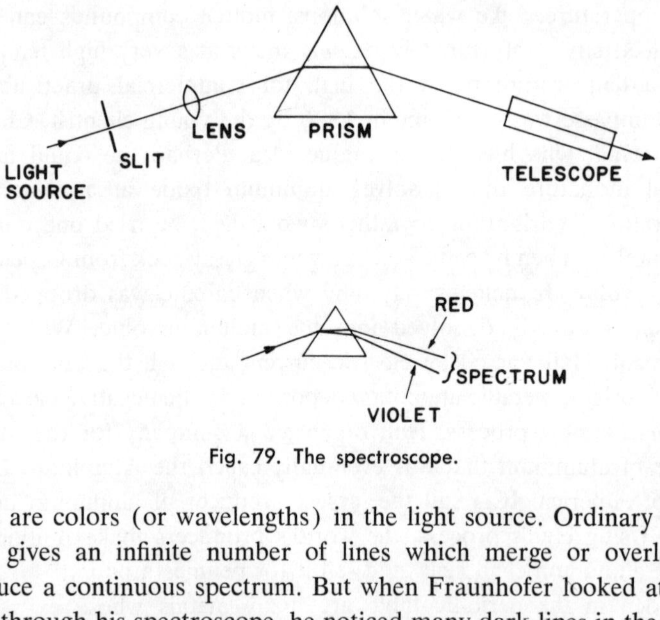

Fig. 79. The spectroscope.

there are colors (or wavelengths) in the light source. Ordinary white light gives an infinite number of lines which merge or overlap to produce a continuous spectrum. But when Fraunhofer looked at sunlight through his spectroscope, he noticed many dark lines in the spectrum, as though certain colors were missing. These lines always kept the same position in the spectrum—each corresponding to a specific wavelength that had somehow disappeared from sunlight. He found the same dark lines in all forms of sunlight, whether direct, or reflected from clouds, moon or planets. Strangely, though, the spectra from the stars had different distributions of lines.

In 1859, Gustav Kirchhoff (1824–1887), working with Robert Bunsen (1811–1899), showed that there is a connection between these lines and certain kinds of matter. Kirchhoff vaporized various elements in a nonluminous flame and found that each element gives off a characteristic series of bright lines when viewed through a spectroscope. While an incandescent solid produces a continuous spectrum of *all* colors, a vaporized element produces only a series of discrete bright lines. The location of these bright lines in the spectrum—their color pattern—is different for each element. Here, then, was a wonderful new way to detect the presence of an element by the telltale bright lines it emits. Because the bright lines are emitted by vaporized substances, they are called "bright line" or "emission spectra."

Kirchhoff also showed that the bright lines were related to the dark lines noticed by Wollaston and Fraunhofer in the solar spectrum. He discovered that the elements, in gaseous form, are able to absorb the same wavelengths (colors) that they are able to emit, and only those wavelengths. As sunlight passes through the sun's atmosphere, the gaseous elements in that atmosphere absorb certain specific wavelengths. This produces corresponding dark lines in the spectroscope against the brighter background of the continuous spectrum. For example, sodium produces two closely spaced yellow lines when it is vaporized in a flame. But if incandescent light (which contains all wavelengths) passes through sodium vapor, light corresponding to the two yellow lines is absorbed and cannot reach the spectroscope. The resulting spectrum is called a "dark line" or "absorption spectrum." With the discovery of dark-line and bright-line spectra, it became possible to chemically analyze the atmospheres of the sun and stars, as well as substances on earth.

The Unsociable Elements

Spectral analysis of the sun's light showed that the chromosphere, an envelope of glowing gas around the sun, consists mainly of hydrogen, calcium and something giving a yellow spectral line not previously observed on earth. Joseph Lockyer (1836–1920) concluded in 1868 that the line was caused by a new element, which he named *helium* from the Greek word meaning "sun." Nevertheless, helium was not accepted immediately as a new element. Mendeléev did not even include it in his table because it did not fit in any column. Later, Hillebrand described a gas expelled by cleveite. William Ramsay (1852–1916) obtained a sample of the gas from the same mineral, vaporized it, and detected the spectral line of Lockyer's helium. He had discovered a new element; one that would not combine with other elements to form compounds; an element that did not fit anywhere in Mendeléev's periodic table. This was the first of a strange new series of elements, undreamed of in all of history; an element first detected by spectroscopic analysis of the sun.

In 1892 Lord Rayleigh (1842–1919) pointed out that nitrogen from the air is more dense than nitrogen from chemical compounds. Perhaps the atmospheric gas that remains when oxygen and carbon dioxide are removed from air is not pure nitrogen. Perhaps this gas is a mixture of nitrogen and one or more other gases as yet undiscovered. Lord Rayleigh and Ramsay planned a series of investigations to isolate the unknown gas. They passed nitrogen from the air over hot magnesium which combines with nitrogen to form magnesium nitride, Mg_2N_2. When all of the nitrogen had been used up, a small amount of gas remained. This gas, which would not combine with any known element, was called argon, from the Greek word *argos:* inert or idle. Argon was about ten times heavier than helium.

After the discovery of helium and argon, Ramsay concluded that a whole new family of elements must exist. He predicted and later discovered (with Travers) three new gases: neon, krypton, and xenon. All three were obtained from liquid air in extremely minute quantities. In 1909, Ramsay and Gray discovered a sixth gas in the family, radioactive radon, which is given off by radium. Like the other inert gases, radon turned out to be chemically inactive.

The six so-called noble gases—undreamed of even by Mendeléev—constituted a strange new group of elements. It was later called the zero group of the periodic table. These unsociable elements refused to unite with the extremely active metals sodium and potassium. Even fluorine, the most violent of the nonmetals, could not induce them to form compounds. In addition to being chemical hermits,* these zero-

* In 1962 Neil Bartlett, at the University of British Columbia, produced a compound of xenon, platinum, and flourine. Noble gas chemistry now includes many compounds and promises to shed new light on the nature of the chemical bond.

group elements are colorless and odorless. No wonder their presence so long eluded detection.

Despite their seeming inertness, the zero-group elements are doing important work. Several are used in electric signs; helium provides a safe and effective substitute for explosive hydrogen in balloons; and argon is used in electric light bulbs to increase their life.

Since the publication of Mendeléev's Periodic Table of the Elements, twenty-three new elements had been discovered, bringing the total to eighty-six. The atomic theory of Dalton had been vindicated in the laboratory and had led to a profound new understanding of the world's very substance. Wonderful new materials had evolved from test tubes to completely change an ancient way of life. Chemistry had assumed its rightful place beside physics, astronomy and mathematics as a dynamic and fruitful pursuit of the scientific mind. But there were still many unanswered questions. The supposedly immutable atom showed unmistakable evidence of change in transforming itself from atom to ion. How could the properties of chlorine change so drastically between the two states? How does electricity play its role in this mystery? Why does electricity move about on ions but not on atoms?

Questions of this sort perplexed the greatest minds of the time. Their solution led eventually to the great explosion of scientific knowledge that the world is now witnessing. But before we discuss the relation between electricity and matter, it is important to pick up the thread of our story that was broken with the downfall of phlogiston. The fire element was an element no more; but it still required an explanation, for life and all human activity proceed only with its blessing. Let's return, therefore, to the study of that mysterious and important quantity, heat.

Ice, Water and Steam Are Put to Work

11

The modern factory worker commands the output of about 250 unseen assistants. At home, his household staff amounts to some forty invisible servants catering to his needs and wishes. His automobile transports him swiftly, with the power of 150 horses, and the commercial airplane he uses generates more power than was available to the greatest Roman legion. The operating engineer of a modern power plant routinely controls the power of 1 million horses, or 20 million men, merely by operating a few switches. This tremendous capacity comes from our ability to convert the heat from fuels into useful work that would otherwise have to be done by human or animal muscles. From this interplay of heat and work has evolved a new scientific discipline—thermodynamics—that dominates the enormous power resources used by man today.

Although heat always played an indispensable role in chemical reactions, chemists initiated a revolution in new materials without really understanding heat. In much the same way, the steam engine created an industrial revolution long before scientists understood its principles of operation. In fact, many new sciences were anticipated by important practical applications—applications that man had blundered into with his genius for innovation by trial and error. Eyeglasses led to the first primitive telescope and eventually to the modern science of optics. The early calendar and the needs of agriculture, animal husbandry and religious worship led to astronomy. Primitive metallurgy led to chemistry. And in the early part of the eighteenth century, Newcomen's steam engine provided the impetus for a full assault on the riddle of heat.

The Logical—but Erroneous—Caloric Theory

It's difficult for us today to understand the hold of the "caloric theory" on scientists of the eighteenth century. The term "caloric" still survives in our word, "calorie"—both words deriving from the Latin word for heat.

Caloric was thought of as a kind of all-pervading fluid. Its particles were attracted by all matter and repelled by one another. When two objects of different temperature came together, caloric was supposed to flow from the hotter to the colder body until equilibrium was reached. In the process of heating, caloric particles entered an object. The mutual repulsion of these particles then caused the body to expand. When two objects were rubbed together, some of the caloric was squeezed out causing them to appear hot.

Although the caloric theory of the physicists was completely erroneous, it outlived the phlogiston theory of the chemists and dominated science until the middle of the nineteenth century. Fortunately, much important experimental work was carried on, unhindered by theoretical errors. In the hands of ingenious scientists, any working theory—even a false one—is often better than none.

Pumping with Steam

By the seventeenth century, the seepage of water into deep mines had become a serious economic and technological problem, threatening to bankrupt the coal and tin mines of England. Then, suddenly, it was found that steam could be used as a source of power for pumping.

The first step in that direction was taken in 1650 by von Guericke who found that a vacuum could be used to transmit power over a considerable distance. His device is illustrated in Figure 80. When the small piston of part (*a*) is pushed down, valve *A* allows some of the air in the cylinder to escape to the atmosphere. When the piston is then raised, as in part (*b*), valve *A* rises to seal the tube from the atmosphere. A partial vacuum is thereby produced in the two cylinders and connecting tube. This means that the pressure under the large piston is lower than the atmospheric pressure acting on the top of the piston. This difference in pressure forces the large piston down in the

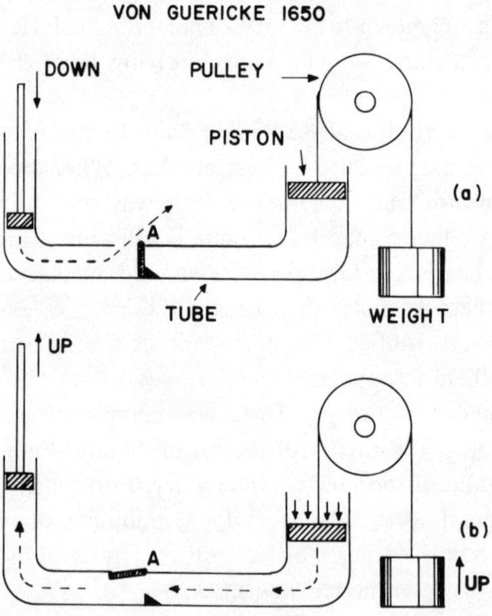

Fig. 80. Von Guericke's pneumatic machine.

cylinder, thereby raising the weight hanging over the pulley. Much later, the principle was used in millions of vacuum brakes on railway cars all over the world. Nevertheless, von Guericke's device was only a means of transmitting power, not of producing useful work.

The next step was taken by the Marquis of Worcester in 1663 and perfected by Savery in 1698. These early heat engines were devices in which a vacuum is created by filling a closed space with steam and then causing the steam to condense by cooling it. You can easily convince yourself that great forces can be developed by a vacuum. Merely place a cup of water in an empty gallon can and start the water boiling. As soon as steam fills the can, remove it from the stove and quickly seal the opening. As the steam cools, it will change back to water and create a partial vacuum in the can. Atmospheric pressure will do the rest, slowly crushing the can into a mass of wrinkled metal.

Savery's "fire engine" made use of these enormous forces created by the weight of the atmosphere. The basic principle is illustrated in Figure 81. All four valves are initially closed and steam is generated

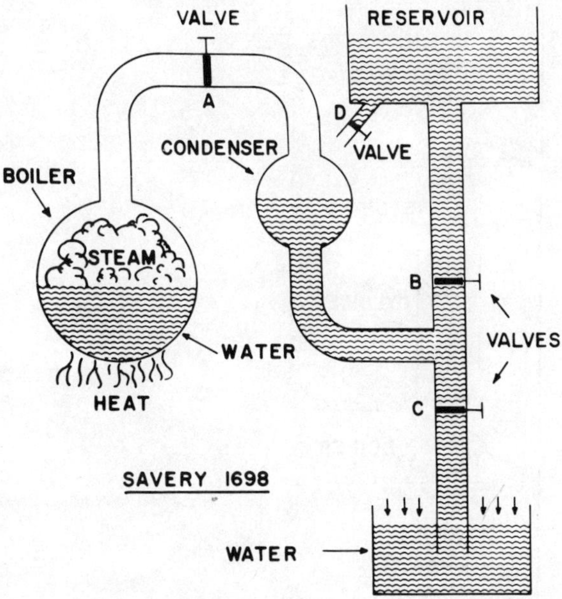

Fig. 81. Savery's "fire engine."

in the boiler. Valves *A* and *B* are then opened, and steam enters the condenser forcing its water up into the reservoir. Valves *A* and *B* are then closed, and valves *C* and *D* are opened. The latter squirts cold water on the condenser, and cools the steam, causing it to condense. This creates a vacuum in the condenser, and atmospheric pressure forces water from the lower tank into the condenser. The cycle is repeated over and over, in order to lift great volumes of water into the reservoir.

Thomas Newcomen's stroke of genius in 1712 was to combine the piston with a way of quickly producing a vacuum by condensing steam. The Newcomen engine is illustrated in Figure 82. The piston is connected to an overhead beam, which also carries a pump rod on the other side. Steam from the boiler is admitted to the cylinder and raises the piston to the top of its stroke. A jet of cold water then enters the cylinder, condenses the steam, and creates a vacuum. At-

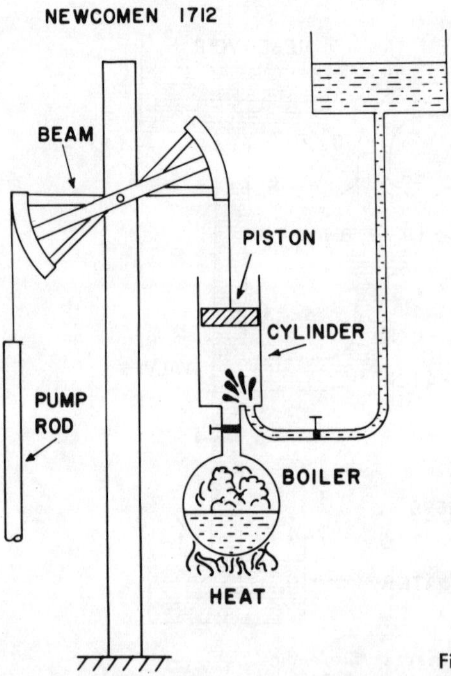

Fig. 82. Newcomen's steam engine.

mospheric pressure then forces the piston to the bottom of its stroke, thereby doing work on the pump attached to the left arm of the beam. Newcomen's ingenuity established the steam engine as a practical machine, sufficiently powerful and reliable to save many English mines from flooding. For the next fifty years his engine was used for pumping, wherever water power was not available.

The main disadvantage of Newcomen's engine for other purposes was its great hunger for fuel. For one bushel of coal, the engine could raise 5 or 6 million pounds of water through a distance of one foot. (Or a smaller weight of water through a correspondingly greater distance.) Much later, Watt built steam engines that produced three or four times as much useful work for the same amount of fuel.

It was not understood in Newcomen's time that the efficiency of a steam engine could be improved by scientific principles of design. No balance sheet was available with which to compare the work accomplished by a machine against the fuel it consumed. There was no way of knowing whether his engine had achieved the limit of efficiency or whether, in fact, any such limit existed. Steam engine performance could not be dealt with effectively until much more was known about the nature of heat.

The First Thermometers

The development of the thermometer as a scientific instrument began with Galileo in the seventeenth century. His early temperature-indicating instrument was really a *thermoscope* rather than a thermometer, because it had no scale upon which to obtain a numerical reading. A simple thermoscope is merely a flask inverted in a dish of water, as shown in Figure 83. Changes in the temperature of a room are indicated by the rise or fall of the water level in the neck of the flask. As the temperature rises, the air in the flask expands, forcing the water to a lower point in the flask. On cooling, the air contracts and atmospheric pressure causes a rise in the water level.

It was soon found that barometric pressure variations from day to day cause corresponding changes in the volume of trapped air. A liquid, on the other hand, does not undergo a significant change in volume as the pressure changes. Liquids are said to be "incompressible." And while the volume of a liquid changes only slightly as the temperature varies, the changes are easily measured in a narrow tube. So liquids—such as water and alcohol—came to be used as the thermometric fluid instead of air.

The thermoscope became a *thermometer* when a scale was attached so that the height of the liquid column could be read. These numerical readings represented so many "degrees" of heat or cold. At first, these "degrees" were different for each thermometer—the maker having selected a convenient numbering scheme that suited his fancy. In order to compare the "degrees of heat" read from different thermometers, it became desirable to adopt a standard scale of temperature.

While Newcomen was making his first engines, Daniel Fahrenheit (1686–1736) was working on improved thermometers. He established a standard temperature scale and made general use of mercury in his instruments. By 1721, at the latest, he had constructed his first

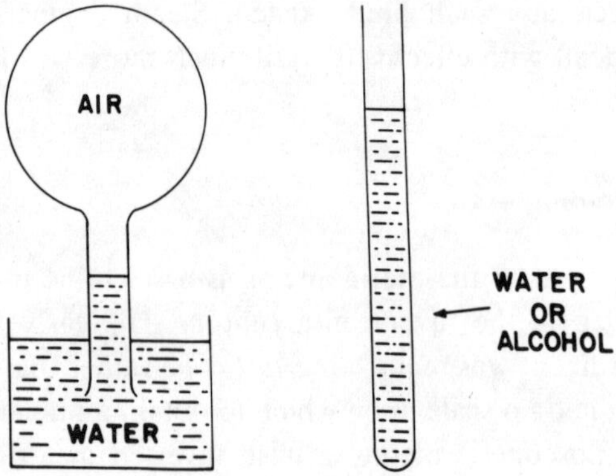

Fig. 83. Two early forms of the thermoscope.

mercury thermometer and had confirmed the fact that water boils at a nearly constant temperature. He correctly attributed the slight variations in the boiling point to changes in atmospheric pressure.

Fahrenheit used the temperature of a mixture of water, ice and salt for the zero point on his scale. A second calibration point on his scale was 96°, which he selected as the temperature of the healthy human body. When this scale was extended, it read 212° for the boiling point of water. Strangely, Fahrenheit did not use the boiling temperature of water as a "fixed point" on his scale.

That step was taken in 1742 by the Swedish astronomer Anders Celsius (1701–1744) of Upsala. The scale of his thermometer had two fixed "degrees of heat": the temperature of a mixture of water and ice; and the temperature of boiling water. The instrument was first placed in ice water and a mark made at the level to which the mercury sank. It was then placed in boiling water and a second mark made. The scale between the two points was then divided into one hundred equal parts—each corresponding to one "degree of heat." But Celsius put his zero at the boiling point of water, and his 100° mark at the melting point of ice. This upside down arrangement avoided the use of negative numbers for temperatures below freezing. The present centigrade scale is just the opposite, increasing *upward* instead of downward in temperature.

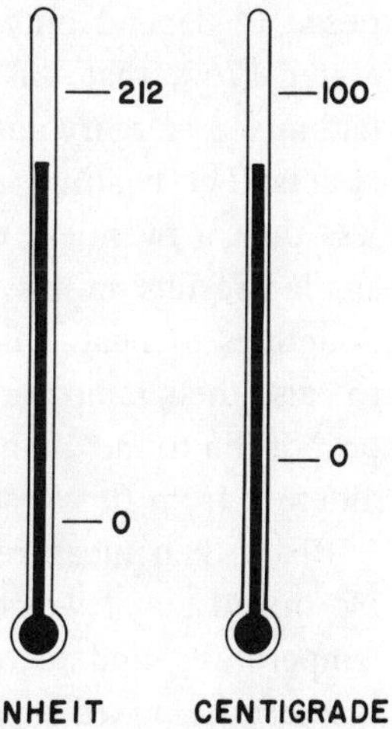

Fig. 84. The Fahrenheit and Centigrade thermometers.

Figure 84 shows the centigrade and Fahrenheit scales side by side for comparison. Temperatures can be converted from one scale to the other by the following relations:

$$F = \frac{9}{5}C + 32$$

$$C = \frac{5}{9}(F - 32)$$

where F refers to Fahrenheit temperatures and C to centigrade temperatures.

Finding a Balance Sheet for Heat

Joseph Black, the Scot, who did so much to start chemistry in the right direction, provided much the same guidance for thermodynamics. Before Black's time, it was generally thought that the amount of heat necessary to raise the temperature of an object depended only on its weight—being the same for equal weights of different substances. An object's "thermal capacity" was the amount of heat it could contain at a certain temperature. The thermal capacity, therefore, was supposed to depend only on the weight of the body. Black concluded, about 1760, that this belief was wrong. He knew that Fahrenheit had mixed mercury and water at different temperatures with surprising results. The heating (or cooling) effect of mercury was found to be less than a twentieth of that of an equal weight of water. He found similar results in his own experiments. Black concluded that "the quantities of heat which different kinds of matter must receive . . . to raise their temperatures by an equal number of degrees are not in proportion to the quantity of matter in each, but in proportions widely different from this."

Black then measured the thermal capacities of various materials. He mixed equal weights of two substances, each having a different temperature, and recorded the new temperature of the mixture. In one test, he mixed a pound of gold at 150° with a pound of water at 50° and found the temperature of the mixture to be 55°. The gold lost 95° and the water gained 5°, so the thermal capacity of gold was found to be $\frac{5}{95}$, or $\frac{1}{19}$ times that of water. (The correct figure is closer to $\frac{1}{30}$.) Black's method is still used today, with appropriate safeguards to prevent the incidental loss of heat. In modern terminology, the ratio of the thermal capacity of a substance to the thermal capacity of water is called "specific heat." The specific heats of a few substances at ordinary temperatures are given below:

SPECIFIC HEATS

Substance	Specific Heat
Alcohol (ethyl)	.58
Aluminum	.21
Copper	.092
Glass	.14
Gold	.031
Ice	.50
Iron	.11
Lead	.031
Mercury	.033
Platinum	.032
Tin	.054
Water	1.000

Hidden Heat

Science also owes to Black its first inkling that large amounts of heat lie hidden in liquid materials. This, his best known discovery, is called "latent heat." He was led to its discovery by a study of the melting of solids. It had long been known that the temperature of a solid, such as ice, does not change perceptively during the process of melting. It was believed, however, that the temperature really did change slightly, by an amount too small to be detected. Most scientists assumed that the addition of a very small "degree of heat" would melt a solid while the removal of the same amount of heat would cause it to solidify. Black was convinced, nevertheless, that ice absorbs a *great* quantity of heat on melting. If the opposite were true, ice would not last all summer in icehouses, and mountain snow would melt in a flash when the temperature rose above freezing.

He settled the matter decisively with many simple experiments. We can understand his reasoning with the help of a numerical example. Suppose we mix 1 gram of ice at 0°C with 1 gram of water at 100°C. What will be the temperature when all of the ice is melted? If no latent heat were absorbed by the melting ice, the new temperature would be simply 50°C, half-way between the initial temperatures. But actual measurements show the resulting temperature to be only 10°C. Each of the two grams of water has lost 40 "degrees of heat." The total amount absorbed by the melting gram of ice, therefore, is 80 "degrees of heat." This is an amount of heat capable of raising *one* gram of water from 0°C to 80°C. Black concluded that "A great quantity of the heat . . . which enters into the melting ice produces no other effect but to give it fluidity, without augmenting its sensible heat (temperature); it appears to be absorbed and concealed within the water, so as not to be discoverable by the application of a thermometer." Black had demonstrated that a melting solid absorbs great quantities of heat with no change in temperature—heat which is "rendered latent."

Black also suspected that a large amount of heat is absorbed by a substance when it changes from a liquid to a vapor at constant temperature. He poured some water at 50°F into a pan and placed the pan on a red hot stove. Because the stove was so much hotter than the water, he assumed that heat would flow into the pan at a constant rate as the water became heated. He found that the water began to boil in 4 minutes, and had completely boiled away after 20 minutes. He reasoned that a certain amount of heat was absorbed by the water during the 4 minutes required to heat it from 50°F to 212°F—a difference of 162°. But the water required 20 minutes to boil away at a constant temperature of 212°F. The latter time interval is five times the former. So the amount of heat absorbed in the process of vaporization is five times the amount required to raise the water through 162°. Thus, the latent heat of vaporization was calculated to be $5 \times 162°$ or 810 "degrees of heat." The correct figure would be 970 "degrees of heat" in Black's system. This "hidden heat" is now called the "heat of vaporization."

In modern units, the *calorie* is the quantity of heat required to raise the temperature of 1 gram of water through 1 centrigrade degree. The BTU (British thermal unit) is the quantity of heat required to raise the temperature of 1 pound of water through 1 fahrenheit degree. The hidden quantities of heat discovered by Black are given below in these units:

	Calories per gram	BTU per pound
Latent heat of melting ice	80	144
Heat of vaporization of water	539	970

Black's researches into the nature of heat were made at the University of Glasgow between 1757 and 1764. In 1757 a young engineer, James Watt (1736–1819), obtained an appointment as an instrument maker at the same university. This was a fortuniate appointment, for both Watt and all mankind. His natural talents and the influence of Black's discoveries combined to produce the most important improvement ever made in the steam engine. In 1764, Black was experimenting with steam flowing through a condenser—a metal chamber kept cool by a surrounding bath of water. From this work Watt got an important idea that would result in a dramatic improvement in the steam engine. For this first time in its history, the operation of the steam engine would be subjected to a scientific analysis.

Watt's Steam Engine

When Watt was twenty-eight years old, he was asked to repair a Newcomen engine that belonged to the university. The engine had been checked by factory experts, but it still failed to work satisfactorily. After some study, Watt concluded that the boiler was too small for the particular application of the engine. When provided with a larger boiler the engine functioned normally. His success with the Newcomen engine induced Watt to investigate the possibility of further improvements.

The Newcomen engine was being made in larger and larger sizes in those days. Consequently, the thin brass cylinders of the smaller engines had been replaced by thick, heavy cast-iron cylinders. The great mass of these cylinders required more and more steam just to heat the cast-iron up to steam temperature. Worse still, more and more cooling water was needed to cool the cylinder so the steam would condense during the downward stroke of the piston. Watt wondered whether this alternate heating and cooling of great masses of metal was reducing the potential work of the engine. He decided to find out.

Watt reasoned that the amount of steam used by the engine went for two purposes: (1) to fill the cylinder and; (2) to heat the cooled cylinder walls, the piston, and any water that was left in the cylinder from previous strokes. It was obvious that any steam used for the latter purposes was wasted. His problem was to find out whether the amount of wasted steam was significant.

From Black's work, he knew that the thermal capacity (or specific heat) of cast-iron and other materials could easily be measured. He also knew the amount of heat (Black's "degrees of heat") liberated by steam when it condenses. On the basis of such data, he calculated that only one-fifth of the boiler's steam was actually used to fill the cylinder. The other four-fifths was used up merely to heat the engine on each stroke following the injection of cold water.

Watt then put all of his ideas together and saw a way to avoid wasting so much steam. He used a *separate chamber*—a condenser—to condense the steam during the downstroke of the piston. The cylinder could be kept at a temperature close to that of the incoming steam at all times. It would not be necessary to reheat the cylinder during each stroke.

The basic principle of Watt's engine is illustrated in Figure 85. When valve *A* is opened, steam enters the cylinder and pushes the piston to the top of its stroke. Valve *A* is then closed and valve *B* is opened. The steam rushes into the cold condenser where it is changed into water, creating a vacuum under the piston. On the next cycle, steam enters a cylinder that is still hot and little heat is wasted in merely heating the metal parts. Using this new principle, Watt was soon making engines that were three times as efficient as the Newcomen engine in terms of useful work performed per bushel of coal used.

In 1781 Watt added a flywheel mechanism to his engines, thereby changing the back and forth motion of the piston into the rotary motion of a shaft. With this addition, the steam engine was no longer restricted to pumping purposes, but could be used to drive all kinds of industrial machinery. In 1782, he disclosed a further improvement, the double acting piston with the steam pressing on either side alternately. The cylinder was sealed from the atmosphere at the top as well as the bottom, the piston rod going through a "stuffing box" to prevent leakage of steam. While one side of the cylinder was connected to the condenser, steam was admitted to the other side, and vice versa. Thus, steam power provided both motions of the piston.

Although Watt's engines were only one-tenth as efficient as modern steam engines, they paved the way for increased demand for steam power and helped usher in the technical age. By 1800, over 500 Watt

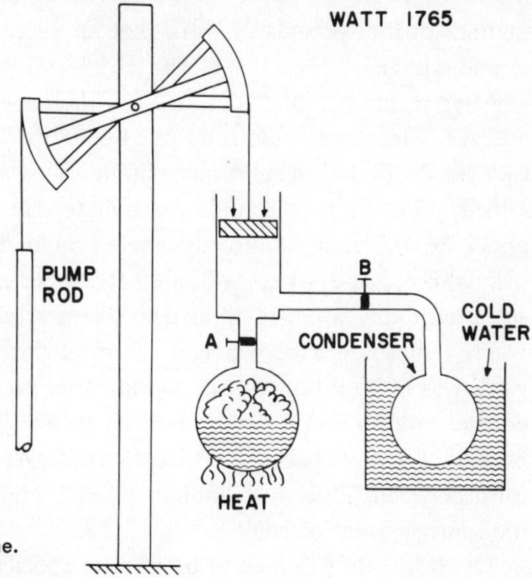

Fig. 85. Watt's steam engine.

engines had been produced, most of these supplying rotary power to factories and mills. Steam power had become a commercial success and the industrial revolution was now in high gear.

Horsepower and Duty

Along with the widespread industrial use of the steam engine came a need to specify the useful work that various engines could be expected to perform. It is only natural that the term "horsepower" should have come into use for this purpose. Savery had used the term to rate his early pumping engines, but there was no general agreement on just how much work the average horse could perform.

The earliest measure of an engine's performance was the number of pounds of water it could lift out of a mine in one minute. It was assumed, correctly, that the product of the weight of water and height lifted was the crux of the matter. A horse did the same amount of useful work whether it lifted 1,000 pounds through a height of 1 foot, or 1 pound through a height of 1,000 feet. In either example, the useful work is $1000 \times 1 = 1000$ foot-pounds. The term "power" meant that *rate of doing work,* so the horsepower was merely the number of foot-pounds of work that an average horse could perform in one minute.

After a number of experiments with horses, Watt found that the average workhorse could raise one hundredweight (112 lbs.) of water to a height of 196 feet in one minute. This amounted to 112 lbs. × 196 ft. = 21,952 foot-pounds per minute. He increased this figure by about 50 per cent so his customers would have no complaint, and defined one horsepower as 33,000 ft-lbs per minute. His definition is still used today whether applied to steam or electrical engines.

The efficiency with which a steam engine changed fuel to useful work was of vital importance to industrial users. The best Newcomen engines could raise 7 million pounds to a height of one foot for one bushel (84 pounds) of Newcastle coal. Watt's best engines could raise between 30 and 40 million pounds of water one foot high for the same amount of coal.

The term "duty" came to be used to specify the number of pounds of water raised one foot high by one bushel of good coal. The duty of steam engines continued to rise as higher steam pressures and other refinements crept into their design. By 1838 an engine was built by Taylor having a duty of 95 million. Its cylinder was 85 inches in diameter and accommodated an 11-foot stroke. It developed 220 horsepower and pumped 500 gallons of water a minute. It was one of many fine machines that came to be known as Cornish steam engines. This engine consumed over four tons of coal per day, and its duty was thus a most important consideration. It was still too early, however, for engineers to know whether the fuel consumption of steam engines could be further reduced. Thermodynamics was still struggling to be born. It would be some years before scientists would discover a numerical correspondence between the heat contained in fuel and the useful work that can be extracted from it.

Invisible Heat Rays

It has long been known that heat can be transferred from place to place in three ways: by *conduction*, when a cold object touches a hot one; by *convection*, when warm air rises; and by a mysterious form of invisible, *radiant heat*. During the latter part of the eighteenth century, scientists studied radiant or "obscure" heat with the help of mirrors and lenses. One of the earliest to make systematic experiments was the Swedish chemist, Carl Wilhelm Scheele (1742–1786).

Scheele used an open stove as a source of heat and light. He found, first of all, that a strong draft of air across the opening had no effect on the passage of heat from the stove to his thermometer in the room. He then put a large glass plate between his face and the stove and could not feel any heat. Whatever this radiant heat was, most or all of it was stopped by the glass. Others later found that glass absorbs radiant heat, and soon gets quite warm as a result. Scheele then studied the reflection of radiant heat by flat and concave mirrors of silvered glass and of metal. With silvered glass he noticed that light was reflected but heat was absorbed by the glass. With metal mirrors, both heat and light followed the same law of reflection as light from the sun. Using the metal mirror, he focused the heat and light on a piece of sulfur and set it afire. He noted also that the metal mirror did not get warm. But if it was blackened with soot from a burning candle, it became too hot to hold in a few minutes. Bright metals apparently reflect radiant heat while glass and dark materials absorb this strange heat. Although light from the stove was mixed with radiant heat, Scheele was convinced that the presence of the light was only incidental. The light could be separated from the heat by a plate of glass, and the heat from the stove was greatest when the wood had burned to red hot coals—and the fire light was at a minimum.

The matter was cleared up decisively by De Saussure in 1786. He used a hot, but nonluminous iron ball as his source of heat. Two tin concave mirrors were placed about 12 feet apart as shown in Figure 86. The iron bullet was heated red hot and allowed to cool until it was invisible in the dark. It was then placed at the focus of one mirror. Thermometer, *A*, placed at the focus of the other mirror, showed a temperature 8° higher than thermometer, *B*, placed outside the focal point. When thermometer *A* was moved slightly from the focus, its temperature fell immediately to the lower reading. De Saussure concluded that radiant heat is given off by hot objects and that these emanations are reflected like light rays.

James Hutton repeated Scheele's experiment with a sheet of glass in 1794, and found that the radiant heat was not absorbed completely by the glass, but merely diminished in intensity. He was the first to

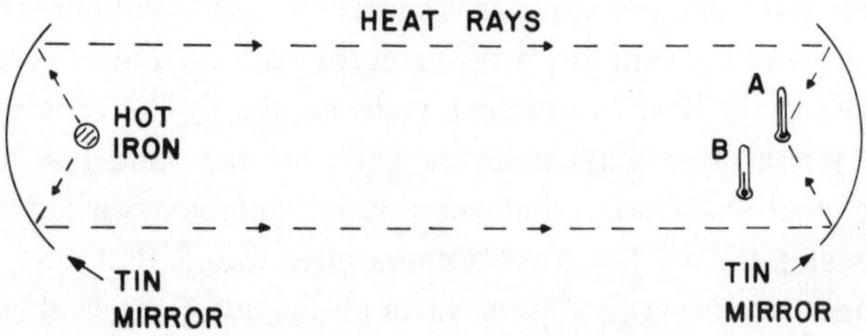

Fig. 86. Heat rays obey the same law of reflection as light rays. The thermometer reaches the highest temperature when placed at the focus of the tin mirror on the right.

regard radiant heat as an invisible kind of light which, though unable to affect the eye, was powerful enough to communicate heat. Radiant heat, then, is not a material emanation, but merely a kind of invisible light that is able to transfer heat from place to place. Thus by 1800 the existence of heat rays was generally accepted, and they were thought to be connected, somehow, with light.

The Weight of Heat

The caloric theory could not explain how invisible rays can transfer heat from one place to another. This raised a doubt about the theory in some minds, and other inconsistencies soon began to crop up. For example, why do gases get hot when they are compressed, and why do bearings overheat when not properly lubricated? The caloric theorists insisted that friction causes heat by squeezing caloric out of a body. The loss of caloric to nearby objects made them hot. Here was another weak point in the caloric theory armor. It would later be attacked by a famous cannon-boring experiment in a Bavarian gun factory. Count Rumford would direct the strategy. But in the mid-eighteenth century, the assault was in another direction.

According to the caloric theory, heat was some kind of material substance that caused bodies to expand when heated. It was natural, under the circumstances, to suppose that heat had weight. So, many attempts were made to measure a change in weight when bodies were heated. The results were inconclusive at first. In 1732, Boerhaave weighed a piece of iron when cold and hot and found no change in weight. Buffon weighed a piece of white hot iron in 1775 and measured 49 lbs. 9 ozs. On cooling, it weighed only 49 lbs. 7 ozs. Roebuck repeated Buffon's experiments in 1776 with smaller masses and more sensitive balances with conflicting results. Some pieces of metal gained weight; others lost.

The most accurate measurements were made by Benjamin Thompson (1753–1814) of Woburn, Massachusetts. Thompson left Massachusetts during revolutionary times because his sympathies were with the Tories. He later went to Bavaria where he served as Minister of War, and Minister of Police. As a reward for his services he was given the title of Count Rumford.

In an attempt to weigh heat, he sealed three flasks containing equal weights of water, spirits of wine, and mercury, respectively. They were put in a warm room for twenty-four hours at 61°F. After carefully weighing the flasks they were placed in a cold room at 30°F for forty-eight hours. A second weighing showed that their weights had not changed. Several repetitions of the experiment gave the same result.

Rumford calculated that a great amount of heat had been given off by the water in his experiment, because of its high thermal capacity and latent heat. Enough heat was liberated to bring an equal weight of gold from the freezing point to a bright red heat. (Gold has a much lower thermal capacity than water.) Furthermore, his balance was sensitive enough to detect a change in weight as small as one part in a million. He concluded, therefore, "that all attempts to discover any effect of heat upon the apparent weights of bodies, will be fruitless."

At about this time, Rumford "was struck with the very considerable degree of heat which a brass gun acquires in a short time, on being bored; and with the still more intense heat . . . of the metallic chips separated from it by the borer." According to the caloric theory, the deformed chips became hot because the squeezing had reduced their thermal capacity. Since they could no longer contain their original amount of heat, the excess was given off as sensible heat. But Rumford measured the thermal capacity of the chips and found it to be exactly the same as for the original bar of metal. Here, then, was another loophole in the caloric theory.

Rumford also noted that the supply of heat provided by friction seemed to be inexhaustible. The longer he continued to bore the bar of brass, the more heat was given off. He concluded:

> It is hardly necessary to add, that anything which any insulated body, or system of bodies, can continue to furnish without limitation, cannot possibly be a material substance. It appears to me to be extremely difficult, if not quite impossible, to form any distinct idea of anything capable of being excited and communicated in the manner in which heat was excited and communicated in these experiments, *except it be motion.* I am very far from pretending to know how or by what means or mechanical contrivance that particular kind of motion in bodies which is supposed to constitute heat is excited, continued, and propagated.

Rumford's experiments had shaken the caloric theory's foundations, but no other theory seemed any better. Half a century would pass before scientists in general would begin to understand how heat can be a form of motion.

The Perfect Engine

The great breakthrough in our understanding of the steam engine occurred in 1824 when Sadi Carnot (1796–1832), a young French scientist, published the pamphlet, "Reflections on the Motive Power of Heat." This mounmental work received little notice at the time and lay virtually unnoticed for a quarter of a century. It was "discovered" about 1847 by W. Thompson (Lord Kelvin) and has since become a scientific classic.

Carnot's unit of thought was a "reversible engine" which implies a measurable relation between heat and useful work. To illustrate, imagine a steam engine that produces 10 million foot-pounds of work upon burning one bushel of coal. Now suppose that engine to be run backwards; no coal is burned, but the shaft of the engine is turned by some external agency. Suppose also that the amount of work done in rotating the shaft is 10 million foot-pounds. If the engine were reversible, heat would be pumped from condenser to boiler in an amount equivalent to the burning of one bushel of coal. In Carnot's view, perfect efficiency meant perfect reversibility. The heat produced in compressing the steam when running backwards is equal to the heat used in driving the engine when running normally. If less heat is produced when working backwards, the engine is not perfectly efficient and some heat is wasted in normal operation. The idea of a reversible engine is of great importance even today in the theory of all heat engines, whether of steam, gasoline, or any fuel whatever.

Although Carnot's article was written in *caloric* form, it is possible that others may have tampered with the manuscript after his death. In any event, we know from his notebooks that Carnot had a modern view of the nature of heat. In the following excerpt, Carnot points out the equivalence of heat and motion:

> Heat is nothing else than motive power, or rather, motion which has changed its form. It is a motion of the particles of a body. Wherever motive power is destroyed, there is, at the same time, a production of heat in a quantity precisely proportional to the quantity of motive power destroyed. Conversely, wherever there is a destruction of heat, there is a production of motive power. One can state as a general rule, therefore, that there is a constant quantity of motive power in nature and that it is never, to speak correctly, produced or destroyed. In truth, it changes form —that is, at times it produces one kind of motion and at other times another—but it is never destroyed.

Carnot pointed out that the steam engine, though extremely successful, was little understood in theory. Attempts to improve it in his day were "still directed almost by chance." In order to learn its secrets, he imagined a perfect reversible engine, and speculated on its performance. Such an engine would be perfectly frictionless, and perfectly insulated against the accidental loss of heat. Heat is added to the engine in the boiler and removed in the condenser, and motion is imparted only to the crankshaft. For such an engine Carnot showed that the useful work done per unit of heat transferred—its efficiency—depends only upon the temperatures of the boiler and condenser. It is completely independent of the fluid used (steam, in the practical engine), and the design of the engine. It can be shown, in fact, that the efficiency is simply:

$$\text{efficiency} = \frac{T - t}{T}$$

where T is the temperature of the steam entering the cylinder and t is the temperature of the steam leaving the cylinder.*

In order to improve the "duty" of an engine, it is necessary either to increase the steam temperature or reduce the condenser temperature. This explains why Watt's invention of the condenser was so phenomenally successful. When exhaust steam is released directly to the atmosphere, its temperature can be no lower than 100°C or it would previously have condensed to water. Watt's water-cooled condenser reduced t to approximately 30°C, a significant improvement which increased the numerator in the equation for efficiency.

Carnot thought of a heat engine as a machine that transfers heat from a high temperature to a low temperature, doing work in the process. The greater the temperature range the greater the amount of work performed. He also thought of heat as a form of motion—a random motion of the molecules of the substance. As the temperature of steam is reduced in an engine, part of its molecular motion is converted to the motion of a crankshaft. These ideas were picked up and extended by Joule, Lord Kelvin, Clausius and others at the middle of the nineteenth century.

The Mechanical Equivalent of Heat

Although Count Rumford's experiments had clearly demolished the caloric theory, it stoutly refused to die. The intellectual inertia of the

* These temperatures are measured on the absolute temperature scale discussed later in the chapter.

time resisted all efforts to lay it to rest. Still, Rumford did make a few converts. One of these was young Humphry Davy who was destined to become distinguished later. Davy was convinced that "heat cannot be considered as matter," but as a "peculiar motion, probably a vibration of the corpuscles of bodies." Carnot held similar views. In 1842, J. R. Mayer of Heilbron, Germany, calculated the so-called mechanical equivalent of heat. He found that the heat used in raising the temperature of a given weight of water from 0°C to 1°C corresponds to the fall of an equal weight of water from a height of 365 meters (1,198 feet).

Although Mayer's result was based on sound reasoning, it fell on deaf ears. The world of science would not be convinced until positive experimental evidence became available to prove Mayer's preposterous hypothesis. This evidence was provided in 1843 by James Prescott Joule (1818–1889).

Joule was born in Salford, England, and became a student of the great Dalton. He was renowned as an experimenter because of his precise measurements and fine experimental technique. His great contribution to science was the proof that heat and work were different forms of the same thing.

As a physicist, Joule was aware of the concept of *kinetic energy*. It was a mathematical expression equal to one-half the mass (m) of a moving object multiplied by the square of its velocity (v).

$$\text{energy} = \tfrac{1}{2}\ \text{mass} \times (\text{velocity})^2$$
$$\text{energy} = \tfrac{1}{2}\ mv^2$$

Another quantity in Newtonian mechanics was called *work*, the product of force (f) and distance (d)

$$\text{work} = \text{force} \times \text{distance}$$
$$\text{work} = fd$$

He knew that these concepts were related and he wondered how heat fit into the picture. He believed firmly in the mechanical theory of heat and started a long series of experiments to show that energy, work and heat had a great deal in common.

The most famous of Joule's experiments was the use of falling weights to turn a paddle wheel immersed in a tub of water. Figure 87 illustrates his apparatus. The weight *W* was allowed to fall a distance *d*, thereby turning the paddle wheel and churning the water. Joule knew that the work expended by the falling weight would show up as a temperature rise of the water. His apparatus was able to convert a measurable amount of work into a measurable amount of heat.

The work performed by the falling weight was simply the product

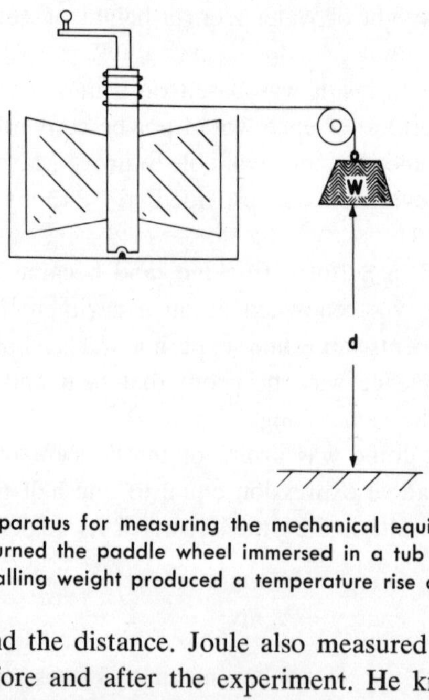

Fig. 87. Joule's apparatus for measuring the mechanical equivalent of heat. The falling weight *W* turned the paddle wheel immersed in a tub of water. The work expended by the falling weight produced a temperature rise of the water.

of the weight and the distance. Joule also measured the temperature of the water before and after the experiment. He knew the mass of water and other materials that were heated, as well as their specific heats. From this information he was able to calculate the amount of heat generated by friction between the paddle wheel and the water. Joule showed that work was converted to heat at a rate of 772 foot-pounds of work for each British thermal unit of heat. The present figure is 778 foot-pounds per BTU. In the metric system, the unit of work is the "erg." One erg of work is done on an object when a force of 1 dyne is exerted on it through a distance of 1 centimeter. Lifting a half-dollar through a height of 100 centimeters (about 3 feet) requires the expenditure of 1,196,000 ergs of work. Because the unit is so small, a larger unit has been defined in honor of James Joule; 1 joule = 10,000,000 ergs. Using these units, the mechanical equivalent of heat is 4.186 joules of work per calorie of heat.

The scientific world became convinced during the following decade that all forms of energy are equivalent, and energy can be converted from one form to another. Kinetic energy, or energy of motion; work, the product of force and the distance through which it acts; electrical energy; the potential energy of water impounded by a dam—all were different manifestations of the same thing, energy. It was also concluded that energy can neither be created nor destroyed—the all important Law of Conservation of Energy. Thermodynamics had finally become a science.

Joule's mechanical equivalent of heat received a cold reception at first. Then, at a scientific meeting at Oxford in 1847, he was scheduled to discuss his latest experiments. In the audience was a brilliant young physicist, William Thompson, later Lord Kelvin (1824–1907). Although only twenty-three, Thompson was a respected scientist and Professor of Natural Philosophy (science) at the University of Glasgow.

Thompson had come to Oxford expressly to criticize Joule's work. He had studied Carnot's pamphlet in its "caloric" form and had been struck by its brilliant reasoning. But along with its greatness was an error. Carnot's paper stated that the amount of caloric that enters the condenser of a steam engine is precisely the same as the amount of caloric fed to the steam by the boiler. This implied that none of the heat was converted to useful work. There was no equivalence, therefore, between heat and work.

After listening to Joule's talk, Thompson became convinced that there was much truth in what he had to say. The two men became lifelong friends and Joule often acted as a catalyst for Thompson's brilliant mind. Though he was impressed with Joule's work, Thompson could not reject Carnot's ideas. Neither could he reconcile Joule's convincing research with Carnot's theory. As we shall see, the answer came just three years later. But before that could happen, Thompson made a discovery of fundamental importance to thermodynamics.

Absolute Temperature, or How Cold Can It Get?

In 1848, Thompson concluded that there must be an absolute scale of temperature. Boiling water, for example, has a temperature of 100° or 212° depending upon which thermometer is in use—Fahrenheit, or centigrade. Each of these temperature readings is purely arbitrary, depending upon the temperature scale selected by the inventor. On the basis of thermodynamic reasoning, Thompson proved that the absolute zero of temperature was approximately −460°F, or −273°C.

The thermodynamic proof of Thompson's discovery is beyond the scope of this book, but we can easily follow a line of reasoning that confirms his conclusion. In 1660 Robert Boyle had discovered how pressure and volume are related in gases. Boyle's law states that pressure, *P*, and volume, *V*, are inversely proportional if the temperature is kept constant:

$$\text{Volume} \propto \frac{1}{\text{Pressure}}$$

$$V \propto \frac{1}{P} \qquad (1)$$

In 1801, the French Chemist Gay-Lussac announced another law that related the volume and temperature of a gas. Gay-Lussac's law states that the volume, *V*, of a gas is directly proportional to the temperature, *T*:

$$\text{Volume} \propto \text{Temperature}$$

$$V \propto T \qquad (2)$$

The law tells us that each degree of temperature change of a gas is accompanied by an identical change in volume, if the pressure is kept constant. Gay-Lussac indicated that Jacques Charles (1746–1823), another French scientist, had preceded him in the discovery of the law, but had published no results. For that reason, the law is also known as Charles' law.

Gay-Lussac's law can be combined with Boyle's law to give a single important relationship:

$$\text{Velocity} \propto \frac{\text{Temperature}}{\text{Pressure}} \tag{3}$$

$$V \propto \frac{T}{P}$$

This expression tells us that the volume of a gas depends directly on its temperature and inversely on its pressure. It can be changed into the "equation of state" for gases by introducing a constant, *K*, that depends on the units used to measure *V*, *T*, and *P*:

$$\frac{PV}{T} = K \tag{4}$$

We can use this equation to determine the absolute zero of temperature.

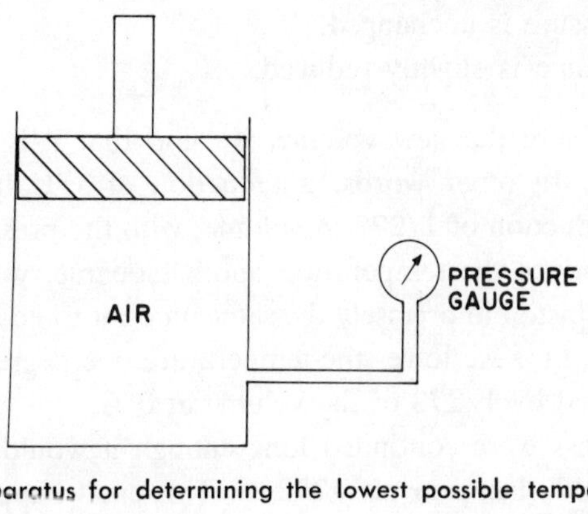

Fig. 88. An apparatus for determining the lowest possible temperature, absolute zero.

To do so, let's imagine that we have a cylinder full of air as shown in Figure 88. The cylinder is closed by a movable airtight piston, and a pressure gauge measures the pressure of the air in the cylinder. A thermometer, not shown, indicates that the temperature of the air in the cylinder is 0°C. The initial volume of the gas is also known. Now let's suppose that we reduce the temperature of the gas precisely 1° centigrade. What happens to the volume and pressure? First of all, the volume remains unchanged because we have not moved the piston. Only the pressure could have changed, as we can see from equation four. The equation states that $\frac{PV}{T}$ is equal to a constant number, *K*, so $\frac{PV}{T}$ must remain unchanged. We have already noted that *V* has not changed. So *P* must have been reduced along with *T* in order to keep $\frac{PV}{T}$ unchanged.

Now let's regain the initial pressure by compressing the air in the cylinder slightly while maintaining the temperature at −1°C. We can do that merely by pushing the piston down into the cylinder a bit until the pressure gauge indicates the initial pressure reading. The condition of the air in the cylinder is now as follows:

1. The temperature is one degree lower than at the start, −1°C.
2. The pressure is unchanged.
3. The volume is slightly reduced.

When we measure the new volume, we find that it is 1/273 of the initial volume. In other words, a reduction of 1°C in temperature produced a reduction of 1/273 in volume, with the pressure held constant. If we reduce the temperature another degree, we find that we must push the piston in precisely the same amount to keep the pressure constant. Each time we lower the temperature one degree, the volume must be reduced by 1/273 of the volume at 0°C.

If this process were continued long enough it would seem that all of the gas would disappear at −273°C. This can't happen, of course, because all gases change into liquids before this low temperature is reached. But the temperature at which gases *theoretically* cease to have a volume is called "absolute zero." Helium liquifies at −268.9°C and freezes at somewhat less than −272°C. Temperatures have been achieved in the laboratory as low as a few tenths of a degree above absolute zero. On the Fahrenheit scale, the absolute zero is at

$-460°F$. According to modern theory, no substance can be colder than $-273°C$ or $-460°F$.

The Two Laws of Thermodynamics

In 1849, Thompson published a scientific paper entitled, "Carnot's Theory of the Motive Power of Heat with Numerical Results Deduced from Regnault's Experiments on Steam." It was this paper that introduced Carnot's ideas to the scientific world. The paper also analyzed the performance of the Cornish steam engine at Fowey Consols, one of the most efficient engines then in existence. Using Carnot's theory, he proved that this engine was producing power at only a little more than half of its theoretical capabilities. Unfortunately, the paper still used the caloric theory. Not until the following year did he resolve his great dilemma and become convinced that Joule's mechanical equivalent of heat could be reconciled with Carnot's work. Before he could publish his ideas, however, the solution appeared in a paper by Rudolf Clausius (1822–1888).

Clausius was a renowned German scientist, a professor of physics at Berlin. He showed that there was no real conflict between the theories of Carnot and the experiments of Joule. Here was the difficulty. If the caloric theory were true, then the amount of caloric entering a steam engine ought to be the same as the amount leaving it. This concept is analogous to the principle of the water wheel—there is no loss in water as it causes the wheel to turn. This principle was advocated in Carnot's paper—at least in the version available to Thompson. But if heat is a form of motion, as Joule maintained, then heat is itself changed into the useful work of the machine.

Clausius pointed out that the idea of heat as a form of motion "is not opposed to the real fundamental principle of Carnot, but rather to the addition, 'no heat is lost,' for it is quite possible that in the production of work, both may take place at the same time; a certain portion of the heat may be consumed and the other portion transmitted from a warm body (the boiler) to a cold one (the condenser), and both portions may stand in a certain definite relation to the amount of work produced." He showed, in other words, that part of the boiler's heat was *converted* into work, and the remainder sent to the condenser.

On the basis of his new interpretation of Carnot's work, Clausius was also able to derive a new and important principle: *It is impossible for any self-acting machine, unaided by any external agency, to transfer heat from one body to another at a higher temperature.* This principle has become known as the Second Law of Thermodynamics. Applied to the steam engine, it tells us that the condenser cannot be hotter than the boiler. Any alternative would imply the possibility of a "perptual motion machine." To illustrate, imagine a steam engine with a boiler temperature of 100°C and a condenser temperature of 500°C. In addition to providing mechanical work, the engine actually supplies steam to the condenser at a very high temperature. Why not feed some of the hot exhaust steam back into the boiler? The engine will then operate indefinitely without the need for fuel.

Thompson discovered the Second Law of Thermodynamics independently and stated it in a different but equivalent form. In addition, he stated that the whole theory of the motive power of heat depends upon another principle, now called the First Law of Thermodynamics: *Whenever heat is changed into work, or* vice versa, *the quantity of energy which disappears in one form is exactly equivalent to the quantity which is produced in the other form.* The First Law is merely the Law of Conservation of Energy as applied to thermodynamic systems.

The Kinetic Theory of Heat

By the middle of the nineteenth century, heat was known to be a kind of motion of the molecules of matter. In solids, the molecules were thought to vibrate or oscillate about fixed places in the substance. In liquids, their motion was less restricted as molecules were able to move past each other with relative ease. Molecular motion was greatest of all in gases. Because gaseous molecules were so far apart, their motion was essentially unaffected by any attractions between molecules. Each molecule merely moved in a straight line until it bounced off another molecule, either of the gas itself or of another object in its path.

In considering the motion of countless tiny molecules of a substance, it is obviously hopeless to try to calculate the path and collisions of each individual particle. The best we can do is to learn the *average* behavior of the molecules under different conditions. This brings the laws of statistics into physics. Insurance companies use statistical methods successfully because they deal with hundreds of thousands of individual policies. But the number of molecules in a cubic centimeter of air is 20,000,000,000,000,000,000. It's easy to see that statistical methods ought to be much more accurate when applied to molecules than to people.

Statistical methods are most easily applied to gases because their molecular motion is relatively free of intermolecular forces of attraction. The molecules rush freely through empty space, colliding with each other and with the walls of the container. The walls receive a continuous bombardment as molecules strike them and bounce off. Each such impact produces a tiny force on the wall and all add up to a steady force—the pressure exerted by the gas.

This concept enables us to understand the basis of Boyle's law. Suppose we compress the volume of a gas to one-half its original volume. The number of molecules per unit volume is now twice as great. Twice as many molecules will bounce off a given area of the wall every second, and the pressure of the gas will be twice the initial pressure. Halving the volume doubles the pressure. This inverse proportionality of pressure and volume is precisely that measured by Boyle.

Now let's consider what happens if the molecules move faster. Two effects are noticed: (1) a greater number of molecules hit the wall each second; and (2) the force of each impact is greater. Both of these effects depend directly on the velocity of the molecules, and both effects cause an increase in gas pressure. The pressure, therefore, depends upon the *square* of the molecular velocity. But the kinetic energy—or energy of motion—of a molecule also depends upon the square of the velocity. So pressure is nothing more than a measure of the *average kinetic energy possessed by the gas molecules.* The pressure rises in direct proportion to the kinetic energy.

$$\text{Pressure} \propto \text{Kinetic energy}$$

But we have already noticed from the combined Boyle-Gay-Lussac Law that

$$\text{Pressure} \times \text{Volume} \propto \text{Temperature}$$

If the volume is kept the same, then

$$\text{Pressure} \propto \text{Temperature}$$

Comparing these relations, we find that the kinetic energy of the molecules is proportional to their absolute temperature (because both are proportional to pressure):

$$\text{Kinetic Energy} \propto \text{Temperature}$$

In other words, the absolute temperature of a gas is simply a measure of the average kinetic energy of the molecules. For that reason, we speak of the *thermal motion* of molecules. The higher the temperature the faster they move.

The kinetic theory helps us understand how a steam engine works. High temperature steam molecules enter the cylinder and impart some of their kinetic energy to the piston, causing it to move. Because the steam has given up some of its energy, its temperature is reduced. The amount of useful work generated by the piston depends upon the change in temperature of the steam as it goes through the engine.

The kinetic theory also accounts for the Second Law of Thermodynamics—the fact that heat will not move of its own accord from a cool body to a hotter one. Imagine a container divided into two parts by a partition. One-half is filled with a hot gas, the other with a cool gas. Now remove the partition. What will happen? The hot, fast-moving molecules lose some of their energy in collisions with the cool, slower molecules. Eventually, the molecules in both halves of the container attain the same average kinetic energy and, therefore, the same temperature. The same argument applies to any object at all. Heat always flows from a high temperature body to a cooler one.

With man's conquest of heat, vast new sources of energy became available to him. Cheap motive power was now available in seemingly unlimited amount. With this new power he was able to fashion the fruits of chemical research into new and better products at prices that increasing numbers of people could afford. But still greater benefits were on the horizon. Electricity, a new and completely unexpected form of energy, was destined to play an amazing role in man's conquest of nature. Let's return, therefore, to the days of vital spirits and a stationary earth, in order to trace the story of electricity.

Magnets and Lightning Rods

12

By the middle of the eighteenth century, the heat engine had established itself as an essential element in man's mastery of his environment. It was the success of the steam engine, in fact, that attracted the attention of men like Carnot, Watt and Thompson to the study of heat. The chemical energy contained in organic or fossil fuels had supplemented the energy of horses and flowing water long before any real progress was made in our knowledge of thermodynamics. The field of electricity and magnetism was quite different in that it received little encouragement or stimulus from an existing technology.

This does not mean that man knew nothing of electricity and magnetism prior to modern times. A magnetic oxide of iron can be found in many parts of the world, and Greek literature contains a number of references to lodestones. The ancients knew that the lodestone can attract pieces of iron and that ordinary iron can be magnetized by rubbing it against a lodestone. The Chinese are thought to have known, furthermore, that a magnet points north and south when allowed to turn freely. In the twelfth century, this remarkable property became known in Europe and the compass became a novel and important instrument of navigation. The Greeks were also familiar with a curious property of amber, a natural resin of vegetable origin. When amber is rubbed, it attracts light bits of straw and dust. Beyond these elementary observations, however, nothing was known about electricity and magnetism until the subjects came under scientific study.

The first experimental investigation of magnetism was reported by Picard Peregrinus in 1269. Peregrinus used a spherical lodestone and located two poles where its magnetic attraction was especially strong. He also recognized the tendency of one pole to seek the north direction while the other seeks the south direction. He found that like poles repel, and unlike poles attract each other; and that a lodestone can be broken to form two magnets in place of one. Peregrinus also noticed the magnetic similarity between a spherical lodestone and the earth. Both influenced a small compass needle in much the same way. Nevertheless, writers in succeeding centuries continued to suppose that the compass needle points toward the North Star, the Great Bear, or to some mysterious mountain in the north.

Early forms of the compass were usually nothing more than a small magnetzied piece of iron floating on wood in a wooden dish of water. Later came the pivoted needle, and the compass card—a disk divided into thirty-two equal "points" of the compass. Compasses were so mounted as to be virtually unaffected by a ship's motion.

Aside from its obvious utility as a north-seeking instrument, the compass had another characteristic that aroused the interest of navigators. During the fifteenth century, it was discovered that the compass needle does not point precisely at the true geographic poles, but is inclined to the true north-south direction by a small angle. This angle varies from place to place and occasionally becomes zero, as Columbus noticed on his voyage of 1492. The angular deviation of the compass needle from true north was called the "magnetic declination," or "variation of the compass."

The compass variation was measured in various parts of the world during much of the sixteenth century. The measurements had a direct practical motive, of course. A navigator wanted to know, and allow for, the variation that existed wherever he might be. But there was also a hope that magnetic variation might provide a solution to the vexing problem of longitude determination at sea. To illustrate, imagine a line (called a "isogonic") drawn on a map through all places having a magnetic declination of say 5° west. Now, imagine isogonics drawn for many other magnetic declinations. These form a network of lines that intersect the parallels of latitude. Having this information, a navigator might find his location at sea by measuring his latitude and the magnetic declination. His position would be the intersection of the two lines. But it was soon observed that the magnetic declination changes slowly from year to year so the system would not be of permanent value.

In 1544, another unexpected property of the magnetic needle was discovered by Georg Hartmann, a German clergyman. It was rediscovered independently in 1576 by Robert Norman, an English compass maker. Both men discovered that if a compass needle is suspended at its center of gravity, so that it is free to move vertically as well as horizontally, it does not come to rest in a horizontal position. It tilts downward, in general, at an angle called the "dip" or "inclination," which also changes from place to place. Norman's book, *The Newe Attractive* (1581), contained the first suggestion that the place toward which the compass points is really within the earth. In 1608, Henry Hudson found that the compass needle was almost vertical at a place on the 75th parallel of north latitude. But, like the variation of the compass, the magnetic dip also changes from year to year as well as from place to place. Neither could provide a reliable solution to the problem of longitude determination.

Gilbert's Terrella, a Model of the Earth

The modern study of magnetism and electricity began with the work of William Gilbert (1540–1603). Gilbert, the greatest English experimental scientist of his time, was a practicing London doctor. He accomplished much of his important work in electricity and magnetism while he was the physician to Queen Elizabeth. His results were recorded in a great book, *De magnete* ("On the Magnet"), published in 1600.

The book begins by refuting the old superstitions about alleged curative powers of the lodestone. Gilbert then shows how to locate the poles of a powerful spherical lodestone using a minute compass needle on a pivot. The compass needle is used as a sensitive indicator to map the magnetic properties of the lodestone. Gilbert placed it on the surface of the lodestone and marked its direction by a chalk line which was extended to form a great circle on the sphere. It was then moved and another great circle drawn, and so on. All of the circles passed through two opposite points on the stone—the two magnetic poles at *A* and *B* in Figure 89.

When the compass needle was placed at any point equidistant between *A* and *B*—the magnetic equator—the needle was found to lie parallel to the lodestone's surface, as at *C*. The magnetic dip at the

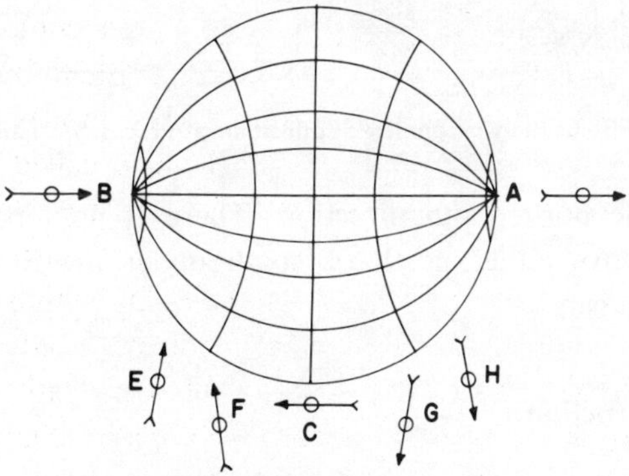

Fig. 89. Gilbert's "little earth."

equator was always zero. When placed at the poles, however, the needle was always perpendicular to the surface. At intermediate points, the needle's inclination to the surface varied with its distance from the poles, as at points *E, F, G* and *H*. He then drew circles parallel to the equator corresponding to places of equal dip. The magnetic behavior of his globular lodestone led Gilbert to conceive of the earth as a hugh spherical magnet. His experimental lodestone was merely a miniature of the earth and, therefore, a *terrella*.

Gilbert also investigated the effect of dividing a magnet in two. He began with an elongated lodestone having its north and south poles as shown in Figure 90. The lodestone was then cut into two approximately equal parts producing two separate magnets. The original poles remained unchanged, but a new pole appeared on each of the newly cut faces.

It was also pointed out in *De Magnete,* that although the earth causes a magnet to rotate, it does not displace the magnet as a whole. If a magnet is floating on a piece of wood, it merely rotates to align

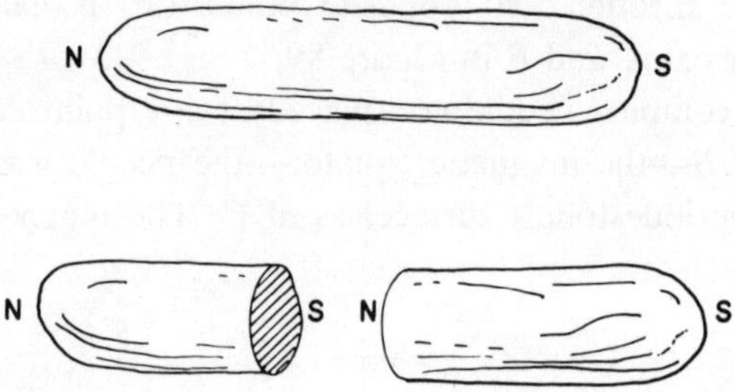

Fig. 90. Cutting magnetized lodestone produces two magnets.

itself with the north-south direction. There is no tendency for the magnet to move either north or south under the influence of the earth's magnetism.

Electrical Attraction

Electrical phenomena were of less practical value than magnetic effects, and little had been learned about them before Gilbert. Certain fish were known to give shocks, lightning was a familiar subject, and rubbed amber was known to attract light objects, but there was no general understanding that these observations were related. It was Gilbert who found that the attractive power of amber was shared by many other substances. His discoveries established the science of frictional electricity.

He made an elementary *electroscope* and used it to detect the presence of electric charge. His instrument was merely a metal pointer balanced to turn easily about its center on a pointed support. He rubbed the substance under study and brought it near one end of the pointer. If the rubbing had produced an electric charge, the pointer would be attracted and would rotate toward the electrically charged body. He found that many substances other than amber produce a deflection of the electroscope including gems, glass, sulfur, crystals and resins. He called them "electrics" because they could develop electrical attraction. The metals were an exception, and he called them "nonelectrics" because he could not electrify them by rubbing. We know, today, that the metals must have lost their frictional electricity to Gilbert's hand as they were being rubbed. It did not occur to him to insulate his metals in order to prevent the leakage of electricity to the earth. For that reason, he missed discovering electrical repulsion.

Electrical Repulsion

The only significant electrical discovery made during the century following Gilbert was that of electrical repulsion. It was first clearly described by Otto von Guericke, the inventor of the air pump. You

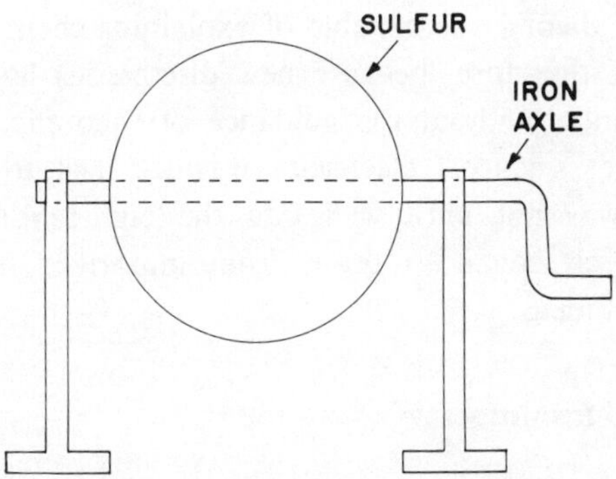

Fig. 91. Von Guericke's electric machine.

can demonstrate electrical repulsion easily in your own home. Merely place some metal filings on a piece of paper and bring a rubbed piece of plastic material near them. The filings will first be attracted to the plastic and then repelled, some "jumping" as far as several inches away.

Guericke discovered this effect with his frictional electric machine. The machine consisted of a sphere of sulfur that could be rotated on an iron axle, as shown in Figure 91. He touched his hand to the rotating sulfur sphere, and friction generated an electric charge on its surface. The charge attracted paper, feathers and many other light objects. While working with the machine, he noticed that a body is sometimes attracted to the sphere and then repelled from it violently. Such bodies are then able to attract other light objects as though they had acquired a charge.

In one experiment, a feather placed between the electried sulfur and the floor jumped up and down between the two. It was first attracted to the sphere where it acquired a quantity of electricity. It was then repelled to the floor where it remained until the electricity leaked off to the ground. Then it was attracted to the sphere once again, starting a new cycle. Guericke also discovered that electricity can travel to the end of a long linen thread—the first example of electricity moving along an electrical conductor. He later found that objects can be electrified merely by bringing them *near* his rubbed sulfur sphere.

All of these observations were mysterious to scientists of the time and no existing theory was capable of explaining them. Progress was extremely slow, therefore, because new discoveries had to be made entirely by chance, without the guidance of theoretical speculation. Of all the exact sciences, electricity was the last to outgrow this deficiency. It was not until well into the eighteenth century that electrical research began to profit from important hints extracted from theoretical ideas.

Conductors and Insulators

Guericke's discovery of electrical conduction was followed up by Stephen Gray in 1729. His apparatus consisted of a glass tube corked at each end as shown in Figure 92. A wire fastened to one cork extended various distances to an ivory ball. When he rubbed the glass tube, he found that the electrical effects were transmitted from the tube through the cork and wire to the ivory ball which would attract a feather. The maximum distance used in the experiment was 765 feet. The wire was kept off the ground by supporting lines of silk thread. The use of silk was necessary because ordinary packing thread would allow the electricity to leak off to ground and no charge could be transmitted to the ivory ball.

Gray's experiments drew a sharp distinction between those materials that can convey electricity from one place to another (electrical *conductors*), and those which cannot (electrical *insulators*). He found that some of the best insulators are hair, silk, resin and glass,

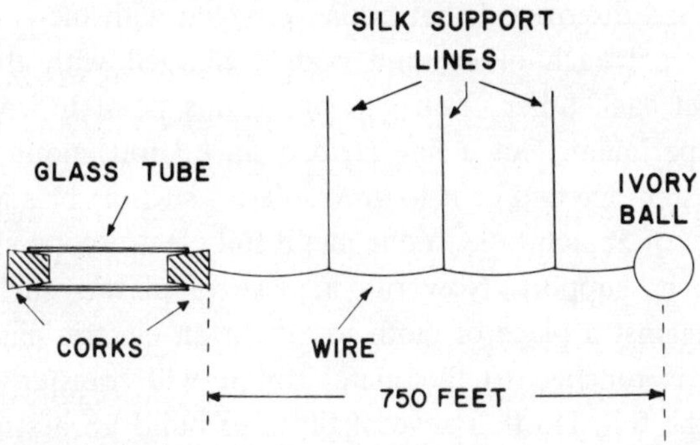

Fig. 92. The transmission of electric charge over a wire.

while the metals are good conductors. It was now clear that Gilbert's "nonelectrics" were merely good conductors—substances that allowed electricity to leak off to the hand or to ground as fast as it was generated. Gilbert's "electrics" were good insulators—materials that impeded the flow of electricity and prevented it form reaching the ground.

Electrical Fluids

A large step forward in our knowledge of electricity was taken by the French scientist Charles Du Fay (1698–1739). His experimental discoveries were two: (1) An electrically charged insulator attracts all conducting bodies, transfers electricity to them upon contact, and thereupon repels them. (2) There are two opposite kinds of electricity, *vitrous* and *resinous*. The latter discovery was extremely important and was published in 1734. Du Fay announced that "there are two distinct electricities, very different from one another. . . . The first (vitreous) is that of glass, rock-crystal, precious stones, animal hair, wool, and many other bodies. The second (resinous) is that of amber, copal, gum-lac, silk, thread, paper, and a vast number of other substances. The nature of these two electricities is that a body of the *vitreous electricity,* for example, repels all that are of the same electricity; and, conversely, attracts all those of the resinous electricity."

Du Fay had discovered that bodies charged with the same kind of electricity repel each other, and bodies charged with the opposite kinds attract each other. You can prove this point to yourself with a single experiment. Cut a fine strip of metal foil about two inches long and fasten one end of it to an insulator, such as plastic or paper, with a piece of Scotch tape. Your metal foil electroscope should hang freely from its support. Now rub a piece of plastic, such as a pen or comb, against a piece of cloth to give it an electric charge. When the plastic is touched to the metal foil it will transfer part of its charge to the foil. Do that several times to build up a strong charge on the electroscope. If the charged plastic is now brought near the metal foil, the latter will be repelled because both objects contain the same kind of electricity. Next, electrify a piece of glass by rubbing it briskly, and bring it *near* the electroscope without touching it. The metal foil will be attracted by the glass because the two objects have opposite kinds of electricity. If you repeat the experiment with various substances and rubbing materials you will have no difficulty in repeating Du Fay's observations.

It's hard to overestimate the importance of Du Fay's discovery. Glass, plastics and metal had been available for centuries, but this great truth about electricity had evaded discovery. With the new knowledge of electrical attraction and repulsion firmly established, Du Fay developed the first important theory of electricity—the two-fluid theory. Although the theory was eventually overturned by Franklin, it served the important purpose of stimulating interest in electricity and guiding experimental research.

Electrical Machines

Following Guericke's lead, electric generating machines were gradually improved as more was learned about frictional electricity.

They eventually superseded the old practice of generating electricity by rubbing various substances by hand. A machine built by Priestley incorporated most of the improvements known in his time. As shown in Figure 93, it consisted of a glass flask attached to a metal axle by cement. The flask was rotated swiftly by means of a crank, two

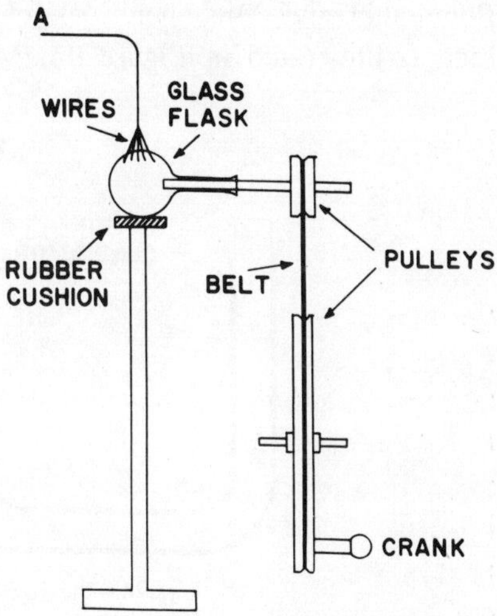

Fig. 93. Priestley's electric machine.

pulleys and a belt. A rubber cushion was placed against the flask where it produced electricity by friction. The electric charge was collected by several wires lightly sweeping the top surface of the globe. Objects could be electrified by touching them to the "prime conductor" at *A*.

Electrical machines became extremely fashionable at the middle of the eighteenth century and scientists began to electrify everything in sight. It was found that electric sparks could ignite gunpowder, ether and alcohol. Children were electrified routinely and safely by suspending them from insulating ropes and having them touch the prime conductor of an electrical machine. Sparks were then drawn from their bodies, to everyone's amusement. Candles were blown out and quickly rekindled by electric sparks, and alcohol was even ignited by an electrified jet of water.

Electricity in a Jar

Perhaps the most astonishing discovery of the century was the famous Leyden jar—a jar or bottle that had the amazing property of storing up large amounts of electricity. It was discovered independently and almost simultaneously in 1745 by E. G. von Kleist of Pomerania and Musschenbroek of Leyden. The form it eventually took is illustrated in Figure 94. Metal foil conductors were placed

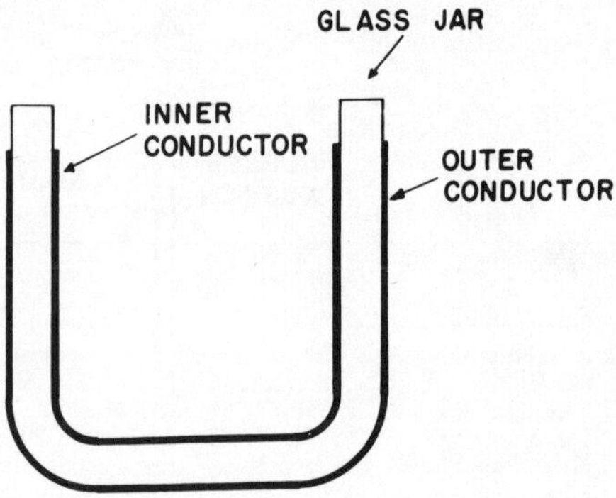

Fig. 94. The famous Leyden jar.

around the inside and outside of a dry glass container, being careful to keep the two conductors from touching. The jar was electrified by connecting the inner conductor to the prime conductor of an electrical machine while the outer conductor was grounded, either through the experimenter's hand or by direct connection to the earth. When disconnected from the machine, the jar was highly charged with electricity. It would then give the experimenter a powerful electric shock as soon as he touched the inner and outer conductors at the same time. Touching one or the other separately produced no shock.

Here was a strange new electrical effect. In the past, sparks and electric shocks had been produced by allowing an accumulation of electricity to rush to the earth through some sort of conductor such as the human body. But the earth played no part in the electrical discharge of the Leyden jar. Its electricity moved only from one of its conductors to the other. A typical experiment of the day was to discharge the jar through a circuit or loop made up of many people holding hands. As soon as the connection was complete from one conductor to the other through the human circuit, each individual received a shock at the same instant.

The principle of the Leyden jar was first explained by Benjamin Franklin (1706–1790), the first great American scientist. He was born in Boston, the tenth son of a soap-maker who had emigrated from England. He became a successful publisher in Philadelphia and played an active part in politics. We know him primarily as a signer of the Declaration of Independence and for his part in the struggle for independence by the American colonies.

Franklin was able to explain all known electrical facts on the basis of a single kind of electrical "fluid." If an object contained the normal amount, it was electrically neutral. If it contained more or less than the normal amount, the object would be electrified one way or the other—*positively* for an excess, and *negatively* for a deficiency of the fluid. Particles of the electrical fluid repel each other. When a body has a large electric charge, the fluid attempts to equalize itself by flowing through a conductor to other bodies.

The Leyden jar consists of two conductors of equal but opposite charge. When the inner conductor is electrified positively, the outer conductor is charged negatively in the same amount. Franklin explained how this condition arises. "Whatever quantity of electrical fire is thrown in at the top (the inner conductor), an equal quantity goes out at the bottom (the outer conductor)." This principle is illustrated in Figure 95. As an electrical charge (Franklin's fluid) flows into the inner conductor, an equal amount flows from the outer conductor to ground. The flow stops when the jar is charged to its full electrical capacity. But if the outer conductor is insulated from the ground, the jar cannot be electrified. Franklin explained that the fluid that enters the inner conductor produces a force of repulsion on the fluid contained in the outer conductor. Similarly, the fluild already existing in the outer conductor repels the fluid entering the inner conductor. If the electricity cannot leak to ground from the outer conductor, very little can enter the inner conductor from the frictional machine. So the jar cannot accept a charge of electricity if its outer conductor is insulated from the ground. This explains why the early experimenters found it necessary to hold the jar in their hands. Their bodies formed the path to ground for the escaping electricity.

Franklin and others found that the shape of the Leyden jar was not

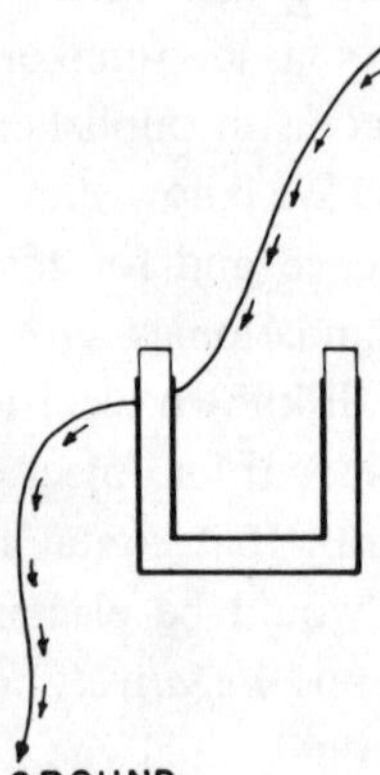

Fig. 95. Franklin's explanation of the charging of a Leyden jar.

important. They obtained the same storage of electricity with a flat piece of glass covered on both surfaces with lead. Such "jars" are still known as "Franklin's panes." Today these reservoirs of electricity are called "capacitors" or, less frequently, "condensers." They owe their large electrical capacity to the relatively large areas and close spacings of the conductors. Modern capacitors are made of metal foil sheets separated by thin sheets of paper or mica. They are an indispensable part of every radio and television set.

Electricity from the Sky

Franklin's greatest experimental feat was the discovery that lightning is merely an enormous electric spark. As late as the seventeenth century, it was generally believed that sulfurous vapors accumulate in thunderstorms and burst into flame in the form of lightning. Franklin used his famous kite experiment of 1752 to show that thunderclouds contain electricity. He used a silk kite in order "to bear the wet and wind of a thunder-gust without tearing." As shown in Figure 96, the wet twine was connected to a dry piece of silk ribbon for the experimenter to hold. The silk prevented leakage of electricity to ground. As soon as a thunder cloud came over the kite, electricity was conducted down the twine and into the metal key. Franklin placed the key in a Leyden jar and thereby charged the jar with electricity. He wrote that

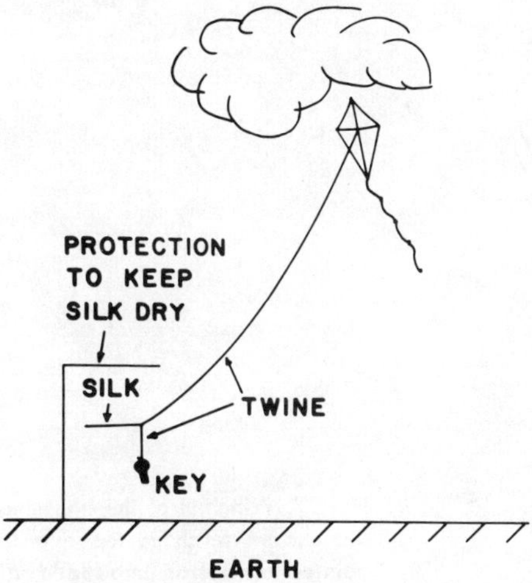

Fig. 96. Franklin's famous kite experiment to capture atmospheric electricity.

"from electric fire thus obtained, spirits may be kindled, and all the other electric experiments be performed, which are usually done by the help of a rubbed glass globe or tube, and thereby the sameness of the electric matter with that of lightning completely demonstrated." The experiment caused great excitement among scientists and was soon repeated many times in Europe. Its great danger to life was not yet suspected, however. In 1753, Richmann of St. Petersburg was killed instantly while experimenting with atmospheric electricity.

The Taming of Lightning

Franklin's discovery that lightning is a sudden flow of electricity between a cloud and the earth led him to the idea of using lightning rods to protect buildings and ships. As shown in Figure 97, Franklin knew that a charged conductor could be discharged by bringing a pointed conductor nearby. The presence of a sharp point seemed to attract the electricity out of the charged body and carry it off to ground. Since lightning is electrical in nature, a cloud could take the place of the charged conductor in the experiment. He wondered if a

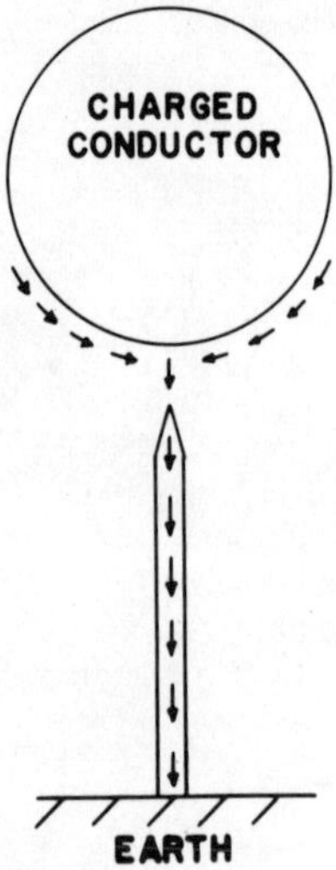

Fig. 97. Principle of the lightning rod. Excess electric charge tends to leave a body and enter a pointed conductor embedded in the earth.

pointed rod high above the top of a building would "draw the electrical fire silently out of a cloud before it became nigh enough to strike, and thereby secure us from that sudden and terrible mischief?"

A great controversy arose about whether the rods should be blunt or pointed, or whether they might attract lightning that would otherwise have passed harmlessly by. Franklin's arguments won out, however, and pointed lightning rods were soon used to protect buildings all around the world.

Following Franklin's work, a number of important discoveries were made in Europe. Johan Carl Wilcke (1732–96), the secretary of the Swedish Academy of Sciences, discovered that the two kinds of electricity are invariably produced when two objects are rubbed together. If one becomes positively charged, the other is always found to be negatively charged to precisely the same degree. It appeared as though a quantity of electricity had been transferred from one body to the other by friction. Wilcke arranged a number of substances in an electrical series such that each substance, when rubbed with a substance lower in the series, became electrified with a positive charge. His series included the following materials in the order given: glass, wool, wood, sealing wax, metal, sulfur. It was later discovered, however, that the position of a substance in the list also depends on the nature of the surfaces being rubbed.

Induced Charges

About 1753 John Canton (1718–1772) discovered a process called "electrical induction." His apparatus is shown in Figure 98. An electrically neutral conductor was insulated from ground to prevent the leakage of electricity during his experiments. A charged rod was then brought near the conductor and Canton found that two opposite charges were developed, or *induced,* as shown in part (*b*) of the diagram. The induced charge nearest the rod had a sign opposite to that of the rod; the induced charge farthest away had the same sign as the rod. Clearly, the electricity in the neutral conductor had merely been induced to move and change its position. When the rod was removed, the conductor reverted to its initial uncharged condition.

Wilcke and his colleague Franz Aepinus (1724–1802) repeated Canton's experiment and added a new twist. They grounded the conductor momentarily while the charged rod was in its vicinity. As shown in part (*c*), the connection to the earth enabled a flow of electricity to neutralize the induced charge on the end of the conductor opposite the rod. The ground connection was then removed as in part (*d*). Lastly, the charged rod was taken away. As shown in part (*e*), the conductor had acquired a negative charge which became evenly distributed over its surface. This induced charge is always opposite in sign to that of the influencing charge.

Aepinus also pointed out that the difference between conductors and insulators is merely one of degree. The difference depends upon the relative *resistance* offered by each to the flow of electricity. Conductors offer very little resistance and insulators offer appreciable

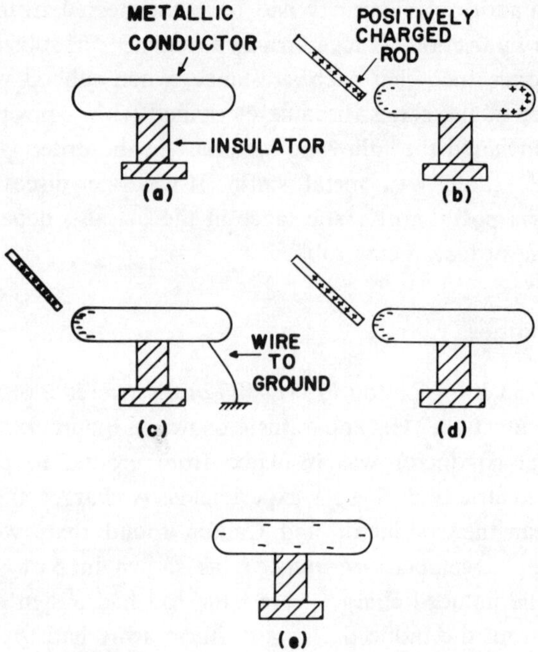

Fig. 98. Charging a conductor by induction.

resistance. The concept of electrical resistance later became extremely important in connection with Ohm's law.

Electricity's Inverse-Square Law

It was generally believed, toward the end of the eighteenth century, that the force of electrical attraction or repulsion falls off in proportion to the square of the distance separating the charges. Priestley and Cavendish had inferred the truth of the *inverse-square law* indirectly from experiments they had performed. But Charles Coulomb (1736–1806), a French military engineer, was the first to derive the law from direct experimental evidence in 1784 and 1785.

To prove the truth of the law he measured the minute forces acting between two charged bodies as their distance of separation was varied. He made these measurements, using the principle of the torsion balance that he had previously discovered. Imagine a horizontal bar hanging from a fine wire or thread as shown in Figure 99. If the bar is rotated somewhat, the wire becomes twisted and exerts a restoring force tending to return the bar to its initial position. Coulomb had discovered that the restoring force is proportional to the angle through which the bar is turned.

Using this principle, Coulomb was able to measure extremely minute

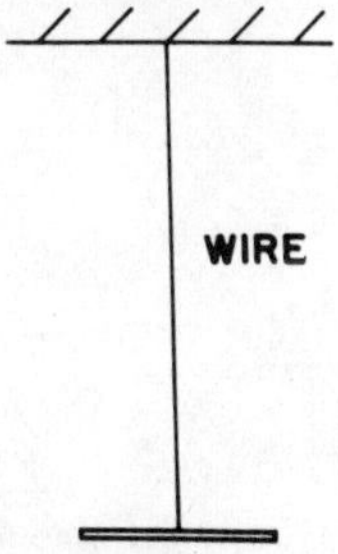

Fig. 99. Coulomb's torsion balance.

forces exerted by small charged bodies. As a result of many experiments he concluded that the force generated by charged bodies varies inversely as the square of the distance apart of their centers. Oppositely charged bodies attract, and similarly charged bodies repel each other. The law is analogous to the inverse square law of gravitational attraction.

Coulomb also verified a finding of Cavendish that electric charge distributes itself on the *surface* of a conductor without penetrating its interior. Scientists of the time proved mathematically that this was an inevitable consequence of the electrical inverse square law.

Measuring Electricity

Most of the discoveries concerning static or stationary electricity were made with the help of extremely crude instruments. These were used to detect the presence and sign of induced charges, to investigate frictional and atmospheric electricity, to standardize the charges of

Leyden jars, and so forth. The first measuring instruments were simple electroscopes which gave no indication of the quantity of electricity involved. They merely indicated the presence of a charge.

The early electroscopes were threads or strips of metal foil that were deflected in the presence of a charge. Later, Du Fay's discovery of electrical repulsion made a better instrument available. In 1753, John Canton used a pair of linen threads about six inches long, as shown in Figure 100. Suspended from each thread was a pea-sized

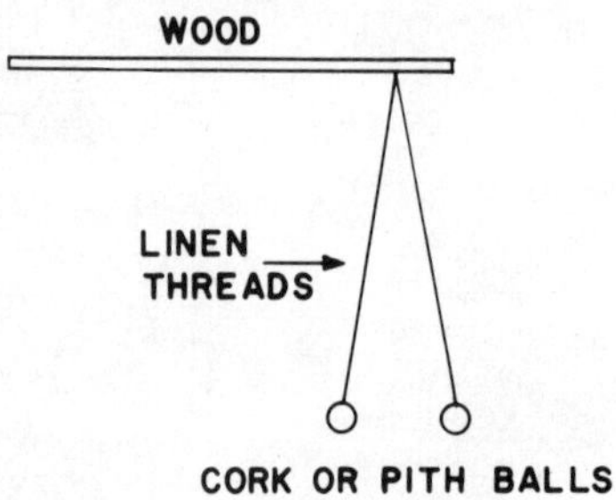

Fig. 100. An electroscope. A charge placed on the balls produces a force of repulsion that separates the two.

ball of cork or pith. When the balls were given a charge, their mutual repulsion caused them to separate as shown. Cavendish used a similar electroscope but added a scale behind it. In this way he was able to estimate the divergence of the balls and the relative amount of charge they carried.

In 1787 Abraham Bennet (1750–1799) built a gold leaf electrometer of the type still used today for teaching and demonstrations. As shown in Figure 101, the electrometer used two strips of gold leaf enclosed in a glass cylinder about 5 inches high. The leaves were connected electrically to a metal cap on top of the cylinder. Two strips of tinfoil were cemented to the sides of the glass cylinder on its inner surface. They served to discharge the leaves to ground when an excessively high charge forced them out far enough to touch the tinfoil. The amount of charge in the electrometer was measured by comparing the divergence of the leaves with a graduated scale (not shown) pasted to the outside of the cylinder.

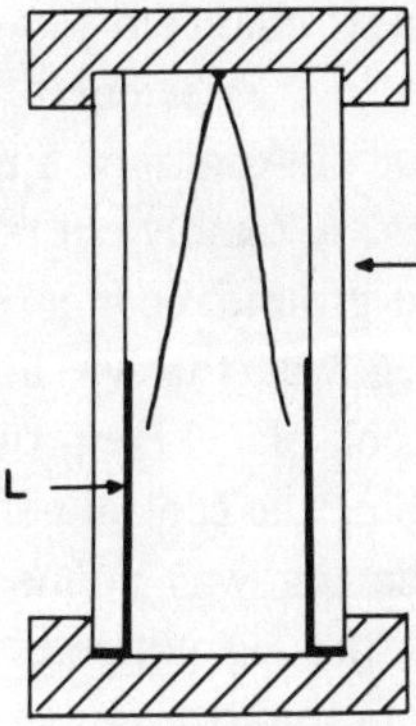

Fig. 101. A gold leaf electrometer.

Electricity from Metals

Scientists of the eighteenth century had learned to generate electricity in several ways: by friction, by heating certain crystals, by drawing electricity from the atmosphere, and by experimenting with electric fishes. Vastly more important than any of these was a source of electricity hidden in Gilbert's "non-electrics," the metals. This implausible source of electricity was destined to generate a revolution in power that rivalled the importance of chemistry to man.

It all began with an unexpected observation made about 1750 by J. G. Sulzer, a German Professor of Mathematics. Sulzer happened to

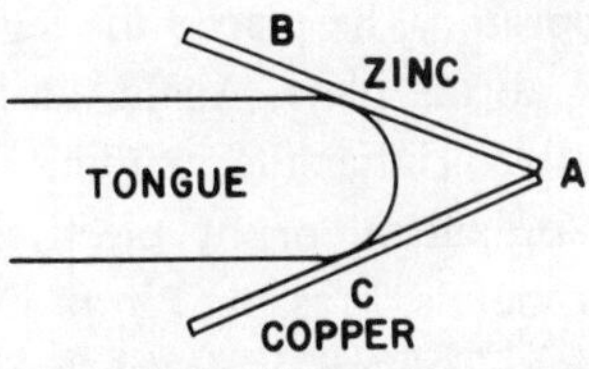

Fig. 102. Sulzer's experiment with contact electricity.

place two pieces of dissimilar metal against his tongue as shown in Figure 102. When the metals were in contact at *A*, and touched the tongue at *B* and *C*, he noticed a strange, pungent taste. When the metals were placed on his tongue separately, no effect was noticed.

Sulzer did not connect his observation with electricity, and it was actually forgotten for half a century.

The next chance discovery of "contact electricity," as it was called, was not lost to science. It happened about 1780 when the Italian Professor of Anatomy, Luigi Galvani (1737–98), noticed a twitching frog's leg. A freshly cut frog's leg happened to be in electrical contact with the ground near an electric machine. Galvani noticed that the frog's leg was thrown into convulsions whenever an electrical discharge took place. He proved that the response was electrical in nature by touching the central nerve with the prime conductor of the machine, while the leg was connected to ground. Each time the nerve was touched, the leg jumped.

Galvani found that electricity from the Leyden jar and from thunderstorms had the same effect on animal nerves. Then he decided to determine whether normal atmospheric electricity could excite the mysterious twitchings. It was this experiment that led to the great discovery of contact electricity.

Galvani had attached the frog's legs to an iron trellis. He had also driven brass hooks into the spinal marrow in order to attract any electricity that the air might contain. He noticed that the legs showed occasional twitching even in fine weather, but the results were sporadic and unpredictable. Finally, tired of waiting, he happened to touch a brass hook to the iron trellis—and the amazing frog's leg went into a convulsion. Each time the brass and iron came into contact the leg twitched and jumped!

Returning to his laboratory, he placed the leg on an iron plate and touched the brass hook to the plate. Again the leg jumped. Only two explanations were possible. Either the jumping was caused by electricity generated in the organism, or by electricity generated by the contact of the different metals, brass and iron. Despite his experiment with the electric machine, Galvani declared the jumping to be caused by animal electricity. Even more strangely, his theory received general acceptance. Frogs were dismembered in vast numbers as scientists repeated his amazing experiments, but only one man thought of contact electricity—Alessandro Volta.

Volta repeated Sulzer's experiment with his tongue as well as Galvani's experiments with frogs. He showed by a number of precise experiments that the nerves were stimulated by an external source of electricity. The metals themselves were the source of that electricity. Galvani and Volta engaged in a violent scientific controversy over the matter. Galvani became depressed and died without ever knowing the magnitude of his discovery, but *metallic electricity* won out over animal electricity. Its conquest became complete in 1800 with Volta's invention of the voltaic pile or electric battery. (This invention was mentioned in Chapter 10.)

A simple voltaic cell is shown in Figure 103. Electrodes of copper

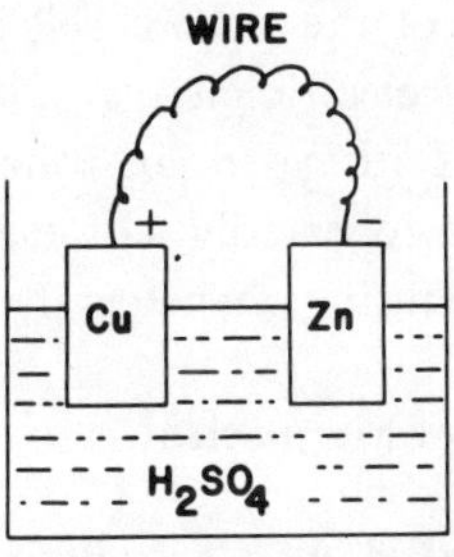

Fig. 103. A voltaic cell.

(Cu) and zinc (Zn) are immersed in a weak solution of sulfuric acid, H_2SO_4. Such a cell sends a continuous flow or current of electricity through the wire connecting the two metallic electrodes. The H_2SO_4 dissociates into electrically charged particles called *ions,* as explained in Chapter 10: The H^+ ion and the SO_4^{--} ion.

$$H_2SO_4 \rightarrow 2H^+ + SO_4^{--}$$

The H^+ ions are deposited on the copper plate, giving it a positive charge. At the zinc plate, some of the zinc atoms dissolve to form Zn^{++} ions leaving a negative charge on the plate.

$$Zn \rightarrow Zn^{++} + \text{negative electrical charge on zinc plate}$$

The negative charge on the zinc plate flows through the wire to the positive plate. This negative charge unites with the H^+ ions forming hydrogen gas.

$$\text{negative charge} + H^+ \rightarrow \text{hydrogen gas}$$

The Zn^{++} and SO_4^{--} ions remain in solution until the water evaporates, when they unite to form zinc sulfate, $ZnSO_4$.

$$Zn^{++} + SO_4^{--} \rightarrow ZnSO_4$$

The electricity flowing in the circuit permits the chemical reactions to take place. If the wire is disconnected from either electrode, the negative charge cannot reach the copper terminal and the chemical changes are effectively stopped. The common battery, then, is merely a device that converts the energy contained in chemical substances to a flow of electricity through an external circuit—to electrical energy.

The great importance of the voltaic cell lay in its ability to produce a *continuous* flow of electric charge. Although its electricity was much weaker than that of a Leyden jar, it had the great advantage of not needing to be charged repeatedly with electricity from outside. Its electricity came spontaneously, whenever it was needed.

The Inverse Square Law of Magnetism

Much had been learned about electricity by 1800, but the knowledge of magnetic effects barely exceeded Gilbert's. The new elements cobalt and nickel were found to be magnetic. Iron, cobalt and nickel came to be called "ferromagnetic" substances because of their relatively strong magnetic effects. On the other hand, the elements bismuth and antimony *repelled* the poles of a magnetic needle—the first example of *diamagnetism*. Instead of being attracted to a magnet, they were repelled by it.

The most important magnetic discovery of the century was Coulomb's inverse square law of magnetic force. Using the principle of the torsion balance, he proved that the force of attraction or repulsion of two magnetic poles varies inversely as the square of the distance between them. Like poles repel, and unlike poles attract each other. Another fundamental law of nature had been found to vary inversely with the square of the distance.

Electromagnetism: The Marriage of Electricity and Magnetism

It had long been suspected that electricity was somehow related to magnetism. Pieces of iron, for example, had often been found magnetized in the vicinity of a recent lightning bolt. But the connection between the two sciences eluded discovery until 1820. In that year a Danish scientist, Hans Christian Oersted (1777–1851), made a great discovery that created the science of electromagnetism. He found that a wire carrying an electric current causes a deflection of a magnetized needle. The wire was first aligned parallel with the compass needle with no current flowing, as shown in Figure 104. When

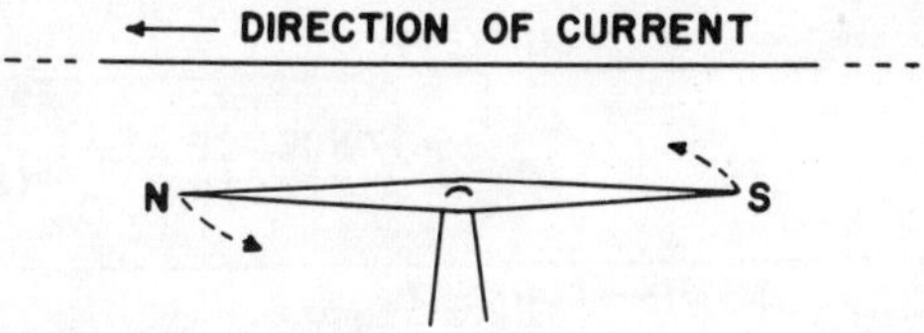

Fig. 104. Oersted's discovery of the interaction between electricity and magnetism. An electric current produces a force on a magnetic needle that was initially lined up with the wire.

the current was caused to flow, the compass needle rotated in the direction indicated until it was at right angles to the wire. When the current was stopped, the needle returned to its normal north-south direction.

Oersted's discovery was truly momentous. Although an electrically charged body has no effect at all on a magnet, a weak current of *moving charges* is able to exert forces on magnetic poles—forces that are transmitted through empty space. The connecting link between electricity and magnetism had turned out to be motion.

The significance of Oersted's discovery was recognized immediately by scientists. Francois Arago (1786–1853), a French experimenter, showed in the same year that a coil of wire acted like a magnet while current was passing through it. Previously unmagnetized iron filings clung to the wire as long as the current flowed, but dropped off when the current ceased. Another great Frenchman, Andre Marie Ampère (1775–1836) determined the laws governing the interaction of electric currents with each other and with magnets. He showed that two parallel wires carrying currents exert forces on each other, as shown in Figure 105. If the currents are in the same direction, (*a*), the wires are attracted to each other; if the currents are in the opposite direction, (*b*), the wires repel each other.

Ampère also showed, following Arago's lead, that a wire coil wound in the shape of a cylinder behaves exactly like a magnet when a current flows through it. It has a north and south pole and attracts magnetic materials. In the same year, 1820, Davy and Arago independently showed that a piece of steel could be permanently magnetized by placing it momentarily in a current carrying coil of wire. But if soft iron were used instead of steel, the iron lost most of its

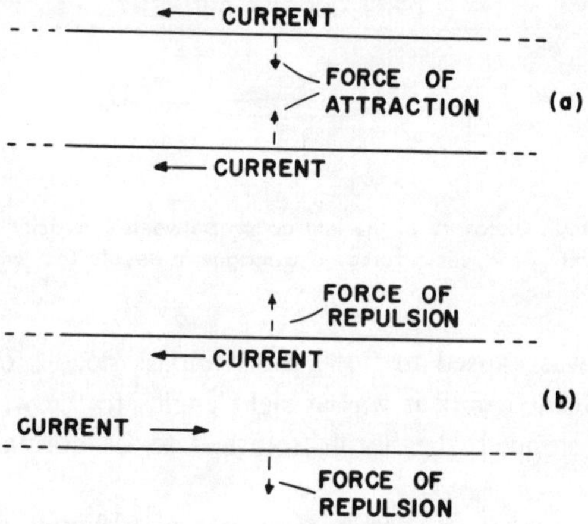

Fig. 105. Ampère's discovery that one electric current can produce a force on another in its vicinity.

magnetism as soon as the current was turned off. These electromagnets, as the latter were called, were extremely powerful and by 1825 they were being made in large numbers.

In 1825 Ampère hit upon the correct explanation of the source of magnetism. He knew that a loop of wire becomes a magnet when current flows through it. Iron is magnetic, therefore, because circular currents of electricity flow in each of its atoms. Each atom or molecule of a magnetic substance has a north and south pole. If these molecular poles point in random directions they cancel and the piece is unmagnetized. If they line up, to some extent, in the same direction, their individual magnetism adds up and the piece is a magnet. The existence of all magnetic effects had been explained on the basis of moving electrical charges.

Soon after the discoveries of Ampère were published in 1820, a number of scientists used electromagnetic forces to detect and measure electric currents. These instruments, called "galvanometers," consisted at first of a coil of wire surrounding a pivoted compass needle. As shown in Figure 106, the instrument was lined up so that the coil windings lay in the north-south direction. When a current was passed through the coil, the compass needle turned away from the north-

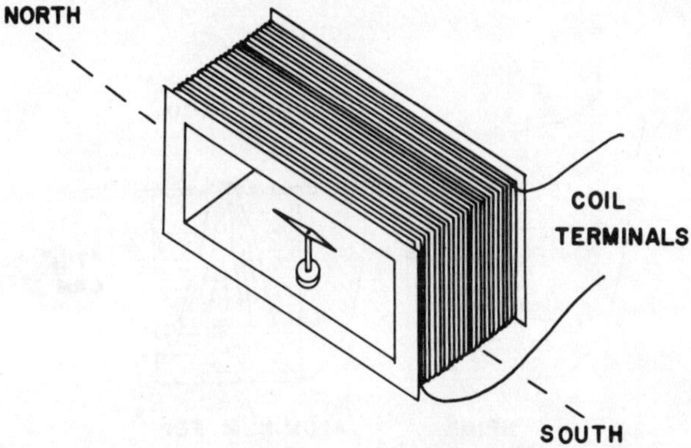

Fig. 106. An early galvanometer to detect the presence of an electric current.

south direction through an angle that depended upon the amount of current flowing. Its deflection was a rough measure of the strength of the electric current.

You can make your own sensitive galvanometer by winding about two hundred turns of insulated copper wire around a cardboard cylinder. Place an inexpensive magnetic compass inside the cylinder and align the galvanometer so that the needle points at right angles to the axis of the cylinder with no current flowing. The needle will be deflected whenever a small battery is connected to the ends of the coil. Or you can make your own "crown of cups" out of "tin" cans and aluminum foil. Select food cans that have a bright tin-plate with no visible coating on the inside. The can is one terminal of each voltaic cell and the aluminum foil is the other. A two-cell battery is illustrated in Figure 107. The electrolyte is brine made by dissolving about four teaspoonfuls of salt per cup of warm water. Use copper wire to connect the terminals as shown. Care must be taken to prevent the aluminum foil from touching the metallic can. Any insulating material, including dry pencils, can be used in place of the glass rods. You can increase the deflection of your galvanometer by using up to ten or twenty voltaic cells in the battery or by increasing the number of turns of wire on the galvanometer.

Although the apparatus described above may seem crude by mod-

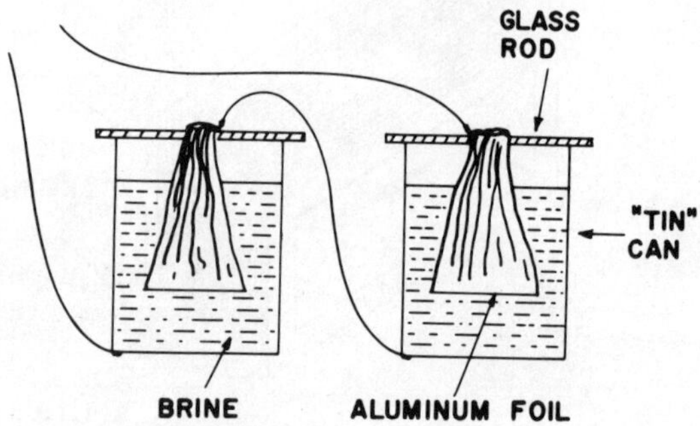

Fig. 107. A homemade voltaic battery.

ern standards, it would rival that found in most electrical laboratories in 1820. With that humble beginning, the electricians of the nineteenth century unravelled the secrets of electricity and magnetism and made possible a marvelous new industry. The discovery that moving charges of electricity can deflect a magnetic needle was a turning point in human history. By 1820, the steam engine had substituted the inanimate energy of coal for the muscle power of man and horse. But there was no convenient way to transmit that new power to every nook and cranny of human activity. Electricity would soon eliminate that deficiency. The theoretical and experimental exploration of electricity and magnetism was the greatest achievement of nineteenth century physics.

Electricity at Work

13

Oersted's discovery of the connection between electricity and magnetism was accepted eagerly by scientists. But few scientific breakthroughs receive such a warm reception. Another great electrical discovery was announced a few years later that received abuse and ridicule for almost twenty years. That discovery, by the great German scientist Georg Simon Ohm (1787–1854), concerned the connection between static or frictional electricity, and current electricity.

Scientists of the eighteen-twenties were convinced that the laws of static electricity were entirely distinct from those of current electricity. And there was considerable evidence to support their view. Nevertheless, Ohm insisted that the two were related by simple scientific laws. It is worth discussing the resulting controversy in detail because of the insight it gives into our modern concept of electricity.

Electrical Resistance

The story begins with the experiments of Sir Humphry Davy, the eminent English chemist. It was well known at the time that frictional charges of electricity can be made to flow along metal conductors. But few investigations had been made of the flow of current electricity. In 1821, Davy began to investigate the ability of metals to conduct current electricity from voltaic cells. His apparatus was simply a voltaic battery whose terminals were connected by two paths: one containing a conducting water solution, as shown in Figure 108 and another consisting of a metallic wire under test. The current from the battery was divided in two parts, therefore, as indicated by the curved arrows. One part flowed through the loop containing the water and the other part flowed through the test wire. The battery current divided unequally, in general, the greater amount flowing in the loop having the lesser resistance to current flow.

When the test wire was relatively long, or disconnected from the battery completely, the water was decomposed by the flowing current

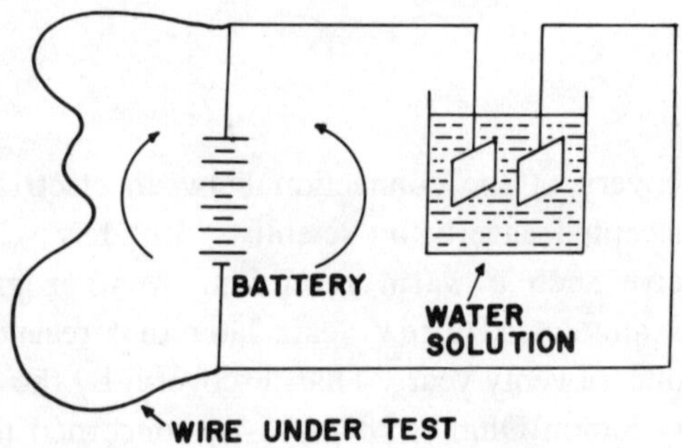

Fig. 108. Davy's apparatus for research on electrical resistance.

and bubbles rose to the surface. As Davy gradually shortened the wire, he noticed a reduction in the number of bubbles coming from the decomposing water. When the length of wire was reduced below a critical length, the water was no longer decomposed, and bubbles stopped forming in the solution. Davy knew that the critical wire length corresponded to a definite amount of electrical *resistance* in the test wire. He also knew that the resistance was reduced as the length of wire was shortened. Here, then, was a way of cutting many pieces of wire so all would have the same amount of resistance. Using his apparatus, he could select wires of different materials and many cross sectional areas and cut each to a length corresponding to precisely the same amount of electrical resistance.

Davy measured the lengths and diameters of different metallic wires under these limiting conditions and showed that the resistance of a wire made of a specific material is directly proportional to its length, and inversely proportional to its cross sectional area. The formula for the resistance of an electrical conductor is:

$$\text{resistance} = K \frac{\text{length}}{\text{cross sectional area}}$$

$$R = K \frac{l}{a}$$

where *K* is a constant for a given material that depends on the units in which the other quantities are measured.

Davy also proved that the resistance of a conductor is independent of the shape of its cross section. Whether rectangular or circular in cross section, two wires have the same resistance per unit length if their areas of cross section are the same. This discovery implied that voltaic currents pass *through the interior* of the conductor and not merely along its surface. A wire of rectangular section has a greater surface area than one of circular section (Figure 109), if both have

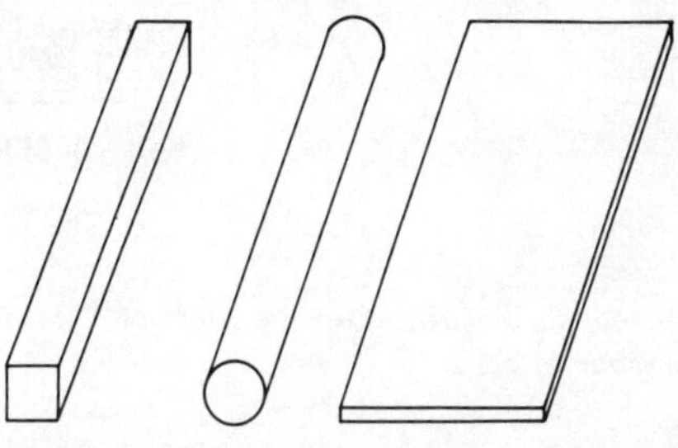

Fig. 109. Wires of the same cross sectional area do not necessarily have the same surface area.

the same cross sectional area. If the voltaic currents flowed only on the surface, the wire of greater surface area would have a lower resistance per unit length. But Davy found that both have the same resistance. Electric current, therefore, is not restricted to the surface of a wire, but flows throughout its interior. This unexpected property of current electricity was clearly different than for static electric charges which were known to remain only on the surface of conductors. Davy's discovery seemed to indicate that static and current electricity were really different in principle.

Electric Tension

Another well known characteristic of voltaic batteries led to the belief that the two kinds of electricity were unrelated. This was the so-called "tension" of the voltaic cell. If one terminal of a cell is grounded, and the other is connected to a gold-leaf electroscope or electrometer, the leaves diverge indicating the presence of an electro-

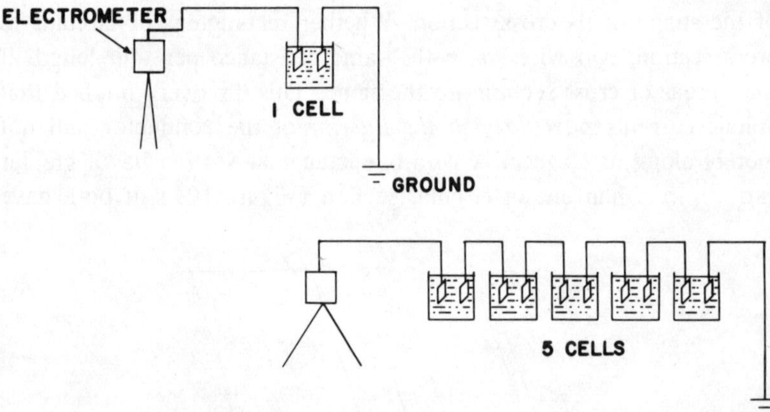

Fig. 110. An experiment to illustrate that the "tension," or voltage of a battery increases with the number of cells.

static charge. The amount of divergence of the leaves is a measure of the amount of charge generated by the cell. Increasing the number of cells increases the amount of charge, as shown in Figure 110. No current flows in the experiment described above, except for the momentary motion of charge necessary to electrify the electroscope. The deflection is a true measure, therefore, of the static electricity generated by the battery. The property of a battery that causes the leaves to diverge was called the "tension" of the battery. Ohm later called it the "electroscopic force."

Scientists found that *tension* depends only on the number of cells, and not upon the size of the plates used in the battery. If a sufficient number of cells were added—even cells having small plates—the battery could produce sparks, shocks and all of the phenomena associated with frictional or static electricity.

It was also known, of course, that a battery can produce current electricity. For this purpose, experiment showed that the areas of the plates were the important factor. Consider the apparatus shown in part (*a*) of Figure 111. A voltaic cell sends a current through a coil

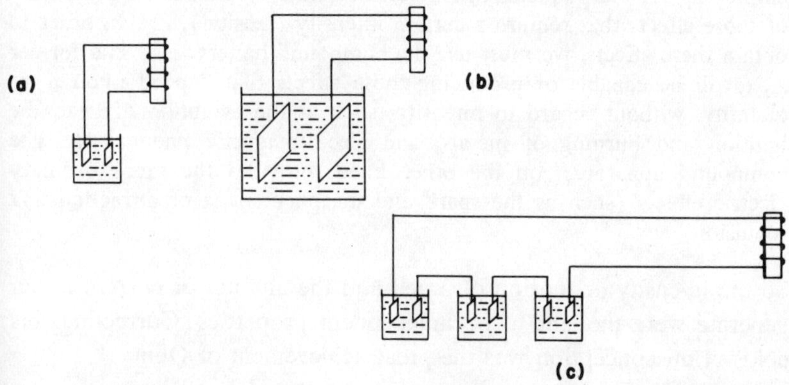

Fig. 111. An electromagnet connected to several sources of elecrticity.

of wire surrounding a bar of iron. The iron is magnetized by the current and the amount of magnetism can be measured by noting its pull on an iron nail. Let's call the degree of magnetism *M*. The apparatus in part (*b*) is similar except for the much larger areas of the electrodes in the cell. When the magnetism was measured for this arrangement, it was found to be much greater than *M*. The battery in part (*b*) produced more current which, in turn, generated a higher degree of magnetism in the iron. In part (*c*) several cells are connected in series to form a compound battery. The extreme terminals of the battery are then connected to the same electromagnet. The size of the electrodes is the same as in part (*a*). A measurement of the magnetism for such an arrangement showed that it was approximately the same as for part (*a*). Increasing the number of cells, and therefore the tension of the battery, did not seem to increase the magnetic effect.

The latter could be increased only by increasing the area of the plates, and therefore the current generated by the battery.

This experiment convinced the greatest scientists of the time that current electricity and tension were not related. Certain kinds of electrical effects came to be associated with the tension of a battery and others with the current it can produce. According to the great Ampère (1820):

> However much we may increase the quantity of electricity (current) by employing very large plates . . . we cannot obtain from such a device any of those effects that require a certain intensity (tension). . . . In order to obtain these effects, we must use the compound battery. . . . The former . . . will be capable of producing those effects that depend upon mere quantity, without regard to intensity; such as the evolution of heat, the ignition and burning of metals, and electromagnetic phenomena. The compound apparatus, on the other hand, provides the more ordinary electric effects (such as the spark and the phenomena of attraction and repulsion).

So the intensity or tension of a cell and the amount of current it can generate were thought to be independent properties. Correcting this colossal misconception was the great achievement of Ohm.

Electromotive Force

Ohm noticed almost immediately that if a number of voltaic cells are connected in series, as in Figure 111(*c*), the current is independent of their number *only if the external resistance is small.* If the external resistance (of the wire coil in Figure 111(*c*), for example), is quite large, the current increases as additional cells are added. The current is then proportional to the number of cells in the battery. Here was a clear connection, at last, between the tension of a battery and the current it can produce. After many related experiments, Ohm attacked the problem of combining all known facts into a consistent theory of electricity.

His analogy was the flow of heat along a metal wire or rod. The theory of heat flow had been familiar to physicists since the publication of Fourier's *Analytic Theory of Heat* in 1822. To illustrate Ohm's ideas, imagine a bar of metal placed in boiling water (100°C) at one end and melting ice (0°C) at the other, as shown in Figure 112. Thermometers record the temperature at equally spaced intervals along the bar. Ohm pointed out that the amount of heat-flow along the bar depends upon the difference in temperature between its ends. Furthermore, the temperature drops uniformly along the bar from right to left, as indicated by the readings of the thermometers.

The similarity between the flow of heat and the flow of electric current suggested the introduction of a quantity whose behavior in

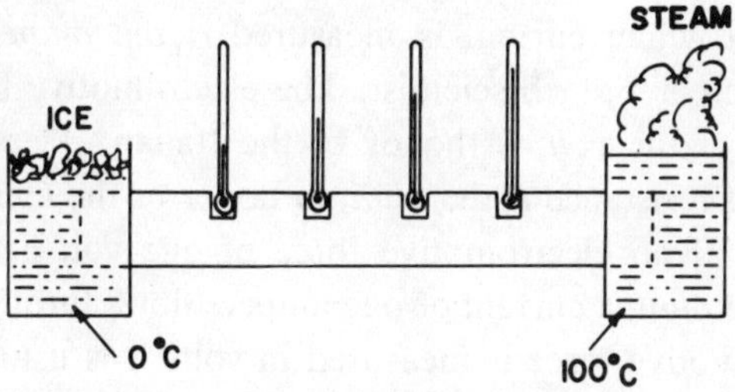

Fig. 112. The flow of heat along a metal rod.

electrical circuits was similar to that of temperature in the theory of heat. That quantity was the tension of a voltaic cell—the physical quantity that causes the leaves of an electroscope to move apart. Ohm called it the *electroscopic force* of a cell, and it is now called an *electromotive force*. It is the driving-power that forces electric charge to move through a wire. Current electricity is merely static electricity that has been set in motion by the electromotive force of a voltaic cell.

Ohm's Law

Ohm published his conclusions in a book and in a series of scientific papers between 1825 and 1827. The rule that he developed is fundamental to all electrical phenomena and is known as Ohm's law. The law states that the voltaic current flowing in any circuit is proportional to the electromotive force in the circuit:

$$\text{voltaic current} \propto \text{electromotive force.}$$

Expressed as an equation, the law becomes:

$$\text{voltaic current} = \frac{\text{electromotive force}}{\text{resistance}}$$

$$I = \frac{E}{R}$$

where the quantity *R* (resistance) is the constant of proportionality. *R* depends upon the electrical resistance through which the current must flow in the circuit.

The unit in which current is measured is the *ampere,* named in honor of the great French scientist. The electromotive force is measured in units of the *volt,* in honor of the Italian, Alessandro Volta, and the unit of resistance is the *ohm,* in honor of the German scientist of that name. If an electromotive force of one volt acts upon a resistance of one *ohm,* a current of one ampere flows through the circuit. When electromotive force is measured in volts it is usually spoken of as *voltage.*

The accompanying diagram (Figure 113) illustrates a simple electric circuit both in pictorial and schematic form. The voltaic cell generates an electromotive force (or voltage) of 1 volt and the resistance around the entire circuit is 10 ohms. Using Ohm's law, the current flowing is simply:

$$I = \frac{E}{R} = \frac{\text{1 volt}}{\text{10 ohms}} = \text{0.1 amperes.}$$

The diagram also depicts a more complicated circuit consisting of several cells connected in series with several pieces of wire of varying resistance. The equivalent voltage generated by the compound battery is the sum of the individual cell voltages, or 4 volts. Similarly, the total resistance in the circuit is the sum of the individual resistances, 8 ohms. The current flowing, therefore, is:

$$I = \frac{E}{R} = \frac{\text{4 volts}}{\text{8 ohms}} = \text{0.5 ampere}$$

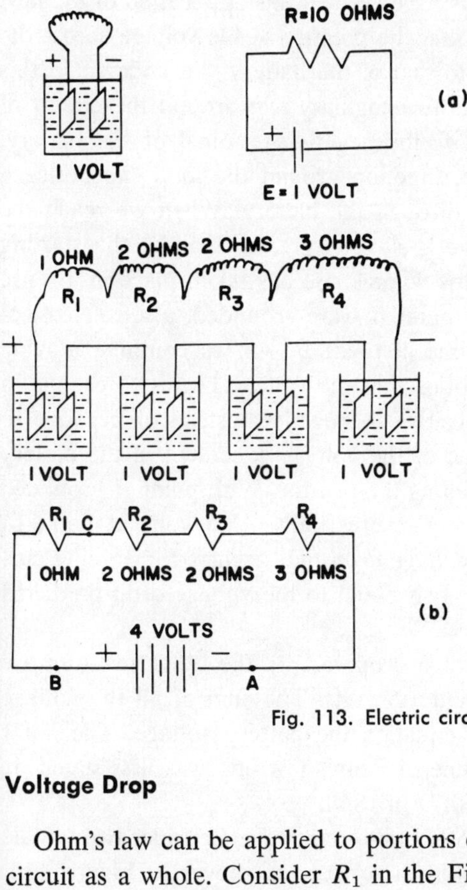

Fig. 113. Electric circuits used to illustrate Ohm's law.

Voltage Drop

Ohm's law can be applied to portions of a circuit as well as to the circuit as a whole. Consider R_1 in the Figure 113, for instance. The current flowing through R_1 produces a measurable voltage across its terminal—a voltage that can be detected by an electroscope and calculated by using Ohm's law. We can rearrange the law in the form:

$$\text{Voltage} = \text{Current} \times \text{Resistance}$$
$$E = IR$$

In that form, the law tells us that the voltage, E, across the terminals of R_1 is the product of the current flowing through the resistor and the magnitude of its resistance:

$$E_1 = \text{Voltage across } R_1 = IR_1 = (0.5)1 = 0.5 \text{ volts}$$

Many accurate measurements have verified this application of the law.

Measurements also show that the polarity of the voltage across the terminals of R_1 is opposite to that of the battery. To understand this polarity difference, let's take an imaginary trip around the circuit of Figure 113(*b*), starting at *A*, the negative terminal of the battery. As we move in a clockwise direction around the loop, we notice a gradual increase in voltage through the battery. When we reach the positive terminal, the voltage is 4 volts greater than at the starting point. We also notice that this *voltage rise* has taken place in a sense from negative to positive. If point *A* were grounded, a sensitive electrometer would read +4 volts at point *B*. As we continue moving clockwise through R_1, we notice that the voltage begins to change in the opposite sense—from positive to negative instead of negative to positive. It is as though some of the voltage generated in the battery is being lost as we progress along the resistor. With point *A* grounded once again, our electrometer measures only +3.5 volts at point *C*. The missing 0.5 volt is the *voltage drop* that occurred in forcing current through the resistor R_1. It is equal to the voltage drop predicted above by Ohm's law.

In a similar way, the voltage drops across the other resistors are found to be 1 volt, 1 volt and 1.5 volts. The sum of all the voltage drops around the circuit is equal to the battery voltage. The latter statement is a natural outcome of Ohm's law and was first stated by Gustav Kirchhoff (1824–1887) in 1848.

Kirchhoff also noticed another important rule. Consider two voltaic cells connected in parallel with a resistor as shown in Figure 114. The two cells are identical and each produces a current that flows through the resistor. The rule states that the sum of the currents entering any junction in the circuit is equal to the sum of the currents leaving that junction. The currents entering and leaving point *A* are indicated by arrows in the diagram. According to Kirchhoff's law, $I_1 + I_2 = I_3$. If each cell supplies 1 ampere, then:

$$I_3 = 1 + 1 = 2 \text{ amperes}$$

Kirchhoff's laws are extremely useful in analyzing the performance of electrical circuits.

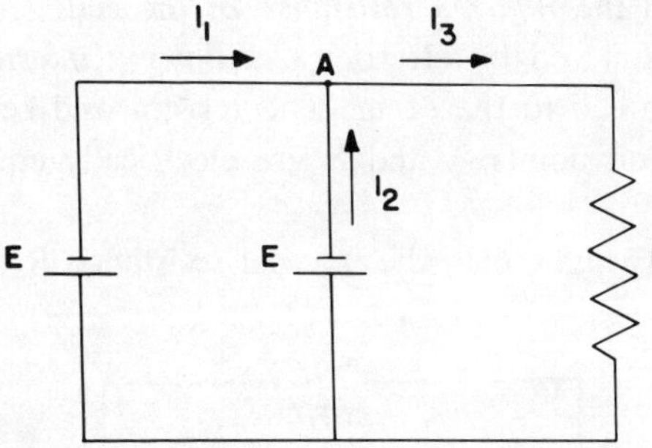

Fig. 114. Kirchhoff's second law. The magnitude of the current leaving point *A* is equal to the sum of the currents entering point *A*.

Internal Resistance

In the examples discussed above, the voltaic cells were assumed to have no internal resistance. This is far from true, especially for the simple cells known in Ohm's time. It was this internal resistance, in fact, that led to the idea that tension and current were unrelated. It was well known, for example, that a voltaic cell "loses" its tension (voltage) when a wire is connected across its terminals. Despite the apparent absence of voltage, the cell sends a current through the wire. Ohm's law enables us to understand why a voltaic cell appears to lose its electromotive force when a short wire is connected across its terminals. Consider the apparatus illustrated in Figure 115. The terminals *A* and *B* of a simple cell are connected in series with a switch and a short piece of wire. When the switch is open, as shown, no current can flow and the presence of an electromotive force can be detected at the terminals of the cell with an electroscope. When the switch is closed, a current flows through the wire and an electroscope fails to detect a measurable electromotive force at the terminals of the cell.

The lower diagram shows the same circuit in schematic form. The connecting wires are now assumed to have zero resistance, and the real resistance of the circuit is lumped in two places, designated R_1 and R_2. The resistor R_2 has a magnitude equal to the resistance of the wire in the part of the circuit external to the cell. The resistor R_1

is equivalent to the *internal resistance of the cell.* It represents the resistance encountered by electricity in flowing *internally* from one electrode of the cell to the other. The resistor and cell drawn schematically between points *A* and *B* are electrically equivalent to the real voltaic cell.

For a simple voltaic cell, the internal resistance R_1 is about 1000

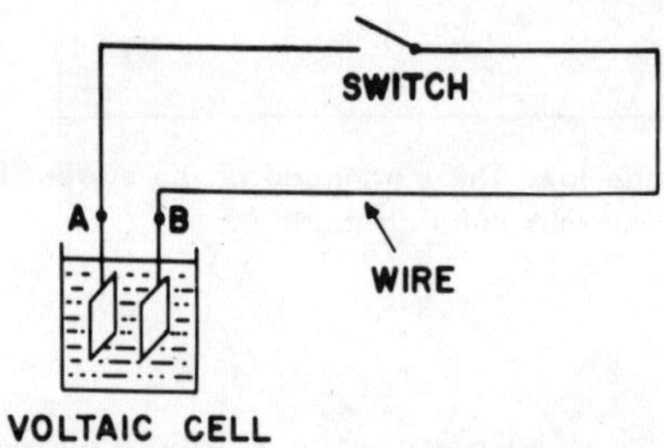

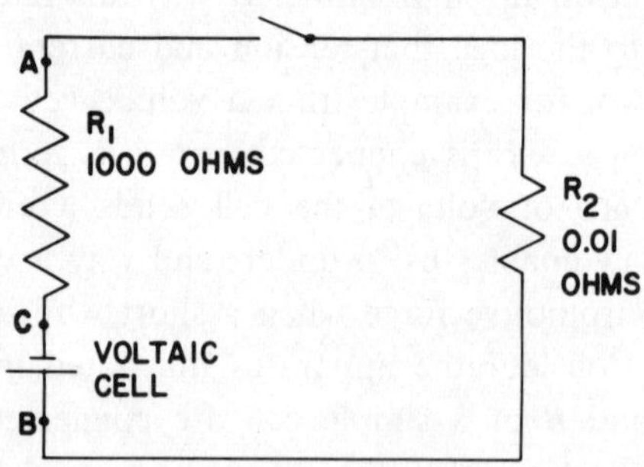

Fig. 115. A simple voltaic cell seems to "lose" its voltage when a wire is connected across its terminals.

ohms. The wire, however, may have a resistance R_2 as low as 0.01 ohms. Using these figures, the total resistance in the circuit is the sum:

$$R = R_1 + R_2$$
$$R = 1000.01 \text{ ohms}$$

If the cell voltage is 1 volt, the current is:

$$I = \frac{E}{R} = \frac{1}{1000.01} = 0.00099999 \text{ amp.}$$

or approximately 0.001 ampere. The resistance of the wire is so small in comparison to the internal resistance of the cell that it has no significant effect on the amount of current flowing. R_2 can be neglected for practical purposes.

Ohm's law can be rearranged in the form E=IR. In that form, it can tell us why the cell voltage disappeared when the switch was closed. The first step is to calculate the voltage drop across the hypothetical terminals *A* and *C* of the internal resistor R_1. That voltage, E_1, is simply the current, *I*, times the resistance, R_1:

$$E_1 = IR_1 = (0.00099999)1000$$
$$E_1 = 0.99999 \text{ volts}$$

The voltage across the terminals of R_1, is almost 1 volt. That voltage is spoken of as a voltage drop because it subtracts from the voltage of the voltaic cell. A large part of the voltage generated by the voltaic cell is "used up" in forcing current to flow through its own internal resistance. The voltage available at the terminals of a battery that is supplying current is always equal to its electromotive force minus the voltage drop caused by its internal resistance. The voltage across the terminals *A* and *B* of Figure 115, when the switch is closed, is the difference $1-0.99999=0.00001$ volts. This voltage is much too small to be detected by means available in the early nineteenth century, so it was natural for scientists to assume that it had disappeared entirely.

Ohm's law also helps us understand why several simple cells might provide no more current than a single cell of the same type. The accompanying schematic diagram (Figure 116) shows several cells connected in series with a short piece of wire. The voltage across the compound battery terminals, *A* and *B*, is the sum of the individual cell voltages, or 5 volts. The total resistance around the circuit is 5000.01 ohms. The current, therefore, is

$$I = \frac{E}{R} = \frac{5}{5000.01} = 0.000999998 \text{ amperes}$$

The current produced by five cells is almost identical to that produced by a single cell of the same type (Figure 115). The current in either example is limited almost entirely by the internal resistance of the battery. Adding an additional cell adds just enough voltage to overcome the added internal resistance and the current remains essentially unchanged.

Batteries manufactured today have much lower internal resistance

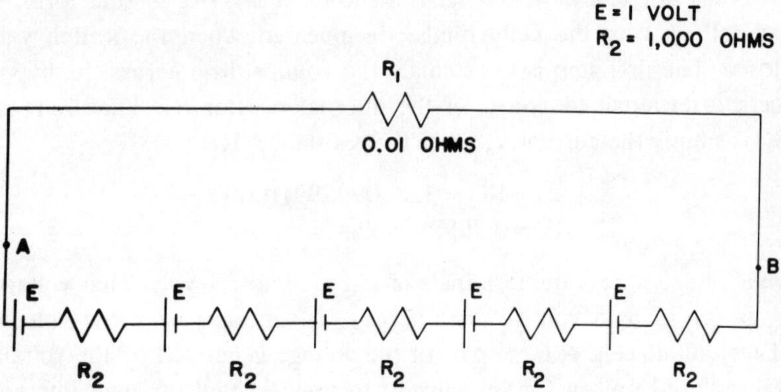

Fig. 116. Five voltaic cells connected in series with a wire.

than the early voltaic cells. If the external resistance is large compared to the battery's internal resistance, the current in the circuit increases as additional cells are added. Figure 117 shows three typical

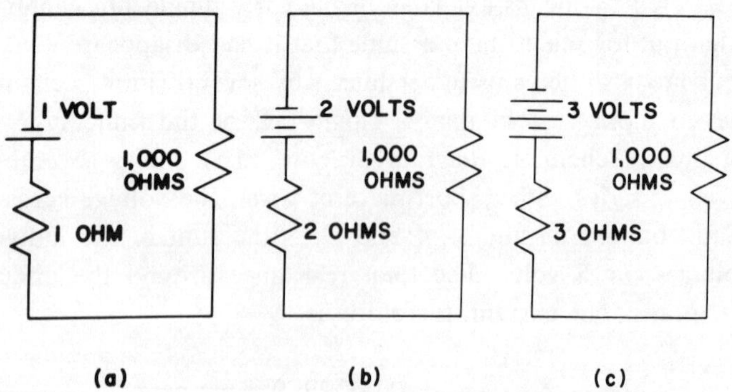

Fig. 117. Batteries of relatively low internal resistance

circuits that illustrate this point. Each battery is connected to a resistance of 1,000 ohms. In part (*a*), the battery is a single cell having an internal resistance of 1 ohm. The two-cell battery of part (*b*) has a voltage of 2 volts and an internal resistance of 2 ohms. The battery of part (*c*) has a voltage of 3 volts and an internal resistance of 3 ohms.

The three electric currents can be calculated using Ohm's law:

$$\text{Circuit } (a) \quad I = \frac{1 \text{ volt}}{1001 \text{ ohms}} = 0.000999 \text{ amperes}$$

$$\text{Circuit } (b) \quad I = \frac{2 \text{ volts}}{1002 \text{ ohms}} = 0.001996 \text{ amperes}$$

$$\text{Circuit } (c) \quad I = \frac{3 \text{ volts}}{1003 \text{ ohms}} = 0.002991 \text{ amperes}$$

The amount of current flowing clearly increases as the number of cells is increased. Ohm's contemporaries failed to notice this truth because of the relatively high internal resistance of the batteries available to them.

Ohm had demonstrated that the current flowing in any conductor depends on two factors—the resistance inherent in the circuit, and the voltage or electromotive force of the source of electricity. The latter is analogous to the role of temperature difference in the transfer of heat. Moreover, he showed that electromotive force is the concept that connects the theory of current electricity with the older theory of static electricity. Much of the work done in the next quarter century was a natural development of Ohm's ideas. Unfortunately, the importance of his discoveries was not fully recognized for many years. Although he received a medal from the Royal Society in 1841, he was not promoted to a full professorship until 1849—twenty-two years after the publication of his book on electricity. He is now recognized as one of the great scientists of his time.

Electric Potential

Electromotive force is often referred to as a difference of *potential*. Potential is a useful concept that evolved in connection with electrostatic theory. Electric charges move in conductors, and jump across air gaps as sparks, because the charges repel each other. They tend to be repelled out of bodies that have large concentrations of charge and to enter bodies that have lower concentrations. If two bodies that have equal concentrations of charge are connected by a wire, no movement of charge—no electric current—takes place. Such bodies are said to have the same potential. If the two bodies have a difference of potential, a corresponding motion of charge will take place. This *difference of potential* is identical in principle to the electromotive force or voltage of a battery.

The First Electric Motor

Immediately after learning of Oersted's discovery of the connection between current electricity and magnetism, Michael Faraday (1791–1867) found a way to put the discovery to work. He knew

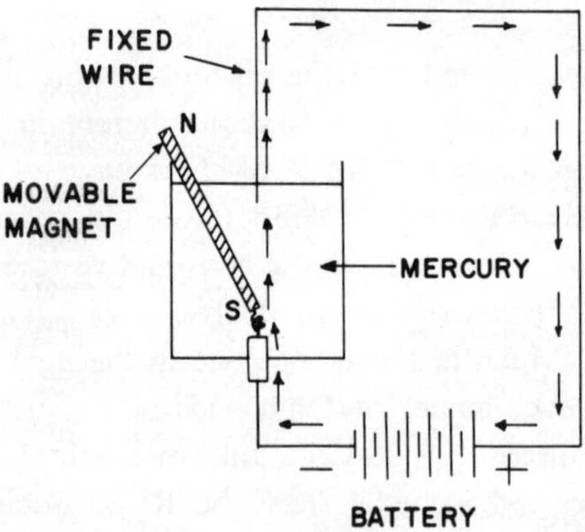

Fig. 118. Faraday's experiment in which a magnetic pole rotates around a current of electricity.

that an electric current exerts a force on a nearby magnetic pole. Perhaps this force might cause a magnetic pole to rotate around a current-carrying wire. Faraday used the apparatus illustrated in Figure 118 to find out. The current flows in a closed circuit as indicated by the arrows.* The mercury, which is a good conductor of electricity

* The current is shown flowing from the negative terminal (−) to the positive terminal (+) through the circuit external to the battery. As we shall see an electric current is really a flow of negatively charged electrons which are repelled out of the battery at the negative terminal and attracted into the battery

forms a part of the path from the bottom of the glass jar to the fixed wire. When a current flows, the north pole of the magnet moves in a circular path around the fixed wire. If the magnetic poles are reversed, or if the direction of current flow is reversed, the direction of rotation is also reversed.

With the success of that experiment, Faraday went on to show that a wire carrying a current can be made to rotate around a magnetic

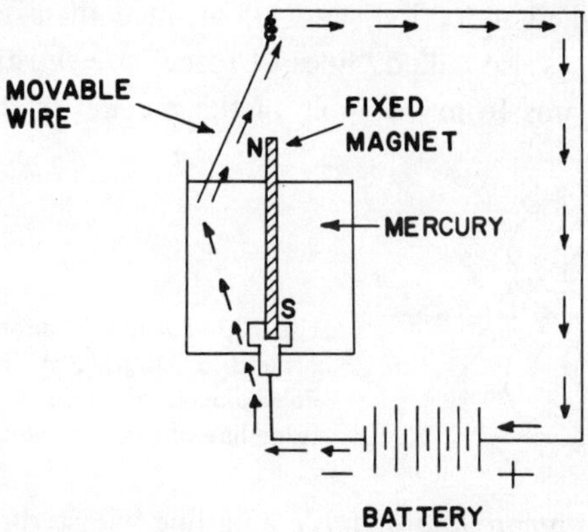

Fig. 119. Faraday's experiment in which a current-carrying wire rotates around a fixed magnetic pole.

pole. The fixed magnet was placed in a socket, as shown in Figure 119, and the wire in the mercury was able to move freely. As soon as the circuit was completed, the wire began to rotate around the magnetic pole. Once again, the direction of rotation could be reversed by reversing either the magnetic poles or the direction of the current.

Magnetic Fields

Faraday explained his results by using the idea of *magnetic lines of force*. Science had long been familiar with the idea of a magnetic field. The forces exerted by a magnet are noticeable not only at its

---

at the positive terminal. Some books show the so-called "conventional current" flowing in the opposite direction. This is a relic of past years when the fluid or particles of electricity were thought to have a positive charge.

poles, but throughout the region of space near the magnet. If this region is explored with a small compass needle, the forces are seen to change in direction and strength depending upon the particular place of measurement. The space around a magnet in which its influence is felt is said to contain a magnetic field.

It is easy to make a "map" of the magnetic field around a magnet. Merely sprinkle iron filings on a large piece of paper and place the paper over a magnet. Each bit of iron becomes a miniature magnet, by a process called induction, and all arrange themselves in curved lines. Some of these so-called "lines of force" are illustrated in Figure 120. Each line runs from one pole of the magnet to the other. Fara-

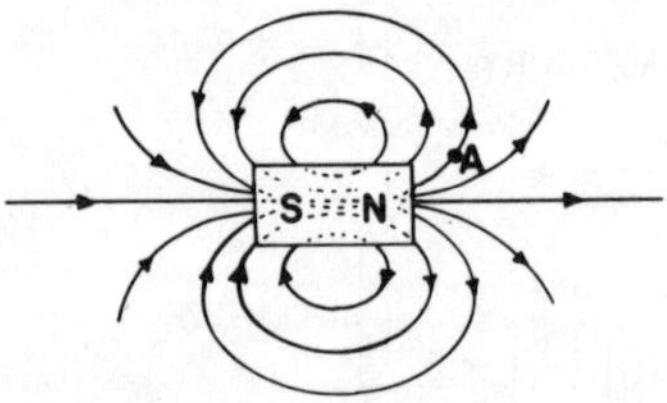

Fig. 120. Faraday's magnetic lines of force around a magnet. A hypothetical North Pole placed at point A would follow the same line of force to the South Pole.

day defined a *magnetic line of force* as a line whose direction at every point coincides with the direction of the magnetic field at that point. If we could somehow isolate a small north pole, and place it at point *A* in the diagram, it would move along the same line of force until it reached the magnet at the south end. Faraday pointed out that lines of force form closed loops that never cross one another. Upon leaving the north pole of a magnet, they move to the south pole and complete their closed path within the magnet.

Oersted had shown, in 1820, that a magnetic field can also be produced by a current of electricity. This can be demonstrated by running a current carrying wire through a card, as shown in Figure 121. The magnetic lines of force are concentric circles at right angles to the current. This can easily be proved either by sprinkling iron filings on the card or by exploring the field with a small compass needle. Just as for a permanent magnet, a line of force gives the path that would be followed by an isolated north pole that is free to move. This explains why the movable magnet rotated around the fixed wire in Figure 118. In Figure 119, the magnet was fixed so the movable wire rotated instead. In each example, the wire and magnet exerted forces on each other. Where one was fixed in place, the other moved in response to the force.

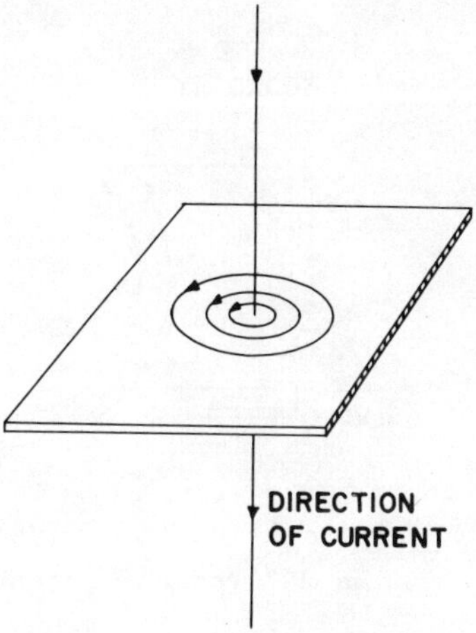

Fig. 121. Lines of force around an electric current.

Faraday's Ring

The year 1831 was one of brilliance for the great Michael Faraday. It was on November 24, 1831 that he outlined the modern concept of magnetic fields in his *Experimental Researches in Electricity*. His hope, he tells us, was to discover analogies between static electricity and current electricity. Static electricity was known to have the power of "induction"—the ability to cause an electrification of opposite polarity on nearby objects. He wondered if electric currents had a similar property. He speculated at first that a current flowing in one circuit might induce a current to flow in a nearby circuit which would continue exactly as long as the first current. Faraday found that his guess was wrong; a current was indeed induced in the second circuit, but it lasted only for an instant. The induced current flowed only when the primary current was started or stopped. It depended not merely upon the *existence* of the primary current, but on its *variation*.

Faraday illustrated his induced currents with a ring of iron upon which two separate coils of wire were wound as shown in Figure 122.

A galvanometer was connected to one coil and a voltaic battery to the other in such a way that the battery circuit could be opened or closed by a switch. On making or breaking the current in the primary circuit, a deflection of the galvanometer pointer showed that a short surge of current also flowed in the secondary circuit. But the currents

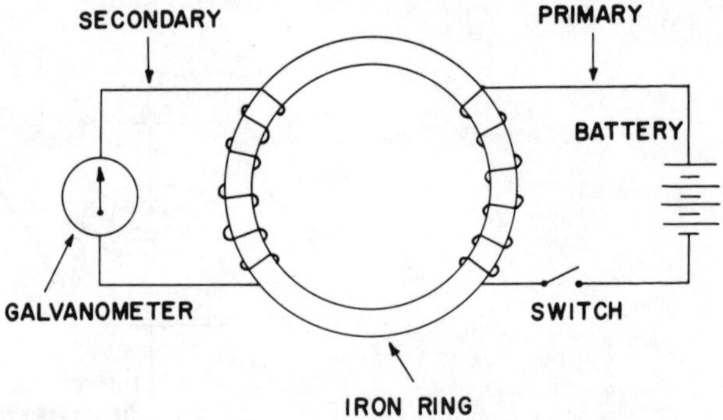

Fig. 122. Faraday's ring, the first electric transformer.

induced on starting and stopping the primary current were in opposite directions—as indicated by opposite deflections of the galvanometer.

As soon as the switch is closed, a current builds up in the primary coil from zero to some steady magnitude depending upon Ohm's law. As it builds up, the changing current produces a changing magnetic field in the iron which, in turn, induces a voltage in the secondary coil. Because the secondary coil forms a closed circuit with the galvanometer, the induced voltage (electromotive force) sends a current through the secondary circuit. As soon as the primary current, and therefore the magnetic field, reaches its steady magnitude, the induced electromotive force and current in the secondary falls to zero.

Faraday soon found that a current can be induced in a test circuit in many ways: by varying the strength of the current in a nearby circuit; by moving a magnet near the test circuit; or by moving the test circuit in the neighborhood of another current or magnet. In all of these examples, the induction of a current depends upon the *relative motion* of the test circuit and the lines of magnetic force in its vicinity. Whenever a conductor is "cut" or crossed by these magnetic lines of force, an electromotive force is induced in the conductor. If the conductor is closed to form a circuit, the electromotive force causes a current to flow. The magnitude of the voltage induced in the conductor depends only upon the number of magnetic lines of force intersected by the conductor in one second.

The Dynamo

In 1821 Faraday had converted electricity to mechanical motion. Ten years later he showed that mechanical motion could be used to generate electricity. His apparatus is illustrated in Figure 123. A

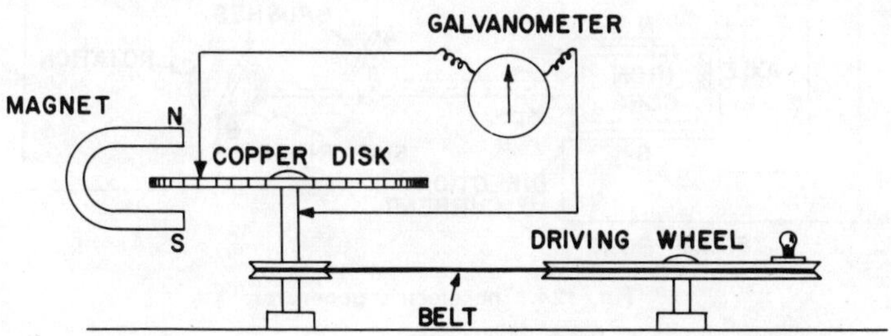

Fig. 123. Faraday's dynamo, or electric generator.

copper disk, supported by a metal axle, was made to rotate between the poles of a horseshoe magnet. The axle and edge of the disk were connected to a galvanometer by wires lightly touching the moving parts. As the disk turned, a deflection of the galvanometer showed that an induced current was generated in the disk.

Figure 124 illustrates the principle of the dynamo or electric generator. A magnet establishes a magnetic field in the region of a movable soft-iron core or armature, around which are wound many turns of copper wire. The armature and its coil of wire are caused to rotate on an axle between the poles of the magnet by a source of mechanical energy, such as a steam engine or water wheel. Each terminal of the armature coil is connected to an insulated metallic cylinder, or slip ring, which is fastened to the armature. Contact is made to the slip rings by fixed metal or carbon brushes, each of which wipes against one of the moving slip rings. The brushes are connected to an electrical load such as the incandescent lamp, *R*.

As the armature rotates, the winding cuts across the magnetic lines of force produced by the field poles. An electromotive force is thereby induced in the winding and transmitted through the slip rings and brushes to the resistance *R*. A current flows through the circuit, as illustrated by the horizontal arrows.

The current described above is an *alternating current,* reversing its direction of flow through the lamp circuit each half revolution of the

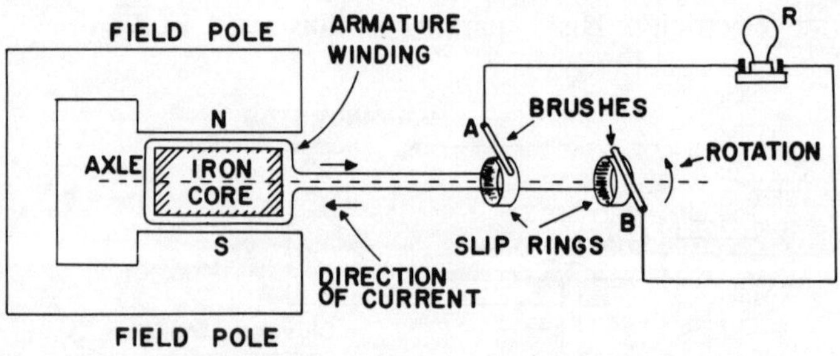

Fig. 124. The electric generator.

armature. Similarly, the voltage at the brushes is called an alternating current voltage. At the instant depicted in the diagram, the current leaves the generator at brush *A* and returns at brush *B*. Now consider the condition one-half revolution later. The top and bottom portions of the armature winding have changed places and the current leaves at brush *B* and returns at brush *A*. The current's direction oscillates back and forth at a rate or frequency equal to the number of revolutions per second of the armature.

An alternating current (A.C.) generator can be changed to an unidirectional or *direct current* (D.C.) generator by replacing the slip rings with a *commutator*. As shown in Figure 125, a simple commutator is a ring split into two insulated halves, called segments, which are mounted on the armature. Each terminal of the armature coil is connected to one segment of the commutator. When the armature rotates, an alternating current is induced in the coil exactly as in the alternating current generator. But before the current enters the external load, *R*, it is changed into a direct current by the commutator. The segments change brushes every half turn of the armature, thereby maintaining a unidirectional current through the external portion of the circuit.

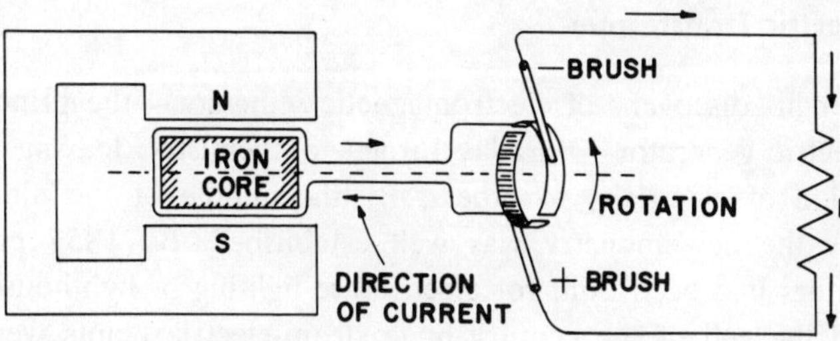

Fig. 125. A direct current generator.

The direct current generator can be operated as a motor merely by replacing the resistance *R* by a source of electromotive force, such as a battery. Its operation is illustrated in Figure 126. Current flows through the armature winding producing a north pole on its left end

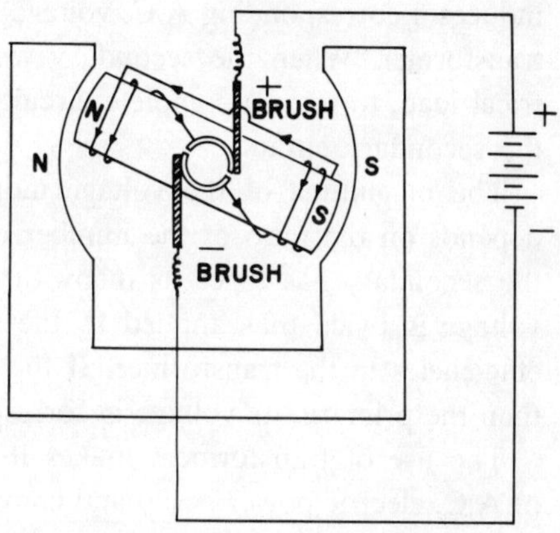

Fig. 126. A direct current motor.

and a south pole on the right. The repulsion between each like armature pole and field pole causes the armature to make a half turn. At that point, the commutator reverses the direction of current flow through the winding, thereby reversing its poles. Again, the left end of the armature has a north pole and the right end a south pole. The repulsion between each like armature pole and field pole causes another half turn of the armature. The process occurs over and over again, and the armature continues to rotate steadily in one direction.

The Electric Transformer

After his discovery of electromagnetic induction—the principle of the electric generator—Faraday turned to chemistry, leaving the exploitation of electricity to others. By the middle of the nineteenth century the new industry was well established. By 1857 powerful generators had been built for electric arc lighting of lighthouses, and toward the end of the century huge steam-electric plants were constructed to supply electricity for public and private use.

Faraday's ring (Figure 122), now called a "transformer," played an important part in the development of the electrical industry. An A.C. voltage is applied to the primary winding sending an A.C. current through that winding. Because the current's magnitude changes constantly in the course of time, the resulting magnetic field also changes in strength from one instant to the next. This varying magnetic field induces a corresponding A.C. voltage in the secondary winding of the transformer. When the secondary winding is connected to an electrical load, to form a complete circuit, an A.C. current flows through the secondary circuit.

The magnitude of the voltage induced in the secondary winding depends on the ratio of the number of turns in the two windings. If the secondary has twice as many turns as the primary, the induced voltage is twice that applied to the primary—except for minor inefficiencies in the transformer. If the secondary contains fewer turns than the primary, its voltage is correspondingly lower.

The use of transformers makes it possible to change the voltage of A.C. electric power easily and conveniently to suit particular applications. They are used at power plants to boost the voltage of A.C. generators to many thousands of volts for efficient transmission over power lines. At local distribution points, they reduce the voltage to relatively safe levels for use in homes and industry. This unique and invaluable property of Faraday's ring is the main reason why all modern power systems use alternating current rather than direct current electricity.

Electric Fields

The idea of an electric field of force around an electric charge grew gradually over the years. It was well known in Franklin's time, for example, that an electric field can exert forces of attraction or repulsion on nearby charged bodies. But just as for magnetic fields, the mechanism by which the electric force was exerted eluded discovery. It was thought, at first, that something—perhaps an "electric effluvia"—emerged from charged bodies to exert the mysterious force. But experiment after experiment proved that electric forces could be exerted through the best insulators—glass, dry air, sulfur—insulators which effectively blocked the movement of charge. Franklin also showed that the electric field in the vicinity of a charged body is not removed when the adjacent air is blown away. Such observations forced Franklin and his contemporaries to drop the idea of electric effluvia as the agent producing the force. No satisfactory mechanical explanation of electric and magnetic fields has ever been devised to replace it.

Our modern concept of the electric field was developed by Michael Faraday in 1837. He had been studying the decomposition of chemical solutions placed between two electrodes that were connected to a source of voltage. He began to wonder what, if anything, would happen if he placed solid insulators between the two flat electrodes instead of a conducting electrolyte. The accompanying diagram (Figure 127) shows two plates connected to a battery. In the upper diagram the space between the plates is filled with air and in the lower diagram with an insulating material such as glass. Both are illustrations of capacitors, identical in principle to the Leyden jar.

Faraday knew that the amount of electric charge that can be placed on the plates of the two capacitors is quite different. If both voltages are equal, and if the areas of the plates are equal, the amount of charge that enters the plates depends on the nature of the insulator.

Suppose that an amount of charge Q enters the plates of the upper capacitor, $+Q$ on the upper plate and $-Q$ on the lower plate. Faraday knew that a greater amount of charge, $+kQ$ and $-kQ$ would enter the plates of the lower capacitor. The letter k stands for a number

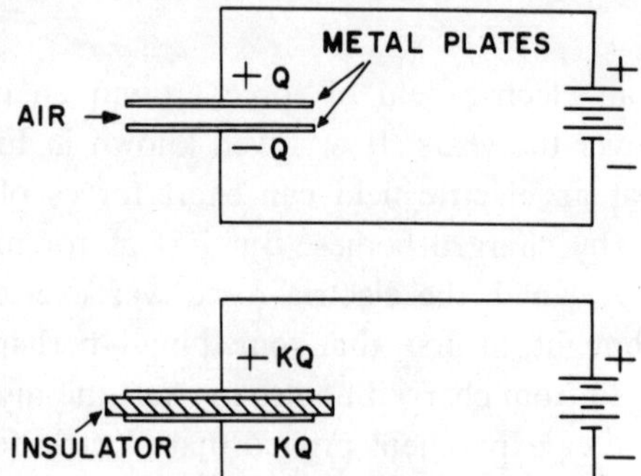

Fig. 127. A capacitor can accept a greater charge from the same battery when air between the plates is replaced by an insulator.

such as two or seven. It tells us that the addition of an insulating substance between the plates increases the charge by a factor, k. For any particular insulator the charge has a magnitude k times the magnitude it would have if the space between the plates were empty. So k is a measure of the influence of the insulator on the electrostatic properties of the capacitor. Faraday called it the "specific inductive capacity" of the insulator; the modern term for it is the "dielectric constant" of a material. Similarly, the insulator between the plates—through which the electrostatic forces act—is called a "dielectric." The dielectric constants of various dielectrics are given below.

Dielectric	Dielectric Constant
vacuum	1.0
air	1.0006
paper	2
paraffin wax	2
rubber	2.2
glass	5 to 10
mica	6 to 7
water	81

Electric Lines of Force

Faraday saw an analogy between the dielectric in a capacitor and the iron core in an electromagnet. The former increases the charge that a capacitor will accept, and the latter increases the magnetism that a coil can produce. He found that he could explain the effect of dielectrics by introducing the concept of *electric lines of force*, similar to the lines of magnetic force that he had used so successfully in his magnetic researches.

Faraday defined a line of electric force as the path followed by a

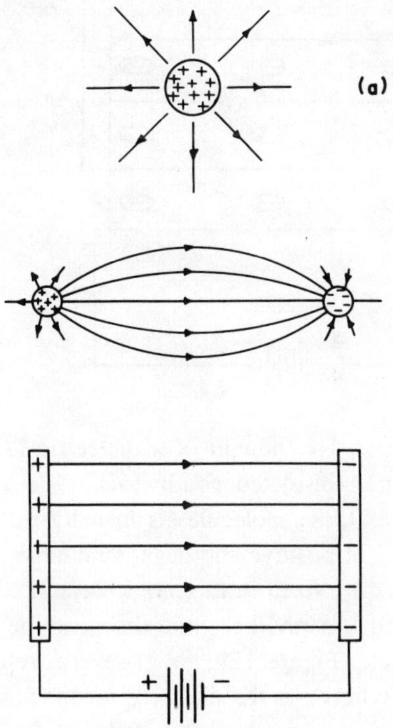

Fig. 128. Faraday's lines of electric force *(a)* around a charged sphere, *(b)* between oppositely charged spheres, and *(c)* between oppositely charged plates.

small positive charge in the vicinity of an electrically charged body. As shown in Figure 128, the electric lines of force around a charged sphere radiate radially from the sphere. Only a few lines are drawn in the diagram, but the entire space should be thought of as filled with such lines of force. The arrows on the lines of force indicate that a positive charge is repelled by the positively charged sphere. The middle diagram illustrates the lines of force between two oppositely charged spheres. Once again, a small positive charge would follow a path indicated by one of the lines of force. The lower diagram shows the lines of force between two parallel plates of a capacitor across which a fixed voltage is maintained. Each line of force can be thought of as originating on a unit positive charge and terminating on a unit negative charge. Also, the strength of the electric field between the plates depends on the number of lines of force between the plates.

Faraday then introduced the idea of induced electric charge in a

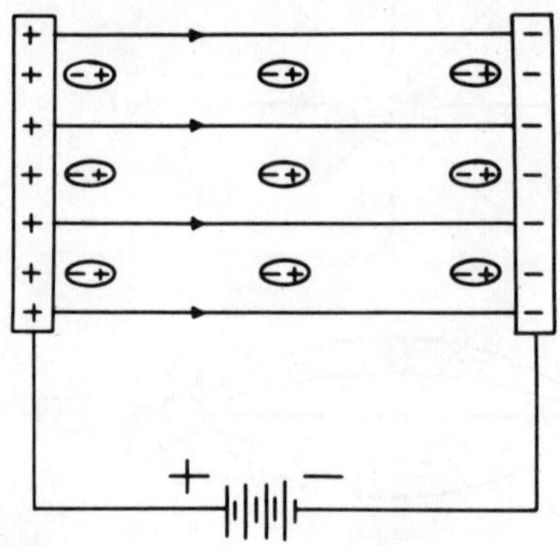

Fig. 129. A polarized dielectric material in an electric field.

dielectric. He thought of a dielectric as consisting of a large number of "small insulated conductors." We would call them "molecules." Each of these molecules is initially uncharged. It contains the same amount of positive and negative charge and is electrically neutral. The charges are fixed in the molecule and cannot be transferred from one molecule to another, but they can be displaced slightly within the molecule. Figure 129 shows several representative molecules of a dielectric between the plates of a capacitor. The electric charges on the plates produce an electric field of force in the region between the plates. Each molecule is *electrically polarized* by the electric field having equal negative and positive charges induced on opposite ends. At the boundaries between the dielectric and each plate, the adjacent molecular charges neutralize the electric field from a part of the charge on the plate. This tends to reduce the number of lines of force extending between the plates, and the battery is able to send additional charge into the plates. The molecular charges cannot leave the dielectric and enter the adjacent plate. They are said to constitute a bound surface charge.

The Relation between Electric and Magnetic Fields

Electric and magnetic fields have become indispensable concepts in many scientific areas. They constitute the intermediary by which electric or magnetic forces can be exerted through otherwise empty space. Scientists think of them as fundamental concepts because they cannot be explained in terms of still more simple ideas. Despite the fact that our senses cannot perceive these fields, they are as real to scientists as the air they breathe.

A review of our knowledge of electricity and magnetism shows an important relation between electric and magnetic fields. An electric current is merely a motion of electric charge. An electric charge has an electric field associated with it. A current of electricity, therefore, carries with it a moving electric field. But Oersted proved that a current of electricity produces a magnetic field. We must conclude, therefore, that *a moving electric field gives rise to a magnetic field.*

Now imagine a magnet moving past a conductor such as a piece of copper wire. The moving magnetic field induces an electromotive force in the wire. This means that one end of the wire obtains a positive charge and the other, a negative charge. But these charges produce an electric field. Therefore, *a moving magnetic field gives rise to an electric field.* As we shall see, Maxwell later incorporated these general ideas into his electromagnetic theory—a theory that united electricity with light, and also predicted the existence of radio waves.

Faraday Rotation

The discovery that light was somehow connected with electricity and magnetism was made by Faraday in 1845. After untangling the hidden relations between electricity and magnetism, he began to look for an interaction between magnetism and light. His apparatus was simply a piece of thick glass and an electromagnet made in the shape of a horseshoe. He placed the piece of glass across the magnetic poles, as shown in Figure 130. Then he passed a beam of polarized light through the glass so that it traveled in essentially the same direction as the magnetic lines of force between the poles. When he studied the light emerging from the glass, he found that it was still polarized, but that its plane of polarization had been changed by the magnetic

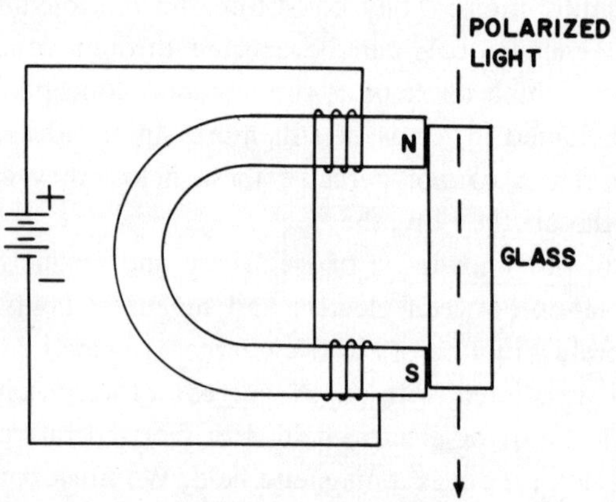

Fig. 130. Faraday's apparatus to show that light is related to electricity and magnetism. Polarized light entering the glass undergoes a change in its plane of polarization while the magnetic field is present.

field. To illustrate, suppose the light entering the glass is horizontally polarized—the light vibrations taking place only in the horizontal plane. The light emerging from the glass would then be vibrating in a different plane—one making an angle with the horizontal plane. This so-called "Faraday effect" occurs whenever polarized light passes through a transparent substance along the lines of force of a strong magnetic field.

The Faraday effect demonstrated the intimate relation between light waves and electromagnetism. We know today that the effect is caused by the tiny electric currents within individual atoms. When placed in a strong magnetic field, these currents are modified slightly within the atom. It was later shown that these changes produce the rotation of the plane of polarization observed by Faraday.

Electrical Energy

As late as 1843, the battle still raged over the principle of the voltaic cell. One group argued that the electricity came from the contact of dissimilar metals, while the other argued that the electricity was produced by chemical reactions in the cell. Faraday pointed out that an electric current can do useful work, as in the deflection of a compass needle. This work must be performed at the expense of chemical energy stored in the cell. He went a step further by showing that the amount of current sent around a circuit is proportional to the quantity of chemicals used up.

The connection between electricity and other forms of energy was cleared up by James Joule. He knew that heat and mechanical work were merely two different forms of energy. In 1843 he proved that electricity was also a form of energy, convertible into the other forms at a definite rate of exchange.

Joule passed a current of known strength through a wire and measured the amount of heat given off by the wire in a unit period of time. The amount of heat energy produced per unit time was proportional to the resistance of the wire multipled by the square of the current strength. In modern units, heat energy is measured in *joules*, current in *amperes*, and resistance in *ohms*. Using these units, Joule's discovery can be expressed as an equation:

Heat energy per unit time = $(\text{Current})^2 \times$ Resistance

$$\frac{E}{t} = I^2R$$

where time, *t*, is measured in seconds. The left hand side of the equation E/t, is defined as "power," the rate at which energy is expended. Power is measured in units called the "watt," in honor of James Watt. One watt is equivalent to an expenditure of 1 joule of energy per second. If *P* stands for power, the equation can be written as follows:

$$P - I^2R$$

To illustrate its use, let us suppose that a current of 7 amperes flows through a resistance of 10 ohms. Joule's relation tells us that electrical energy is changed to heat energy in the resistance at a rate of $P = (7)^2 \times 10 = 490$ watts.

Using this new discovery, Joule was able to show a numerical relation between the chemical energy in a battery, the electrical energy in a circuit, and the mechanical energy or work performed by a motor. He dissolved a measured weight of zinc in an acid, and found that a definite amount of heat was given off. Then he made a voltaic cell using the same weight of zinc and found that considerably *less* heat was given off to the cell as the zinc dissolved. He sent the current produced by the cell through a wire and measured the heat given off by the wire. This amount of heat corresponded, he showed, to the difference between the heat produced by the simple solution of zinc in acid, and that produced when it is dissolved as an element in a voltaic cell. This was clear proof that the electrical energy from a battery comes from the chemical energy contained in its constituents.

Joule also demonstrated that electrical energy is converted to mechanical work by an electric motor. He sent current from a battery through a motor and found that the chemical energy used up in the battery was equal to the sum of two quantities: the I^2R heating of the conducting wires; and the mechanical work done by the motor.

Joule published his revolutionary ideas in the historic paper of 1843, *On the calorific effects of Magneto-Electricity and on the mechanical value of Heat*. Others extended his ideas in the years that followed. The contributions of William Thompson (Lord Kelvin) to thermodynamics have already been described. In 1847, Hermann Helmholtz (1821–1894) published an important pamphlet, *Erhaltung der Kraft* ("Conservation of Energy"), which had been refused for publication by an important scientific journal, *Annalen der Physik*. His ideas were too revolutionary for the scientists of the time. Helmholtz asserted that the conservation of energy is a basic law of nature; that the kinetic energy of moving bodies, or the potential energy of water behind a dam, can be converted into heat according to definite laws as taught by Joule; and that any of these forms of energy can be converted into chemical, electrostatic, magnetic or voltaic forms.

Energy had become a common denominator that united chemistry, heat, mechanics, electricity and magnetism. As a form of energy, electricity offered great promise for the future. It could be generated in large quantities at central power plants and transmitted over wires to the ultimate users. Once there, it could be converted to other forms of energy to do specific jobs. Men like Davy, Ohm, Faraday and Joule had discovered the basic laws of electricity and magnetism. Now others assumed the task of using electricity in the service of man.

Electricity Under the Sea

14

The electrical discoveries of the early 1800's led to the invention of a variety of useful devices. Like the early steam engines, these inventions were based more on observation and practical knowledge than on true theoretical understanding. Theory caught up later, providing a deeper insight that led to vast improvements in electrical devices and to a fantastic increase in the number of tasks that electricity could accomplish.

Long before the importance of Ohm's law was universally acknowledged, industrialists had begun to exploit electrical knowledge with astonishing speed. The *chemical effect* of electricity was put to work in the plating of metals. In this process, the salt of a metal is decomposed by the current so that particles of the metal are deposited on another piece of metal connected to the negative terminal of a battery.

The *heating effect* of electricity also found important applications. As early as 1845, the electric arc lamp was used to light streets in Baltimore. The light of an arc lamp is produced by the intense glow in the gap between two carbon rods connected to a source of D.C. electricity. The arc is started by first touching the ends of the rods together and then separating them slightly. The arc lamp represented the first commercial application of electricity to illumination. By about 1879 Joseph Swan in England and Thomas Edison in America had developed practical incandescent lamps using thin carbon filaments.

Incandescent lamps had been made as early as 1820 using platinum filaments, but their life had been extremely short. This was partly a result of the inability of inventors to achieve a good enough vacuum in the lamp bulb. As soon as the current heated the filament to incandescence, the metal would oxidize and disintegrate. The major difficulty, however, was the narrow margin between the temperature at which plantinum begins to glow and the temperature at which it melts. Modern incandescent lamp filaments are made of the metal tungsten, a so-called refractory metal which glows well below its melting temperature. The heating effect of the electric current was also used in electric welding, electric cooking and the electric furnace of the metallurgist.

The electric current has a third useful property, in addition to electrical heating and chemical decomposition: the ability to produce a magnetic field. The *magnetic effect* was the first to be exploited commercially, as the principle behind magnetic telegraph systems. Telegraphy was introduced in the 1820's and expanded rapidly in the next decade, especially in America.

The deflection of a compass needle by a magnet had been suggested much earlier as a means of signalling. But practical exploitation of the idea was quite out of the question as long as lodestone was the only source of magnetism. With Oersted's discovery of electromagnetism, the ability to create or destroy magnetism at will made long distance telegraphy a distinct possibility. The first electromagnetic telegraph was developed in 1820 by André Marie Ampère, the great French scientist. Ampère used one wire for each letter of the alphabet, the earth serving as a return conductor for each circuit. Small magnetic needles were placed at the distant end of each of the wires. A current through one of the wires (and the ground return) caused the associated needle to deflect, thereby indicating the desired letter.

A number of needle telegraphs soon followed and, in 1831, the American Joseph Henry (1799–1878) built an electromagnetic telegraph using only one wire. His apparatus consisted of a magnetized steel bar that was balanced on a pivot so it could rotate in the horizontal plane. A nearby electromagnet could attract the steel bar.

When a current was sent through the electromagnet, the end of the steel bar would strike a bell. Information was conveyed by coded sound combinations.

The Morse Telegraph

In 1832, Samuel F. B. Morse, an American artist, was returning from Europe on a packet ship. He had read a recent publication by Faraday on electromagnetism and made plans to construct a telegraphic recording instrument. Three years later he constructed an experimental model based on the famous dot-dash code that still bears his name. The dots and dashes were distinguished by short and long

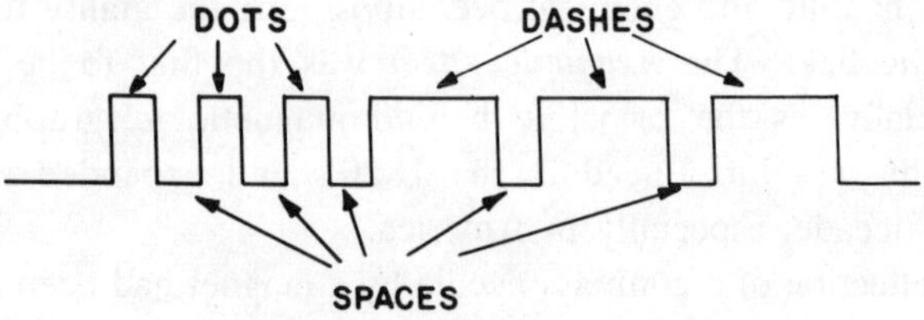

Fig. 131. Dots and dashes of the Morse code.

impulses of current. A dot corresponded to a current duration of about 1/24 of a second. The dashes were about three times as long.

The heart of Morse's receiving apparatus was an electromagnet that caused a pencil to move when energized by an electric current. The pencil was in contact with a moving roll of paper and drew a straight line on the moving paper when no signal was being received. On reception of a current, the pencil moved to the side, returning to its neutral position when the current dropped to zero. In this way, an on-off current sequence drew a rectangular "box" or current waveform, as shown in Figure 131. If the current remained on for 1/24 second, a "dot" was drawn by the pencil. For 3/24 second, a "dash" was drawn. Morse's recording telegraph gave a permanent record of the message that could later be decoded at leisure.

In the course of his experiments, Morse discovered that the electric current became too feeble, after traveling 20 miles or so, to operate the recording apparatus. Working with Joseph Henry, he devised a relay that would automatically repeat the signals into another section of telegraph line. The principle is illustrated in Figure 132. When the telegraph key is closed, a current is sent along the first 20-mile length of line, through the coil *A*, and back to the battery. The current magnetizes the iron frame of the relay, and attracts the hinged iron armature which rotates clockwise against the spring tension. The mo-

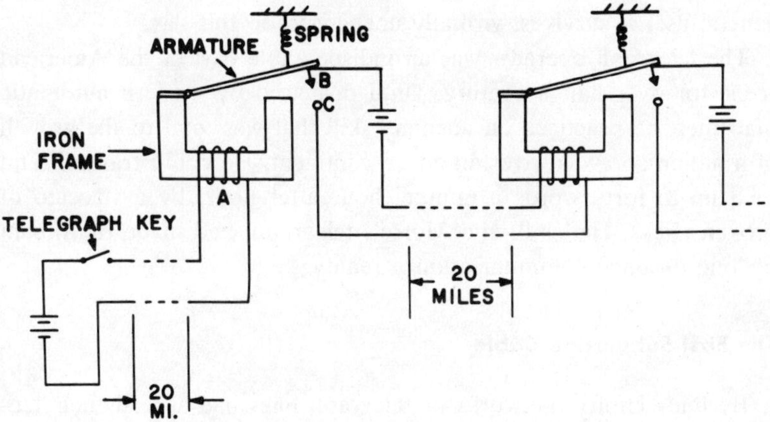

Fig. 132. The Morse relay.

tion of the armature brings point *B* in electrical contact with point *C*, thereby closing the circuit containing the second 20-mile length of line. The battery in this circuit sends a new current through the second section of line, thereby energizing a second relay at the distant end. Circuits could be joined in this way for any desired distance.

In 1838 Morse demonstrated his telegraph before President Van Buren and his cabinet. With characteristic speed, Congress passed a bill in 1843 to build an experimental line between Washington, D. C. and Baltimore, Maryland. On May 24, 1844, Morse sent the first public telegram, "What hath God wrought," over this 40-mile length of line. A short while later, the postmaster general concluded that Morse's toy would never support itself and the government withdrew its assistance. Morse obtained private capital, however, and extended the line to Philadelphia and New York in 1845. By mid-century, fifty companies using Morse's telegraphic patents were operating in the United States.

By that time, Alfred Vail had discovered that telegraph signals could be "read" by the clicks of the relay at the receiving end. He noticed that the armature made two distinct sounds for a dot and a dash. With practice, an operator could decode messages merely by listening to the relay. In time, the Morse sounder or buzzer came into general use. It survives, virtually unchanged, to this day.

The telegraph operator was an indispensable part of the American scene for over half a century. Until displaced by modern automatic machines, he practiced an uncanny skill that was vital to the growth of a nation sprawled over an entire continent. He could transmit and read up to forty words a minute, hour after hour, by a staccato of broken clicks. His skill, and Morse's telegraph, had made the dream of long distance communication a reality.

The First Submarine Cable

By mid-century, networks of telegraph lines had spread their tentacles over much of England, Europe and North America. But the new art of communication still stopped at the edge of the sea. The English channel, for centuries a protective moat between England and the Continent, would soon be attacked by ships carrying miles of copper wire.

The earliest attempts to lay a submarine telegraph cable were made in 1839 by a Dr. O'Shaughnessy of the East India Company. Somewhat later, Morse carried out experiments under New York harbor, sending signals through rubber-covered wires enclosed in lead pipe. The results of these tests led him to conclude that trans-Atlantic telegraphy was completely feasible. At about the same time, Charles Wheatstone carried out successful underwater tests in Swansea Bay, Wales.

The first major submarine cable was laid in 1850 across the Straits of Dover. Two brothers, John and Jacob Brett, secured a ten-year concession for laying the cross-Channel cable, and contracted with the Gutta-Percha Company for its manufacture. The core of the cable was a single copper wire surrounded by a quarter inch of an insulator called gutta-percha.

Finding a suitable insulator in the early days of telegraphy was a far from simple matter. Luckily, gutta-percha turned up at just the right time. Although it has been replaced almost completely by synthetic plastics, gutta-percha was indispensable to the development of telegraphy. In 1843, William Montgomerie, a resident of Singapore, sent a few samples of gutta-percha to the Royal Society. After examining them, Faraday concluded that the material would be quite useful as electrical insulation. The Gutta-Percha Company of London (later the Telegraph Construction and Maintenance Company) was formed in 1845 to market the product.

Gutta-percha comes from the gum of a number of related trees found in the jungles of Malaya, Borneo and Sumatra. The term comes from the Malay words *getah percha,* gum of the percha (tree). Unlike rubber, it does not stretch, being quite hard at ordinary temperatures. It becomes as soft as putty in hot water, reverting to its original hardness when cooled to room temperature. This makes it easy to mold gutta-percha around a wire in the process of cable manufacture.

Such, then, was the construction of the first cable across the English Channel—a thin strand of copper insulated by a single layer of gutta-percha. The Brett brothers assumed that once the cable had been deposited on the floor of the Channel, it would be perfectly safe and so would need no outer protection. Only the ends near the shore were protected by a lead outer covering.

On August 28, 1850 the cable was laid between Dover and Cap Gris Nez in France. Amid great excitement, an automatic printer was connected to the line to receive a message from John Brett to Prince Louis Napoleon Bonaparte. Unfortunately, nothing came out but a jumble of unintelligible chatter. A needle instrument was tried in place of the automatic printer with slightly improved results. A word here, a word there, but no real messages.

Once again, practice had outdistanced theory. The Bretts could not know that a properly insulated wire behaves quite differently in salt water than in air. The conducting salt water had transformed the cable into a gigantic Leyden jar. Surrounded by conducting sea water, the submerged wire became extremely sluggish because of its increased electrical capacitance. If the operators had merely slowed down their rate of transmission, the signals would have got through.

Unfortunately, the Bretts had little time to expermient with the cable. The next morning the cross channel link was completely dead. The fine thread of copper uniting England and Europe had parted somewhere along its 25-mile length. The Bretts had come face-to-face with the natural enemy of all submarine cables—a fisherman with an anchor caught on the submerged line. Pulling it to the surface, the simple fisherman puzzled over this strange seaweed that seemed to grow around a metal core. Suspecting the metal might be gold, he cut out a goodly section for possible profit and the entertainment of his friends. His innocent vandalism touched off an undeclared war between fishermen and cable companies that persists to this day. More damage is done to underseas cables by trawlers and fiishing boats than by any other means.

A New Industry Is Born

Despite their initial failure, the Bretts had shown that electrical signals, such as they were, could be sent across the English Channel. A year later, they obtained the financial and technical support of Thomas Crampton, a railway engineer, and the project went ahead once again. Crampton designed the new cable personally, and this time no fisherman would dare tamper with it. A layer of hemp was wound around the gutta-percha, and an armoring layer of galvanized iron was spun on top of that. It had the strength of many hawsers and weighed thirty times as much as the earlier cable.

On September 25, 1851 the huge cable was paid out across the Channel, straining the inadequate release mechanism aboard the cable ship. Although much cable was wasted, there was just enough on board to reach the French shore. After a few weeks of testing, the first successful undersea cable was opened to the public, and England suddenly found itself a fraction of a second away from Europe.

The success of Crampton's cable touched off a boom in the submarine cable business. Within two years, over fifteen hundred miles of submarine cable had been manufactured. Cables soon spanned the Irish Sea, the Black Sea and portions of the Mediterranean. Inevitably, attention became focused on the most important ocean of all—the Atlantic.

Spanning the Great Atlantic

The driving force behind the first Atlantic cable was Cyrus West Field (1819–1892), an American businessman who had turned eight dollars into $250,000 by the time he retired at thirty-three. Intrigued with the idea of a trans-Atlantic cable, he organized the New York, Newfoundland and London Telegraph Company in 1854. It took two and a half years merely to construct an overland line between the first two cities mentioned in the company's name. The "and London" part of the title involved ten years of frustration and failure before a permanently successful trans-Atlantic cable was laid in 1866.

The first attempt failed, in 1857, when the cable was broken by faulty operation of the paying-out mechanism. Though Field was disappointed, there was considerable reason for optimism. The break had occurred after 335 miles of cable had been laid in water up to two miles deep. Communications had been maintained through the cable between the cable ship, *Niagara,* and the shore terminal at Valentia Bay, Ireland, until the moment of the break. This proved that there was nothing impossible about the project. They would buy more cable and try again.

The delay was put to good use by the great William Thompson who had been a director of Field's company from the start. Thompson had already established his reputation as one of the great scientists of the time. His studies convinced him that if an extremely sensitive detector of current signals were placed at the receiving end of the cable, the rate of signalling could be increased significantly.

Thompson knew that the sharp "square wave" pulses of current fed into a submarine cable suffered severe degradation in tranmission. When the pulses emerged at the distant end, their leading edges were no longer sharp, but rose very gradually to a maximum value of current. The accompanying diagram illustrates the general idea in simplified form. We can understand why the pulses are rounded and "smeared" together in the diagram by applying the knowledge of sine waves gained in connection with sound and light. A sharp pulse of current, such as those illustrated in the part (*a*), really consists of an alternating current at some fundamental frequency accompanied by a large number of harmonics. The frequencies of these harmonics are multiples of the fundamental frequency, just as for complex sound waves. If the fundamental frequency is 10 cycles per second, the

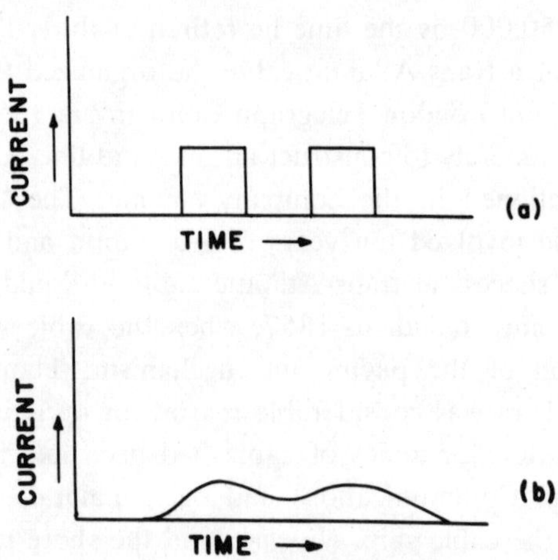

Fig. 133. Current pulses before (*a*) and after (*b*) being sent through an early submarine cable.

harmonic frequencies are 20, 30, 40, 50 c.p.s., etc. The sharper the pulse, the greater the number of harmonics that the pulse contains. When a simple cable is submerged in a conducting medium, such as salt water, it turns out that the higher frequencies travel faster than the lower frequencies. So the higher harmonics tend to "catch up" with the lower harmonics *of the previous pulse*. Individual pulses are no longer rectangular in shape, but tend to drag out. If insufficient time is allowed between pulses, they tend to run together as shown in the lower diagram.

Kelvin proposed to speed up the rate of signalling by using a sensitive pulse detector—one that would operate efficiently on the small variations in current provided by the "smeared" pulses.

Unfortunately, Edward Whitehouse, the official electrician of the company, had other ideas. He decided to solve the problem by another method which sounded just as reasonable at the time. He would merely increase the magnitude of the current pulses fed into the cable. His giant pulses were produced by great spark coils generating at least 2,000 volts of electricity. We shall see how Whitehouse's use of the wrong idea later led to disaster.

In the spring of 1858, the great enterprise was once again under way. The cable had been divided between two ships, the *Niagara* and the *Agamemnon.* After splicing the cable in mid-Atlantic, the two ships began to lay cable in opposite directions. Following a great storm, and after three other mishaps during which the cable parted, much time and cable had been lost. Once again the ships headed back to Ireland in defeat.

In late July of the same year, the *Niagara* and *Agamemnon* were back in mid-Atlantic for another try. By August 5, the mighty Atlantic had been successfully spanned. After some testing, a message was received on August 16 from Queen Victoria to President Buchanan. On September 1, Cyrus Field received a tumultuous public ovation in New York. But on that very day, by a curious twist of irony, the great cable went dead.

As soon as communication had been established using Thompson's sensitive apparatus, Whitehouse at the Valentia end had put his patented automatic recorder on the line. Although the machine worked well over short lines, it could not interpret the weak, distorted pulses emerging from the cable. He thereupon insisted on replacing Thompson's low voltage batteries with his 2,000 volt spark coils. Although the cable was poorly designed, and carelessly put together, there is little doubt that such high voltage hastened its end. The insulation was just not capable of withstanding so high a voltage. The very last message to pass through the cable informed Cyrus Field in New York that the company was now ready to forward U. S. Government messages to England. After a month of union, the continents were once again separated by the Atlantic Ocean.

This time it took Cyrus Field seven years to get his project under way again. A new cable was designed this time with great care.

Scores of samples were tested both electrically and mechanically. The new cable had a conducting wire three times as large as the 1858 cable, and it was much more heavily armored. Over an inch in diameter, it weighed 1¾ tons per mile. The largest ship in the world, the *Great Eastern*, had been outfitted to lay the cable. All seemed ready for a final successful assault on the stubborn Atlantic. But the new, carefully designed, cable contained the seeds of disaster.

On July 23, 1865, the *Great Eastern* began her first cablelaying voyage. After eighty-four miles of cable had been laid, an electrical fault was indicated in a recently submerged portion of the cable. There was no alternative but to pull up the cable and investigate the reason for the fault. After ten miles of cable had been retrieved from the sea, the fault was discovered. A two-inch piece of iron wire had been driven through the cable, producing a short-circuit between the inner conductor and the salt water of the ocean. Those in charge suspected sabotage, but the piece of iron wire turned out to be a piece of the cable's iron armor. The cable had been coiled, layer upon layer, in a great hold of the ship. The enormous weight of the over-laying cable had snapped the brittle armor wire and had forced a piece of it into the cable. The fault was quickly repaired, and the *Great Eastern* returned to its task of laying cable.

By the morning of August 2, the *Great Eastern* had laid almost three-quarters of her cable, when another short circuit showed up in the cable. During the process of retrieving the fault, a weakened section parted and the broken end sank quickly into the depths. Although the *Great Eastern* was not suitably equipped for the job, it was decided to fish for the broken end two and a half miles below. The grapnel did catch onto the cable, but the wire rope parted less than half-way up with its prize. A second attempt also failed because of inadequate grappling equipment. A large buoy was then improvised to mark the location of the cable's end, and the *Great Eastern* sailed away in solitude and frustration.

By early 1866, Field had raised enough capital to try again. He immediately ordered 1990 miles of cable that was designed to prevent a recurrence of the earlier difficulties. The *Great Eastern* set sail once again on June 30, 1866. This time the trip was uneventful, and the cable a complete success. On July 27, 1866, the *Great Eastern* reached Newfoundland with a cable in direct communication with Ireland. The cable could transmit 15 letters per minute and on its first day of operation it earned five thousand dollars. The cold waters of the Atlantic had finally begun to repay the fortunes it had swallowed during the previous decade.

Field set sail immediately in the *Great Eastern* to retrieve the broken end of the 1865 cable, seven hundred miles away. By the end of August, the cable had been caught, hauled aboard, and spliced. For the first time in history, a ship was in contact with both Europe and North America at the same time. The *Great Eastern* was in communication with England through the repaired cable, and England could communicate with America through the cable of 1866. Despite a tremendous storm, the *Great Eastern* brought in her second transatlantic cable just four weeks after her first. Since that time, the United States and Europe have never been out of communication for more than a few hours at a time. The age of rapid global communications had finally arrived.

The Wires Learn to Speak

Many scientific discoveries and important inventions were made by accident, but Alexander Graham Bell knew exactly what he was doing when he designed the first telephone. The idea came to him in the summer of 1874. He later stated, "If I could make a current of electricity vary in intensity precisely as the air varies in density during the production of sound, I should be able to transmit speech telegraphically." Bell knew that the complex sounds of speech produce a rapidly and continually varying pressure pattern that moves away from the speaker in all directions. As these pressure variations fall on an object near the speaker, the object experiences forces that vary in a corresponding way. Bell's novel idea was to use these gentle forces to generate a current that varies in the course of time exactly the way the pressure varies.

Bell's first instrument was based on electromagnetism and was used for both the transmitter and receiver. Its principle of operation is illustrated in Figure 134. A permanent magnet exerts a pull on a thin iron diaphragm. A plastic housing (not shown) prevents the diaphragm from coming in direct contact with the poles of the magnet. When the pressure variations of speech reach the diaphragm, they produce in it a corresponding back and forth motion. These vibrations of the diaphragm produce corresponding variations in the distance between the poles and the diaphragm. This, in turn, produces variations in the magnetic field in the magnet. The changing magnetic

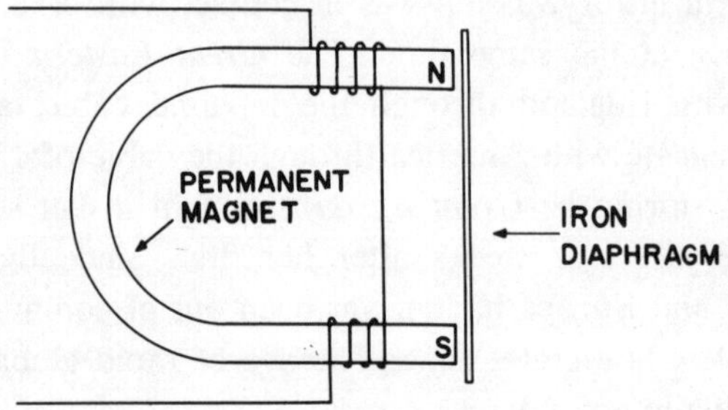

Fig. 134. Bell's early telephone transmitter and receiver.

field then induces a changing electromotive force in the coils that is an exact electrical copy of the original pressure variations.

When used as a receiver, the sequence of operations is reversed. The changing electromotive force from a distant transmitter sends a corresponding current through the telephone line and through the coils of the receiver. This varying current modifies the strength of the magnetic field pulling on the diaphragm. The diaphragm moves to and fro in response to the changing magnetic field, and generates sound waves that reproduce the original sounds. The first sounds of speech were transmitted over wire on June 3, 1875.

Amplifying the Sounds of Speech

Bell's early transmitters were based on correct scientific principles, but they produced electric currents that were too feeble for practical use. In his electromagnetic transmitter, only the small amount of energy in sound waves was available for conversion to electrical energy. What was needed was a method of increasing or amplifying the electric current. Ways were sought, therefore, to control the current from a battery in accordance with the pressure variations of speech. Instead of a mere converter of energy, a transmitter would then be a sort of valve—a device that could use a small amount of acoustical energy to control a relatively large amount of electrical energy from a battery.

The modern telephone transmitter was invented along these lines in 1878 by Henry Hunnings, an English clergyman. The essential feature of the transmitter is a loosely packed quantity of carbon granules as illustrated in the accompanying diagram (Figure 135).

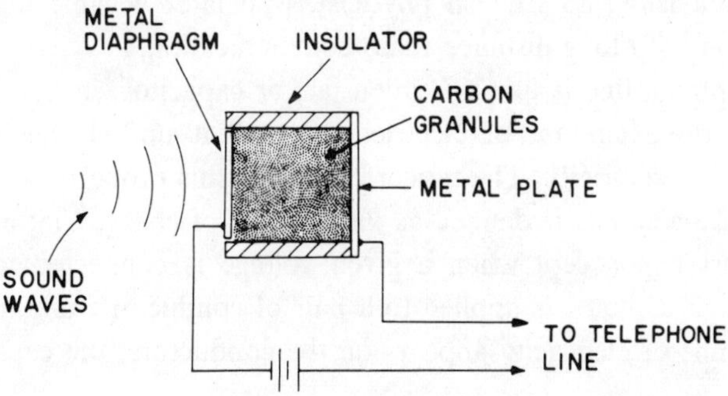

Fig. 135. A carbon granule telephone transmitter.

The telephone line terminals are connected to the metal diaphragm and plate, as shown. A direct current flows normally from the diaphragm to the plate through the loosely packed carbon particles, and through the telephone line to the distant receiver. In normal operation, sound waves strike the transmitter and the diaphragm vibrates back and forth. When it moves to the right, the granules become more tightly packed and the current increases proportionately. When it moves to the left, the granules are packed more loosely and the current is reduced. The small amount of sound energy striking the diaphragm is used merely to control a much larger amount of electrical energy from a battery. Modern transmitters working on this principle produce a thousandfold amplification of the original speech sounds reaching the diaphragm.

Capacitance

As telephone service grew, it became apparent that telephone lines would have to be inproved significantly before efficient service could be rendered. It was found that ordinary lines consisting of two elevated wires produced an unexpectedly great reduction in signal current. The current at the receiving end was much weaker than could be accounted for on the basis of Ohm's law. Also, it was clear that the higher frequency portions of the voice current were reduced, or attenuated, to a greater degree than the low frequency portions. The reason for this behavior of telephone lines was first made clear by Oliver Heaviside, an English physicist, who later became famous in connection with long distance radio communication.

A telephone line is like a Leyden jar, or capacitor, in that it takes a considerable amount of electricity to "fill it up." It has a great *capacity* for electricity. The modern term for this property is "capacitance." Capacitance is defined as the amount of charge that a pair of conductors can accept when a given voltage is connected to them. If a voltage, E volts, is applied to a pair of conductors, and a charge Q coulombs of electricity appears on the conductors, the capacitance C is:

$$C = \frac{Q}{E}$$

Capacitance is measured in *farads* in honor of Michael Faraday, the great English scientist. The amount of capacitance increases when the conductors are brought closer together. Heaviside showed theoretically that the attenuation of a telephone line could be reduced if some way were found to counteract the effect of capacitance inherent in the construction of the line. This led to the study of a different effect, already hinted at, that was discovered independently by Joseph Henry and later by Faraday. That effect is called "self-induction," or "inductance."

Inductance

Henry noticed, in 1829, that extremely high voltages can be produced quite simply with a low-voltage battery and a length of wire wound in the form of a coil. Consider the circuit shown in Figure 136. When the switch is closed, a current builds up through the coil and reaches a steady magnitude determined by Ohm's law. There is then a steady magnetic field inside and around the coil. When the switch is opened, the field begins to fall to zero. In doing so, the lines of magnetic force sweep through the turns of the coil, and induce an

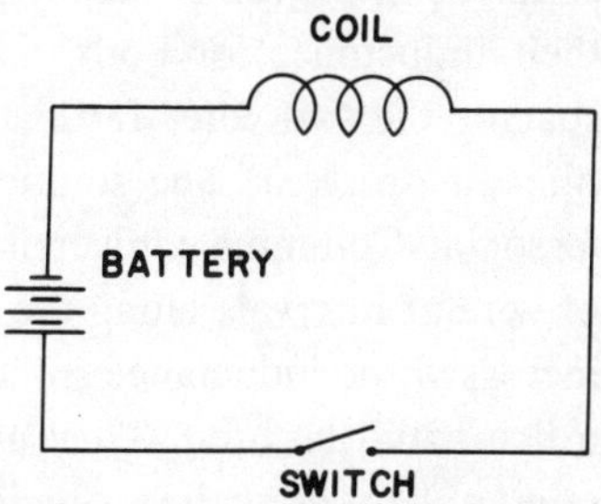

Fig. 136. A circuit to illustrate the effects of self-induction.

electromotive force in it. This voltage is said to be *self-induced*, and tends to keep the current flowing. It is as though an electrical inertia tends to maintain the flow of current for a short time, even though the switch has been opened. If the coil has may turns, the self-induction of the coil produces enough voltage to generate a spark across the switch terminals as they are drawn apart. This property of wires and coils that resembles mechanical inertia is called "inductance," or "self-induction."

A corresponding effect takes place when the switch is closed once again. As current begins to flow through the coil, it induces a voltage in it. As before, this voltage tends to oppose any change in the current flowing through the coil, just as mechanical inertia tends to oppose a change in velocity of a moving automobile. The coil's inductance slows down the rate at which the current builds up in the coil. The greater the inductance, the longer it takes for the current to build up to its steady magnitude.

Inductance is measured in units called "henrys." A conductor or coil has an inductance of 1 henry if a current that is changing at a rate of 1 ampere per second produces a self-induced voltage of 1 volt across its terminals. All conductors have some inductance, but a wire wound into a coil has much more inductance than it would have as a straight wire.

Loaded Lines

Heaviside showed theoretically that telephone lines then in use could be improved either by reducing their capacitance, or increasing their inductance, or both. Capacitance could be reduced merely by spacing the two wires farther apart, but inductance presented a more difficult problem. The solution was found by Michael Pupin, a professor at Columbia University in New York. He merely added coils of wire at intervals along the line. These coils, in series with the line, increased the inductance sufficiently to double the distance over which a line could be used. They are called "loading coils," and lines containing them are called "loaded lines." The loading of telephone lines illustrates dramatically the advantages of having theory and practice work hand in hand.

The amount of excess capacitance is considerably greater in submarine cables than in land lines. To understand why, imagine a single insulated wire suspended above the earth by poles. The capacitance between the wire and the earth is relatively small because of the large spacing between the two. Now imagine the same insulated wire immersed in salt water. The inner conductor can be thought of as one plate of a capacitor, and the conducting water as the other. Because the two are so close together, the cable has a great deal of capacitance and its performance is poor. Nevertheless, Heaviside's equations told engineers just how much inductance to add in order to counteract the capacitance and improve performance. By correct adjustment, in fact, the inductance and capacitance can be balanced perfectly, producing what Heaviside called a "distortionless line"—one in which *all* frequencies travel at the same speed and suffer the same minimal amount of attenuation. Because of the difficulty of inserting coils in a submarine cable, loading was usually accomplished by winding an iron wire or tape around the inner conductor, thereby increasing its inductance.

Heaviside's discovery made submarine telephony possible over distances of a few hundred miles. Inductive loading was also used subsequently on telegraph cables, greatly increasing their speed of signalling and, therefore, the amount of traffic they could carry. A loaded telegraph cable across the Atlantic can operate at five times the speed of an unloaded one, and so has five times the earning power.

With the invention of the telephone, and the success of undersea telegraph cables, it would seem to be a simple matter to lay a telephone cable across the Atlantic. Nevertheless, the first voice communication across the Atlantic took place almost half a century before the first transatlantic telephone cable was completed between Oban, Scotland and Clarenville, Newfoundland in 1956. Three great men of science, Maxwell, Hertz and Marconi, had found a way to project Faraday's electromagnetic fields far into space, carrying with them the tell-tale currents of telegraphy and, later, of human speech. It was radio, in fact, that provided the technological know-how that made long distance submarine telephony practical.

Even the best telephone cables provide feeble currents to a receiver two hundred miles away. It had been a simple matter to boost the weakened pulses of a telegraph signal by means of the Morse relay or repeater, but doing the same thing for voice currents had to await the invention of the radio tube. The first transatlantic telephone cable uses electronic amplifiers spaced about 40 miles apart on the floor of the ocean. Without such amplification, the currents would fade below hearing levels after a few hundred miles.

The telephone cable borrowed another equally important technique from the field of radio. Instead of a single voice conversation, the first cable was designed to carry thirty-six telephone conversations over the same wire at the same time. This electronic magic is merely a simple application of the same tuning principle that enables us to select any radio station we wish from the crowded air waves. All this means is that when you speak into a telephone transmitter, it is not the voice currents from your telephone transmitter that are fed into the cable. Instead, they are used to control one of thirty-six devices which are, in effect, tiny radio transmitters. The output of these transmitters feeds the equivalent of thirty-six conversations into the cable. Each transmitter is tuned to a slightly different frequency so that the thirty-six conversations can be unravelled by an equal number of radio receivers at the other end. The only essential difference between true radio communication and the methods used in the cable, is the medium of transmission. Radio waves are projected into space whereas the telephone cable guides its signals along a wire. The many conversations do not interfere with each other any more than do the multitude of unrelated programs entering your radio set.

To learn how these marvels came to be, we must return to Faraday's electric and magnetic fields and discover the dramatic role they were to play in man's never-ending quest for knowledge. These invisible fields of force have come to be the fundamental concepts upon which much of science rests. They carry the human voice around the world and with their help, scientists have discovered an important and undreamed of belt of electricity in the sky.

Over a Mountain of Water

15

As a young bookbinder's apprentice, Michael Faraday educated himself by reading volumes of the Encyclopedia Britannica and other worthwhile books that came to him for binding. Yet, despite a lack of formal education, he became one of the greatest scientists of all time. A lack of mathematical training did not prevent him from conceiving a theoretical picture of electrical phenomena that is still valid today.

Before Faraday pointed the way, electric and magnetic forces were thought to act across empty space. To his way of thinking, such "action at a distance"—as it was called—did not appear to make sense. When he saw a stone moving across a field, he expected to see a rope pulling it, or a stick pushing it. In order to account for the forces acting between charges or magnets, he had to visualize the space between them being filled with an invisible "something" that can pull or push. This "something" was a magnetic or an electric field, or both, depending upon the circumstances.

He conceived of a field as a system of invisible rubberlike tubes that stretch between a pair of unlike charges or unlike magnetic poles and pull them together. If the charges, or poles, are of the same sign, the elastic tubes take a different configuration and push the charges or poles apart. These rubberlike tubes, or lines of force, tend to shorten their length just as a stretched rubber band does. They also exert a sideways force of repulsion on one another, tending to spread apart as they move from their source.

These revolutionary ideas of Faraday opened a new era of scientific development. Inexplicable forces acting at a distance had been replaced by electric or magnetic fields that fill the space around charges and magnets. The fields could be "mapped" by measuring the direction and magnitude of the force acting upon a small electric charge or magnetic pole anywhere in the field. Using this approach, a scientist could calculate, or at least estimate, the forces between pairs of charged or magnetic objects separated by seemingly empty space. Although the concept of electric and magnetic fields was no easier to comprehend than the older action-at-a-distance ideas, it did provide an intermediary by which such action could be propagated through space.

Faraday did not combine all of his ideas into mathematical laws. That task was accomplished by a brilliant Scottish mathematician and physicist named James Clerk Maxwell (1831–1879).

Electromagnetism

Maxwell was born in Edinburgh in the same year that Faraday published his discovery of electromagnetic induction, the principle of the electric generator. He was an excellent mathematician, and at the age of fourteen won a medal for a paper showing how to draw a perfect ellipse with pins and thread. In 1856 he became Professor of Natural Philosophy at Mareschal College in Aberdeen where he remained until appointed the first director of the new Cavendish Laboratory at Cambridge in 1874.

Maxwell did important work in connection with the kinetic theory of gases, but his greatest contribution to science and to mankind was his theory of electromagnetism. Maxwell combined all knowledge of electricity and magnetism into two mathematical equations which are still fundamental, not only to electricity, but to all of physics. They represent a summit of achievement comparable to Newton's laws of motion, and have received the profound admiration of all succeeding generations of scientists. All later advances, such as quantum mechanics and relativity, presuppose a familiarity both with Newton's laws and Maxwell's equations.

The application of Maxwell's equations requires a knowledge of a higher form of mathematics—the calculus—so a mathematical discussion of them is well beyond the scope of this book. Nevertheless, we can gain an insight into the meaning of Maxwell's wonderful equations without resorting to mathematical methods.

We already have acquired the notion of electric and magnetic fields, and their physical representation by Faraday's lines of force. The direction of each line of force represents the direction of the electric or magnetic field at a particular point. This merely means that the direction of an electric field at a certain place is the direction in which a small positive charge would move if it were located at that place. Similarly, the direction of a magnetic field at any location is the direction in which a small north pole would move at that location. It is also clear that the strength or intensity of a field varies from place to place. The field strength at any point is determined by the density or closeness of the lines of force at that point. If a relatively large number of lines of force are crowded into a small region of the field, the field strength is relatively high in that region. To summarize, each kind of field has three characteristics associated with it: lines of force, field strength and direction.

Before proceeding with Maxwell's equations, we should keep in mind that motion or change of some kind is necessary for the interaction of electric and magnetic phenomena. A stationary electric charge produces no force on a stationary magnet. Only when one or the other is set in motion can a force be developed between them. Carrying this empirical knowledge a step further, we can say that a stationary electric field (produced by the charge) does not interact with a stationary magnetic field (produced by the magnet). They can interact only when one changes in strength with respect to the other.

If a magnet rests on a table, both the magnet and its field are stationary. If an electric charge rests nearby, its field is also stationary. There is no relative motion and no interaction. Now let's suppose that we move the magnet so that the magnetic lines of force cut across the stationary electric charge. During such motion, the magnetic field strength is constantly changing when measured in the vicinity of the charge. A force is then experienced by both the magnet and the charge. The same effect takes place if we move the electric charge so that the electric lines of force cut across the stationary magnet. We say that their fields interact during the period of relative motion so as to generate the observed forces. (Notice that no force is generated when, for example, the charge moves along the axis of the magnet. No lines of force are crossed by such motion and no force is developed.)

By *electromagnetism*, we merely mean the way in which electric and magnetic fields interact. Maxwell's theory of electromagnetism is a mathematical statement of the laws governing the interaction of these fields.

To understand Maxwell's first equation, imagine a magnet and a charged body resting on a table. If the magnet is set in motion so that the magnetic field strength at the charged body varies in the course of time, a force is experienced by the charged body. Maxwell took the position that the force is not produced directly by the varying magnetic field. Instead, *the changing magnetic field produces an electric field* which exerts a force on the charged body. The latter field should not be confused with the original electric field associated with the charged body. It is a new electric field produced by the changing magnetic field. It would exist even if no charged body were nearby to sense its presence.

In the example given above, the changing magnetic field was produced by a moving magnet. A magnetic field can also be produced by a current of electricity—a flow of electric charges. If the flow is constant, that is, if the magnitude of the current is unchanging, the magnetic field is also unchanging. But if the current varies in strength in the course of time, the magnetic field also varies in strength. Imagine a region of space occupied by a magnetic field whose field strength changes in the course of time. Maxwell's first equation tells us that an electric field is produced in such a region by the varying magnetic field. The strength of the induced electric field depends only on the rate at which the magnetic field strength varies.

Maxwell's second equation describes how a magnetic field is created in a region of space. It tells us that a magnetic field can be induced in either of two ways: by a current of electricity, or by a *changing electric field*. We are familiar with the former method; it is the principle of the electromagnet. The latter method is novel with Maxwell. He was the first scientist to realize that a changing electric field induces a magnetic field in a region of space. It was this new idea that led to the theoretical prediction and later discovery of radio waves, and to the discovery that light is electromagnetic in character.

Maxwell was able to deduce from his equations that oscillating electric and magnetic fields should move through space in the form of waves. These electromagnetic waves would carry off the energy originally supplied to the two fields. Maxwell's theory of electromagnetism was published in 1873 in his great *Treatise on Electricity and Magnetism*.

The Velocity of Electromagnetic Waves

An important conclusion drawn from Maxwell's equations was that electromagnetic waves should propagate through space at a velocity equal to that of light. "We can scarcely avoid the inference," he said, "that light consists in the transverse undulations of the same medium which is the cause of electric and magnetic phenomena." His equations predicted that electromagnetic waves should travel at a speed of 3×10^8 meters per second (186,000 miles per second). By that time, the velocity of light was known as a result of Fizeau's measurements of 1849. It agreed almost perfectly with the speed deduced from Maxwell's equations. If his equations were correct, then light must be of electrical origin and must consist of electromagnetic waves.

Through chemistry, electricity had found its way into the very structure of atoms, ions and molecules. It now began to appear that electricity was also bound up in the nature of light. A German scientist named Heinrich Hertz (1857–94) decided to put Maxwell's equations to the test. If Maxwell had been right, it should be possible to generate electromagnetic waves of relatively long wavelength—our radio waves—and detect them at some distance from their source. He decided to try.

The Oscillatory Discharge

In the spring of 1886 Hertz noticed an effect that gave him a start in the right direction. His apparatus is illustrated in Figure 137. A piece of copper wire was bent in the form of a rectangle so that the ends of the wire at *A* were separated by a very short air gap. This rectangle was placed near an induction coil which could be made to produce a spark discharge when the switch was operated. Hertz found that a secondary spark was induced across the air gap, *A*, when-

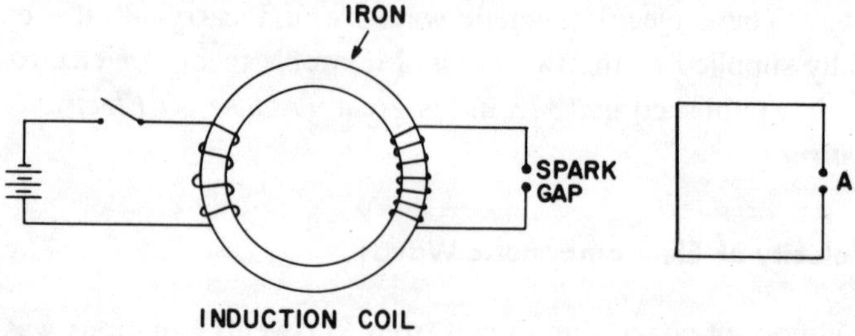

Fig. 137. Hertz's apparatus to induce a spark in a secondary circuit.

ever a spark was produced in the primary circuit containing the induction coil. Hertz had found a way to detect electric phenomena at some distance from its source.

Hertz also noticed that the intensity of the secondary spark depended upon the dimensions of the rectangle. A certain specific size gave the most intense spark. If the rectangle were slightly larger or smaller, the spark became weaker. This is an example of electric resonance, a most important property of alternating current circuits.

The Nature of Sparks

Before we can understand the importance of electrical resonance in Hertz's experiment, we must look into the nature of a spark discharge. Suppose we have a circuit, as shown in Figure 138, consisting of a Leyden jar, or capacitor, an induction coil, and a spark gap consisting of two small spheres. Now let's charge the capacitor by connecting it momentarily to a source of high voltage. As we bring the spheres closer and closer together, the voltage stored on the capacitor plates suddenly causes a spark discharge across the gap. When Henry performed this experiment in 1842, he discovered an amazing property of the spark. The current that produced the visible display did not flow continuously in one direction; it surged to and fro across the gap during the passage of the spark. The current oscillated back

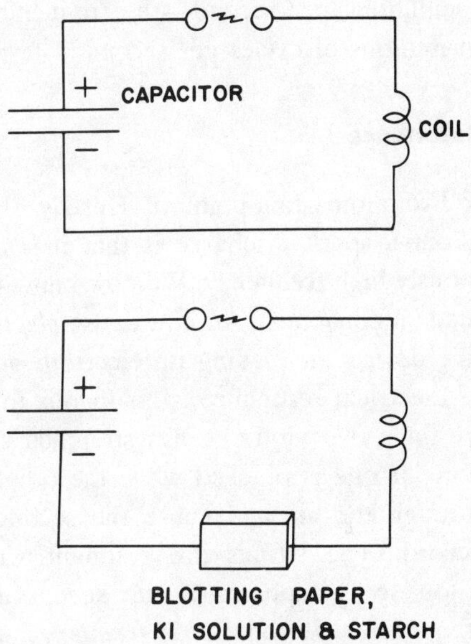

Fig. 138. A circuit to illustrate the oscillatory nature of a spark discharge.

and forth until all of the electrical energy was dissipated. Here was an electrical phenomenon analogous to the motion of a pendulum.

The oscillating nature of the spark discharge can be proved by cutting the wire at any point as shown in part (b) of the diagram, and inserting a piece of blotting paper soaked with a solution of potassium iodide mixed with starch. The oscillating current must now flow through the blotting paper as well as the other portions of the circuit. When current flows through potassium iodide, the resulting electrolysis releases free iodine at the *positive* electrode. Iodine turns starch a deep blue. If the current travels only in one direction, the region of the blotting paper nearest the positive terminal should become blue.

the other end of the blotting paper should experience no change in color. But experiment shows that the starch becomes blue near both terminals connecting it to the circuit. The current, therefore, is not unidirectional, but oscillates back and forth through the circuit and through the spark discharge. A spark is an alternating current accompanied by incandescence. Motion pictures show that there may be many thousands of oscillations in a single spark discharge which itself usually lasts less than a thousandth of a second. The number of oscillations per second—the frequency of the current—is measured in millions of cycles per second.

Resonance

From the standpoint of Hertz's experiment, the important thing about a spark discharge is that it is an alternating current of enormously high frequency. We now know that this fortunate circumstance made it convenient for him to use *electrical resonance* to build a sensitive device for picking up elecrtomagnetic waves.

Electrical resonance is analogous to acoustical resonance. Suppose we have two violin strings stretched so that both produce the same note. If one is plucked while the other is at rest, sound waves move through the air and cause the second string to vibrate of its own accord. The strings are resonant with each other. This state of resonance disappears if the strings are tuned to different natural frequencies. Resonance is *frequency selective.*

Hertz's pickup loop illustrated on the right of Figure 137 was of just the right size—about 2 feet on each side—to resonate with the alternating current through the spark discharge. As the current in the primary circuit surged back and forth, it induced corresponding currents in the secondary circuit. Because of resonance, the oscillating currents in the latter circuit built up rapidly in magnitude until a spark appeared at the air gap of the loop.

The current oscillating in a resonant pickup loop, such as the one used by Hertz, is hundreds of times greater than for a nonresonant loop. Had he not made use of resonance, it is doubtful that he could have proceeded with his later experiments. There would have been no visible spark in the pickup loop and therefore no available sensing device to pick up the weak electromagnetic waves that his spark was able to generate.

The Discovery of Radio Waves

Hertz's discovery that sparks can be produced in the air gap of a secondary circuit, provided it has the dimensions necessary for resonance, was the key to his later success. Having a suitable detector of

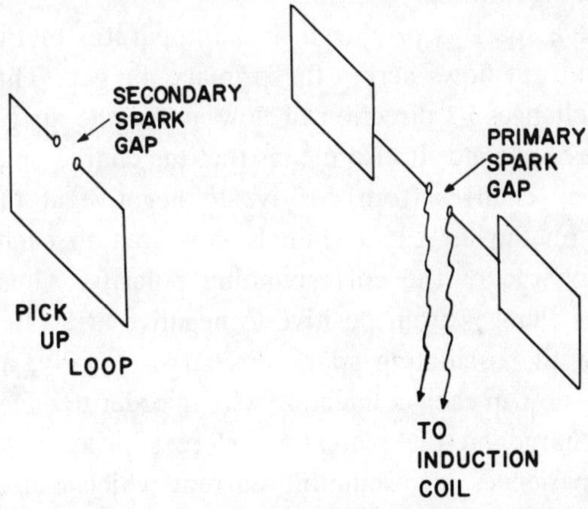

Fig. 139. Hertz's apparatus used to transmit and receive radio waves.

electromagnetic waves, he was now able to test Maxwell's theory by direct means.

His apparatus is illustrated in Figure 139. The "primary conductor" —or antenna as we would say—consisted of two square brass plates, 40 centimeters (about 16 inches) on each side, connected by a copper wire 60 centimeters (about two feet) long. In the middle of the wire was a spark gap in which spark oscillations were produced by powerful discharges of an induction coil. The conductor was set up 1.5 meters (about 5 feet) above the floor with the plane of the plates vertical. The secondary circuit was that already described—a wire of suitable length bent into an incompletely closed loop. Its dimensions were carefully chosen so that its natural frequency of oscillation was the same as that of the primary circuit. The two circuits were in electrical resonance.

The brass plates of the primary circuit are really the two plates of a modified Leyden jar, or capacitor, with air as the dielectric between them. Because of their relatively great spacing, the electric field between them extends throughout a large region, instead of being confined to a narrow space as in an ordinary capacitor. This arrangement insures that the antenna will radiate a large amount of its energy in the form of electromagnetic waves upon each oscillation of current between the plates.

Each time a spark is produced in the apparatus of Figure 139, an oscillating current flows across the primary air gap. This means that the current changes its direction of flow across the air gap many millions of times a second. It also means that the charge on each terminal of the air gap changes from positive to negative at the same rate. The charge on the air gap terminals flows out to each brass plate, giving it a charge of the corresponding polarity. Thus, the charge on the plates changes from positive to negative at a rate equal to the frequency of the oscillating spark discharge. Finally, an alternating current must flow in each connecting wire in order to effect the changes in electric charge on the plates. So Hertz's primary conductor, or antenna, experiences an oscillating current which causes the charge on each brass plate to change continuously from positive to negative at an extremely rapid rate.

This electric charge generates an electric field by its presence on the plates, and a magnetic field by its motion. Both of these fields vary continuously in strength and direction because of the oscillatory motion of the electric charge. During each oscillation, the lines of electric force move outward into the surrounding space, as shown in Figure 140. When the charge on each plate reaches its greatest magnitude, the lines stop moving outward and tend to contract inwards toward the antenna. But the outer lines of force contract in such a way that their parts touch each other at some distance from the antenna. The portion of each of these lines of force most distant from the antenna takes the form of a closed loop. As the rest of the line of force contracts toward the antenna, this loop becomes detached and is propagated outward as an electromagnetic or radio wave. As these fields move away from the antenna they become wider and wider, expanding to fill all of the space through which they propagate.

Radio waves of this kind were detected by Hertz at some distance from the antenna as continuously varying electric and magnetic fields. He proved that they travel at a speed just about equal to that of light. They could be reflected, refracted and polarized. He sent them through

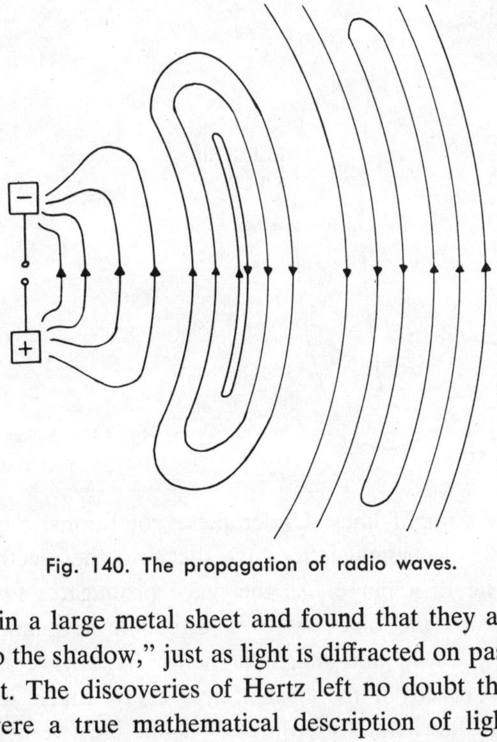

Fig. 140. The propagation of radio waves.

an opening in a large metal sheet and found that they are diffracted, or "bent into the shadow," just as light is diffracted on passing through a narrow slit. The discoveries of Hertz left no doubt that Maxwell's equations were a true mathematical description of light and other forms of electromagnetic radiation. Light and radio waves were shown to be identical in principle, differing only in the vastly greater frequencies that are characteristic of light. Both consist of electric and magnetic fields that are propagated through space in accordance with Maxwell's equations.

Hertz's experiments also made it clear that the directions of the electric and magnetic fields are at right angles to each other, and that both directions are at right angles to the direction of travel of the wave. To understand why, consider the vertical antenna illustrated in Figure 141. An observer to the right of the diagram detects an electric field that would cause a small positive charge to move in the vertical direction. Similarly, the magnetic field would cause a north magnetic pole to move to his left, out of the plane of the paper. If he is far enough away from the antenna, both of these motions would

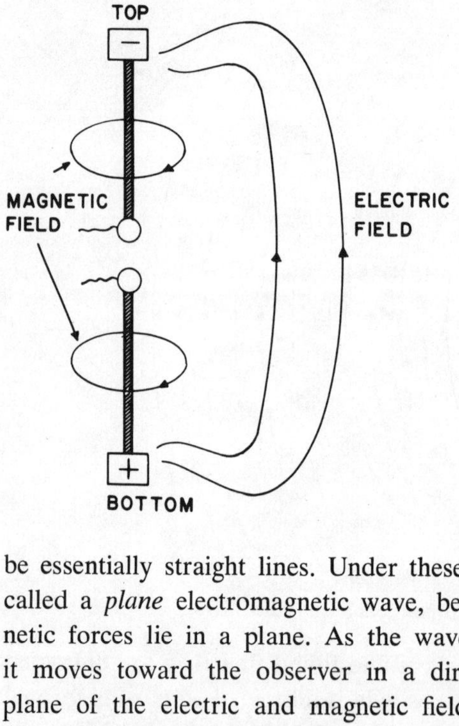

Fig. 141. An antenna radiating electromagnetic waves.

be essentially straight lines. Under these conditions, the radiation is called a *plane* electromagnetic wave, because the electric and magnetic forces lie in a plane. As the wave propagates through space, it moves toward the observer in a direction perpendicular to the plane of the electric and magnetic fields. The wave is said to be "vertically polarized" if the electric field exerts forces in the vertical direction, and "horizontally polarized" if the electric field lies in the horizontal direction.

The Coherer

By 1894, Oliver Lodge (1851–1940) had developed an improved method of detecting radio waves. His device was actually invented as a lightning protector for telegraph lines. It consisted of two closely spaced metal spheres as shown in Figure 142. One was ordinarily connected to the telegraph line and the other to a rod embedded in moist earth. He found that the spheres tended to fuse together after

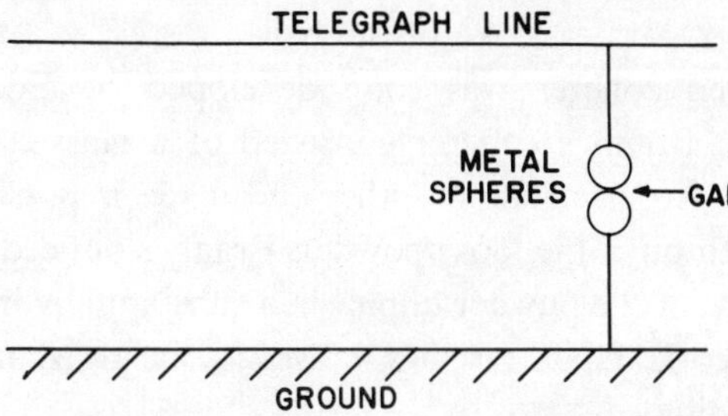

Fig. 142. Lodge's coherer, a more efficient detector of radio waves.

the passage of a spark, but that they could be separated again by a gentle tap. Thus, the electrical resistance from one sphere to the other was very high before the passage of a spark, and very low after the spark.

Lodge explained that the spheres tended to "cohere" after a spark

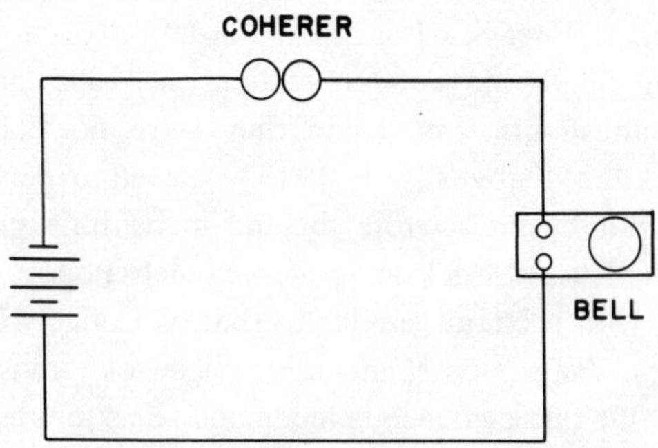

Fig. 143. Lodge's receiving circuit to illustrate the use of radio waves in signaling.

and the device became known as a *coherer*, a name that stuck. Lodge demonstrated his coherer in a lecture given at the Royal Institution in London in 1894. The coherer was connected in series with a battery and an electric bell as shown in Figure 143. When a spark occurred in a nearby circuit, the spheres cohered, the circuit was closed, and the bell began to ring and kept ringing until the spheres were separated by a gentle tap. When Lodge placed the bell on the same stand as the coherer, the mechanical vibration of the first stroke of the bell was enough to tap the spheres apart. Each spark was then signalled by a separate stroke of the bell.

An improved coherer was soon developed by Edouard Branly (1846–1900). Branly's coherer consisted of a glass tube filled with metal powder. An electrode at either end made is possible to send a weak current through the loose powder. Branly's noticed that the electrical resistance of the powder diminished substantially in the presence of a nearby spark. Upon tapping, however, the resistance rose again to a high magnitude.

For some reason, Lodge made no attempt to transmit information using his new techniques. He later wrote in 1923, "I was too busy with teaching work to take up telegraphic or any other development; nor had I the foresight to perceive what has turned out to be its extraordinary importance to the navy, the merchant service, and indeed land and war service, too."

Lodge's early lectures inspired a great deal of interest in radio telegraphy. Henry B. Jackson (1855–1929), a British naval officer who later became First Sea Lord of the Admiralty, carried out secret trials of radiotelegraphic communications in 1895 and 1896. But he did not publish his results and they were not followed up. In Germany, Adolph Slaby (1849–1913) started experimenting with radio waves, but he later wrote that he made little progress in the early years. In Russia, Aleksandr Stepanovich Popov (1859–1905) experimented with a circuit similar to that of Lodge. He was able to register electric disturbances, including those of atmospheric origin, as early as 1895. The evidence is inconclusive as to whether he really used this apparatus to transmit telegraphic signals. In any event, he produced no practical system of radio telegraphy. The use of radio waves as a means of transmitting telegraphic signals had its real beginning in Italy. There, a young man named Guglielmo Marconi (1874–1937) began a career that resulted in three major scientific discoveries in the field of radio transmission and reception.

The Birth of Wireless Communication

In the summer of 1894, while vacationing in the Alps, twenty year old Marconi came across an obituary article describing the experiments of Hertz. He later said:

> It seemed to me that if the radiation could be increased, developed, and controlled, it would be possible to signal across space for considerable distances. My chief trouble was that the idea was so elementary, so simple in its logic, that it seemed difficult for me to believe that no one else had thought to put it into practice. I argued, there must be more mature scientists who had followed the same line of thought and arrived at almost similar conclusions. From the first, the idea was so real to me that I did not realize that to others the theory might appear quite fantastic.

After studying Hertz's experimental results, Marconi rushed home and put together an induction coil, a spark gap, a telegraph key, batteries and a Branly coherer. Following the methods of Hertz and Lodge, his first apparatus sent signals from one end of a table to the other. By improving his apparatus and using more intense sparks, he was able to increase this distance from a matter of a few hundred feet to several miles. All of this was accomplished with devices that in form, at least, were essentially like those of his predecessors. Being an optimist, he predicted that the range would eventually be extended to hundreds of miles.

Most scientists of the time ridiculed Marconi's view. They argued that radio waves have the same properties as light waves and so travel only in straight lines. Once beyond the horizon, they would move off into space and reception would be cut off. Despite the unanimous skepticism, Marconi continued his experiments and discovered a new principle. He learned to "attach" his radio waves to the surface of the earth by connecting his transmitter to two conductors, one elevated high above ground, and the other buried in moist earth. In this way his waves were guided around the earth's surface to distances of about two hundred miles. The vital secret of Marconi's first success was his "grounded wave."

Realizing he had a new force at his disposal, Marconi conceived the bold project of attempting to span the Atlantic. He set up a powerful transmitter at Poldhu, Cornwall, England, and a receiving system at St. John's, Newfoundland, in 1901. (A large antenna system at Cape Cod had just been blown down by a hurricane.) After considerable difficulty with balloon and kite supported antennas at St. John's, a 400-foot wire was kept aloft long enough on December 12 to convince Marconi that he had received the Poldhu signals. The letter *S*, three dots in the Morse code, had flashed from Poldhu to Newfoundland, some 2,000 miles away. In their transmission across the ocean, his radio waves had curved over a moutain of water more than 100 miles high!

Marconi's claim was so incredible that many scientists and laymen just couldn't believe it. The matter was settled in February 1902, however, when Marconi received the Poldhu signals aboard the steamship *Philadelphia* out some two thousand miles. When the ship docked in New York in March of that year, he came down the gangplank and handed news reporters yards of "telegraph" tape showing dots and dashes recorded automatically while enroute across the ocean. "Now," he exclaimed, "will anyone say I was mistaken in Newfoundland!"

On his voyage across the Atlantic, Marconi discovered that the range of reception was greater at night than by day. While signals had been received at 2,000 miles during the night, the daylight range was only one-third as great. His subsequent experiments showed, however, that the gap between day and night ranges could be narrowed by using extremely long wavelengths. This research culminated in the establishment during 1907 of the first transatlantic radio telegraph service between Nova Scotia and Ireland. Marconi's spark transmitter generated radio waves having wavelengths of several thousand meters. At least 50 kilowatts of electric power were used to generate these waves, and gigantic antennas were constructed to transmit and receive them. In succeeding years, wavelengths reached 10,000 meters in length and mile-long antennas were constructed 1,000 feet in the air to receive them. Scientists and engineers were now convinced that reliable long-distance communication depended upon the use of great amounts of power and ever longer wavelengths. As we shall see, radio had turned down a wrong path. But long before the error of using long waves became obvious, science discovered the mechanism by which radio waves encircle the earth.

The Ionosphere

Marconi's achievement of spanning the Atlantic in 1901 could not be explained on the basis of existing knowledge. His earlier experiments, of much shorter range, had not seriously challenged the scientific belief that radio waves travel only in straight lines. But the two thousand mile path across the Atlantic brought to mind the picture of a mountain of sea water interposed between transmitter and receiver. It was clear that something had been left out of the scientific reasoning that had led to the initial pessimism.

The answer was first suggested by A. E. Kennelly in 1902, who concluded that an electrically conducting layer must exist in the earth's upper atmosphere at a height of about 80 kilometers. He calculated that the layer would have an electrical conductivity about twenty times as great as ocean water. It was well known that a large conducting surface is a good reflector of radio waves. Hertz had demonstrated that fact by experiment. Perhaps the upper atmosphere holds a great radio mirror that reflects radio waves back to earth. Sea water and moist earth are also good conductors of electricity, so there are really two *mirrors* involved. In Kennelly's own words:

> It seems reasonable to infer that electromagnetic disturbances emitted from a wireless sending antenna spread horizontally outwards, and also upwards, until the conducting strata of the atmosphere are encountered after which the waves will move horizontally outward in a 50-mile layer between the electrically reflecting surface, or successive series of surfaces, in the rarefied air above.

Shortly after Kennelly's paper was published, Oliver Heaviside made a similar suggestion:

> Sea water, though transparent to light, has quite enough conductivity to make it behave as a conductor for Hertzian waves, and the same is true in a more imperfect manner of the earth. Hence the waves accommodate themselves to the surface of the sea, in the same way as waves follow wires. The irregularities make confusion, no doubt, but the main waves

are pulled round by the curvature of the earth, and do not jump off. There is another consideration. There may possibly be a sufficiently conducting layer in the upper air. If so, the waves will, so to speak, catch on to it more or less. Then the guidance will be by the sea on the one side and the upper layer on the other.

The Kennelly-Heaviside layer, as it was called, met abuse and scorn for almost a generation. Most scientists tried to explain Marconi's results on the basis of diffraction, or bending into the shadow of the earth. They pointed out that the wavelengths were extremely long, and enough diffraction might take place to account for the observed results. Although a small amount of diffractive bending of long radio waves does take place, no completely satisfactory theory was developed to explain ranges of 2,000 miles or more. Nevertheless, the Kennelly-Heaviside layer, or *ionosphere* as it is now called, was not accepted universally until after Marconi made his second great discovery—with the help of radio amateurs.

The Daylight Wave

In 1920, British and American amateurs organized a test to determine whether the "commercially useless" wavelengths of 200 meters and under could span the Atlantic. It failed. Another test was run in 1921, despite the almost universal belief that it, too, would fail. But this time there was success. Twenty U. S. amateur station call letters were identified in the British Isles, and one station even transmitted a complete message. But the signals were only received at night, a fatal defect as far as communication is concerned. Neither the commercial companies nor the amateurs were interested in further investigation. Marconi seems to have been the only man whose imagination was fired by the amazing achievement of the amateurs. He wrote in 1922:

> I have brought these results and ideas to your notice as I feel—and perhaps you will agree with me—that the study of short electric waves, although sadly neglected practically all through the history of wireless, is still likely to develop in many unexpected directions, and open up new fields of profitable research.

During the spring of 1923, Marconi began the first of a series of classic experiments on the "useless" short waves. He set up a transmitter at the historic Poldhu site operating on a wavelength of 97 meters and sailed south on his yacht *Elettra.* He listened to the Poldhu signals on his way south and found them to be exceptionally good. In the Cape Verde Islands, over 2,500 miles from Poldhu, the signals were far clearer than had ever been received over a comparable distance from a long wave station. Even when the transmitted power at Poldhu was reduced to 1 kilowatt—a minute fraction of the power used with long wave stations—the signals were stronger than those received from the transoceanic stations in the British Isles.

As he expected, Marconi found that the short wave signals disappeared during the daylight hours. But there was a curious difference between the 97-meter Poldhu signals and the 200-meter amateur signals of 1921. Although the latter had disappeared at sunrise, the 97-meter signals from Poldhu persisted for a considerable time *after* sunrise. Marconi wondered whether even shorter wavelengths might persist throughout the entire day.

The following year, he laid out a program to test a series of wavelengths from 90 meters down to 30 meters. Cruising through the Mediterranean, he compared the signal strengths received on all of the wavelengths. In Beyrouth harbor in September, 1924, he made the astounding discovery that the 30-meter wavelength from Poldhu held throughout the day. After spanning 2400 miles, these short wave signals were as good as nighttime signals. The 90-meter signals, on the other hand, behaved as at the Cape Verde Islands. Marconi had made his second great discovery. Short waves are more suitable for long-range communication than long waves.

What Marconi had observed was transmission by reflection from an elevated region of the ionosphere that later became known as the F-layer. In order to reach this layer, radio waves had to traverse lower regions of the ionosphere called the D- and E-layers. It turned out that the 90-meter and 200-meter signals were almost completely absorbed by the lower regions of the ionosphere *during the daytime* and so could not reach the F-layer. The shorter 30-meter signals passed through these layers with little loss of energy, were reflected by the

F-layer, and received over 2,400 miles away. Within a month of his discovery, Marconi had arranged to set up world-wide communication links with places as far away as Australia—and all were amazingly successful. His short wave stations handled messages at speeds of 100 words per minute, speeds that no undersea cable or long wave transmitter had ever approached. The long waves were now obsolete and the cables had become a secondary means of communication.

Probing the Ionosphere

Marconi's dramatic discovery aroused considerable interest about the ionosphere. Researches to test the Kennelly-Heaviside theory were carried out in Great Britain in the year 1924 by Appleton and Barnett. They were able to show that the signals received from a broadcasting station arrived by two routes—a direct ground path and an overhead atmospheric path. By measuring the angles of arrival of the signals, they were able to calculate the height of the ionosphere. Nevertheless, many scientists remained unconvinced.

A crucial experiment was performed in 1925 by Gregory Breit and Merle Tuve which established the existence of the ionosphere. The idea arose out of an earlier discussion with two fellow scientists of the University of Minnesota: Professor W. F. G. Swann and Dr. J. G. Frayne. The idea was to send a signal in a nearly vertical direction and measure the round-trip time required for the radio echo to return to earth. Echoes were first observed in July, 1925 from station NKF of the U. S. Naval Research Laboratory, distant only 11 kilometers from the receiver. A direct signal was also received via the ground path and the time-delay between the two corresponded to the height of reflection of the ionosphere. Most of the reflections came from heights in the range from 160 to 200 kilometers. Although the Kennelly-Heaviside layer was first suggested in 1902, it was not until 1925 that its existence was experimentally proved beyond a reasonable doubt. During that year, measurements of its height were made independently by numerous scientists.

The success of long range radio communication illustrates the equal importance of theory and experiment in the progress of science. The point was aptly made in 1938 by Enrico Fermi, the great physicist who played an indispensable role in conceiving and carrying out the wartime atomic bomb project in the United States.

It is well known that the discoveries of Marconi were at first received with a certain skepticism in the scientific community. This skepticism was based on the conviction that transmission of radio waves would not be possible between stations when one is beyond the horizon of the other. The reasoning, in fact, was more or less as follows: The electromagnetic waves used in radio transmissions are substantially analogous to light waves, from which they differ only in wavelength; and the earth, by virtue of its electrical conductivity, acts as an opaque body to them. The radiations emitted by a station, propagating in a straight line, must leave in shadow all of the stations situated below the horizon of the transmitting station; except for a small correction due to the phenomenon of diffraction. It is fortunate for humanity that these arguments, which *a priori* could seem reasonable and well founded, did not deter Marconi from experimenting with transmissions to great distances. The story of this first success of radio transmission constitutes a reconfirmation of the fact that in the study of natural phenomena, *theory and experiment must walk at the same pace.** Rarely can experiment, unguided by a theoretical idea, attain results of great importance; and it is certainly one of the most significant successes for theory that the very existence and the essential properties of electromagnetic waves were predicted mathematically by Maxwell before the experimental verification of their existence, and before these (electromagnetic waves), through the ingenious intuition of Marconi, were to find their field of practical application; on the other hand, an excessive reliance thrust upon theoretical predictions would have discouraged the insistence on experiments which were destined to revolutionize the techniques of communications.

Radio Twilight

In the summer of 1932, Marconi discovered a third new form of radio wave propagation that depends only on the blanket of (unionized) air surrounding the earth. Although short wavelengths—in the vicinity of a few tens of meters—were reflected by the ionosphere, still shorter waves were known to penetrate the ionosphere without significant reflection. The so-called "microwaves"—below a meter or

*Italics mine.

so in length—were therefore useless for long range communication. Or so everyone thought.

Once again, an aging Marconi decided to find out whether the laws of nature were as they were supposed to be. He constructed a transmitter that generated 60-centimeter radio waves and installed it atop Rocca di Papa, an elevation some 2,500 feet above the Mediterranean Sea. Sailing to Sardinia, he installed the receiving equipment at a height of about 1,000 feet above sea level. The distance was 168 miles, and the intervening sea represented a mile-high obstacle between the two stations. Yet Marconi received useful signals from the station in Italy.

Once again, his announcement was met with skepticism. No existing theory could explain the success of his experiment. The "line-of-sight" theory of transmission of centimeter waves was too firmly ingrained to be disturbed by one experiment, even if the experimenter was Marconi. Ill health and death prevented him from pursuing his third discovery, but there is no doubt about his confidence in this new means of propagation. "In regard to the limited range of propagation of these microwaves, the last word has not been said." In August of the following year, a similar experiment was performed in the United States by W. D. Hershberger, a scientist at the Signal Corps Laboratories in New Jersey. Using a 75-centimeter wavelength, Hershberger was able to receive useful telegraphic signals aboard a small boat at a distance of 80 miles from the transmitter. This distance was five times as great as could be accounted for by existing line-of-sight theories. Hershberger concluded, "So we are not dealing with anything remotely resembling rectilinear (line-of-sight) propagation of 75 centimeter waves. These experimental observations are offered without comment."

There the matter rested for nearly twenty years, until the effect was rediscovered independently when television receivers began picking up unexpectedly strong interfering signals from stations in distant cities. The celebrated television allocations freeze followed in 1948, and no new TV stations were authorized until the interference problem was better understood four years later.

Today we know that very high frequency (very short wavelength) radio waves do not disappear completely as a receiver moves below the transmitter's horizon. The received signal strength does drop off considerably just beyond the horizon, but the dropoff at greater distances is much more moderate. The behavior of these waves is similar to ordinary twilight. As the sun sets below the horizon, we notice a great reduction in illumination. Then a period of twilight sets in during which the illumination falls off much more gradually until the sun is about 15 degrees below the horizon. To make use of the weak but ever-present radio "twilight" signals, it is necessary to use extremely high power transmitters and large antennas to compensate for the initial drop in signal strength near the horizon. Called tropospheric scatter systems, communication stations of this kind carry scores of telephone conversations and even television signals for hundreds of miles.

Scientists agree that this new kind of radio wave propagation is somehow made possible by the lower portion of the atmosphere called the troposphere. One theory holds that the radio waves are reflected, or scattered, by turbulent "blobs" or volumes in the air. Another holds that the signals are caused by reflections from layers of air that become less dense as the height increases. A great controversy now rages on the matter among scientific specialists. No matter which of these theories gains ultimate acceptance, radio "twilight" propagation is recognized as the most important discovery in radio wave propagation to come along in nearly thirty years.

In recent times, there has been a flurry of conjecture as to who was really the Father of Radio. The honor was claimed by Lee De Forest, the inventor of the radio tube, in his autobiography bearing that title. Others point out that no progress was made until Lodge gave his important lectures. Still others claim that Popov transmitted Morse code a few months before Marconi did. But all of these claims seem quite irrelevant. Only Marconi had the courage and insight to proceed with his experiments against the scientific beliefs of the time. Even the great Hertz said that long-distance radio telegraphy, beyond the horizon, could not be achieved unless "you could construct a mirror as large as a continent." It could not have occurred to him that nature had provided such a radio mirror in the form of a blanket of electrons high in the atmosphere. If the title Father of Radio can be assigned to any man, it can only belong to Marconi.

The Riddle of the Atom

16

Physics found itself in an enviable position toward the end of the nineteenth century. An imposing scientific edifice had been constructed upon the foundation of Newton's mechanics and Maxwell's electromagnetic theory. The former accurately described the motion of material bodies and the latter united the theories of light, radiant heat and radio waves. Underlying all of physics was the concept of energy which appeared in many forms: kinetic energy of moving bodies, electrical energy, chemical energy, and energy of molecular motion, or heat.

There were many blank spaces to be filled in, of course, but they would surely require no more than a mopping up operation. Classical physics had all the important answers, and no additional facts were expected to turn up that could possibly disturb its basic concepts.

Little did most scientists realize that within a few short years their most fundamental ideas of space, time and matter would be shattered by fact and logic. The indivisible atom of Democritus and Dalton would slowly fall apart in a proliferation of smaller particles. The ether, which provided the medium through which electromagnetic waves were thought to travel would be found nonexistent. Matter and energy would turn out to be convertible, one into the other. And light itself—clearly a form of wave motion—would begin to show tell-tale evidences of behaving as a particle instead.

Out of the changes that followed, there arose a new physics and with it a vastly deeper understanding of the intricacies of nature. But even with that understanding, the modern scientist has grown much more cautious than his Victorian predecessor about the fundamental nature of his knowledge. He is content to fret about his facts, and to invent tentative theories about them. Occasionally he finds a germ of truth upon which to build an intricate superstructure. Every new fact erupts in a series of tremors that threaten the edifice, and he suspects deep in his heart that eventually it must fall. But he also knows that an even stronger structure will rise from the ruins. Solving the riddle of the atom was an adventure of that sort, full of success and failure, chaos and simplification—a major achievement of the human mind.

Cathode Rays

To understand the compelling need for a new kind of physics, we must begin with a string of curious observations concerning the flow of electricity through gases. It was known for centuries that gases, although ordinarily good electrical insulators, can sometimes be broken down by high voltages. Furthermore, it was known that the passage of electricity through gases takes place much more easily if its pressure is reduced to a small fraction of atmospheric pressure. As early as 1744, Sir William Watson (1715–1787) had sent electricity through air at low pressure in a glass tube three feet long. Instead of the usual small spark, his tube was completely filled with a bright silver light "forming a continued arch of lambent flame." Watson had anticipated the everyday "neon" lamp by about two hundred years.

Almost a century later, Faraday noticed a strange feature of Watson's light. A purple haze or stream of light began at the tube's positive terminal—called the "anode"—but stopped short before reaching the negative terminal, or "cathode." But the negative terminal itself was covered with a continuous glow, the two visible discharges being separated by a dark space. Improved air pumps became available in the 1850's, and Julius Plucker (1801–1865) took up the study of the discharge. Placing a magnet near the tube, he noticed that the discharge is deflected by a magnetic field. He also noticed that during the discharge the walls of the tube around the cathode glowed with a phosphorescent light. This strange glow also moved about in response to the location of his magnet. Clearly, something was emanating from the negative cathode and striking the walls of the tube—and that invisible "something" was deflected by a magnetic field. Later, it was discovered that a shadow was cast by an object placed between the cathode and the glass tube. The term "cathode rays" was introduced to describe these mysterious emanations that proceed from the cathode in straight lines and cause phosphorescence when they strike the tube.

In 1870, Cromwell Varley (1828–1883) suggested that cathode rays are composed of "particles of matter, projected from the negative pole by electricity." They were supposed to be negatively charged bits of matter. In motion, they constitute a current of electricity and so are deflected by a magnetic field.

In 1879, Sir William Crookes (1832–1919) suggested that the glowing of the gas is caused by collisions of particles in the tube. He attributed the fluorescence or phosphorescence of the walls of the tube to the impact of the particles on the glass. In 1894, J. J. Thomson measured the velocity of the cathode rays, finding it to be 190,000 meters per second, less than one-thousandth of the speed of light. Thomson's measurement was a severe blow to a rival hypothesis that cathode rays might be a form of electromagnetic radiation traveling at the speed of light. At about the same time, Jean Perrin placed a piece of metal in the path of the cathode rays and found that it became charged with "resinous" or negative electricity.

All of this seemed to indicate that cathode rays must be negatively charged particles flying through rarefield gas just as Faraday's ions move through liquids during electrolysis. But Philipp Lenard, a pupil of Hertz, had shown in 1894 that cathode rays can pass through a thin foil of aluminum without making a hole in it, as any material particle would surely do. This convinced many scientists that the rays were undoubtedly a kind of wave rather than a particle. Of course, we know today that the neutron, a subatomic particle, can pass with little difficulty through concrete walls many feet thick, but Lenard had no such information. In his time, the argument seemed exceptionally strong.

Thomson got around Lenard's objection by pointing out that cathode-ray particles might be very small, and might pass through the spaces between the atoms of a metal. Furthermore, the cathode-ray particles were identical *no matter which materials or gases were used in the tube.* This led to the idea that the cathode-ray particles are a constituent of all atoms. In Thomson's words, "The assumption of a state of matter more finely subdivided than the atom of an element is a somewhat startling one; but a hypothesis that would involve somewhat similar consequences—viz., that the so-called elements are compounds of some primordial element—has been put forward from time to time by various chemists." The term "electron," suggested by G. J. Stoney, now came into use to describe these negatively charged subatomic particles. The assault on the atom had begun.

Thomson made his victory complete in 1897 and 1898 by measuring the mass and amount of charge associated with the electron. Before describing his ingenious methods, it is necessary to go back a few years and pick up the story of a kind of radiation different from cathode rays that also emanated from the cathode-ray tube.

X-Rays

On November 8, 1895, Wilhelm Konrad Röntgen (1845–1923) was experimenting with a cathode-ray tube when a strange event took place. His tube happened to be covered completely by black cardboard on that particular occasion. Near the tube was a piece of paper painted with a fluorescent material. When the discharge was turned on, the fluorescent paint became brilliantly luminous. A new kind of radiation was passing through the walls of the tube, through the opaque cardboard, and causing fluorescence some distance away. Röntgen's new rays were surely not cathode rays which barely make it through a thin sheet of aluminum foil. Furthermore, the rays were not deflected by a magnetic field. Neither could they be refracted by glass prisms or any other material that bends ordinary light.

The new X-rays, as Röntgen called them, travelled in straight lines that originated at the cathode of the tube. They could be stopped by metal plates, but they passed to some extent through human flesh.

Röntgen also discovered that his X-rays were able to affect a photographic plate. The first X-ray photograph taken by Röntgen was the hand of his wife, which clearly showed her wedding ring and the bone structure. The new rays were immediately put to use in surgery and in locating foreign objects in the human body.

The nature of X-rays was not long in doubt. It was soon agreed that they are electromagnetic waves similar to light and Hertz's radio waves, but of exceedingly small wavelength. It was known by the 1890's that the visible spectrum is bounded on one end by infrared (heat) rays of relatively long wavelength and on the other by ultraviolet rays of relatively short wavelength. X-rays are merely much shorter wavelengths beyond the ultraviolet. Because of their minute wavelengths, they are not refracted by prisms, but pass through without bending.

The Wilson Cloud Chamber

In 1896, C. T. R. Wilson (1869–1959), a Scottish physicist of the Cavendish Laboratory at Cambridge, made a discovery of great importance. He found that if dust-free air saturated with water vapor is suddenly cooled by expansion, tiny water droplets are formed on any ions that may be present in it. (The droplets also form on dust particles, so these must be removed to avoid confusion.) For small expansions, below about 30 per cent, only negative ions serve as *condensation centers* as they are called. For larger expansions, both negative and positive ions cause droplets. In the same year, Wilson passed X-rays through his "cloud chamber" and found that X-rays are able to ionize gases. J. J. Thomson saw in Wilson's cloud chamber just the apparatus he needed to measure the charge on an electron.

Measuring the Electron

Thomson's approach involved three separate measurements. The first two required the use of a rather special cathode ray tube as illustrated in Figure 144. Electrons given off by the cathode C are attracted toward the positive anode A. A few pass through the hole in the anode and form a fast moving beam. A voltage between the parallel plates *D* produces an electric field which deflects the beam off course. Thomson measured the amount of the deflection. Then he added a magnetic field (not shown) which was directed perpendicular to the paper in the vicinity of *D*. Its effect was to raise the beam back to its undeflected path. Thomson measured the magnetic field strength just necessary to overcome the original deflection of the beam.

From these measurements, and a knowledge of Maxwell's electromagnetic theory, Thomson calculated the ratio of the electric charge on

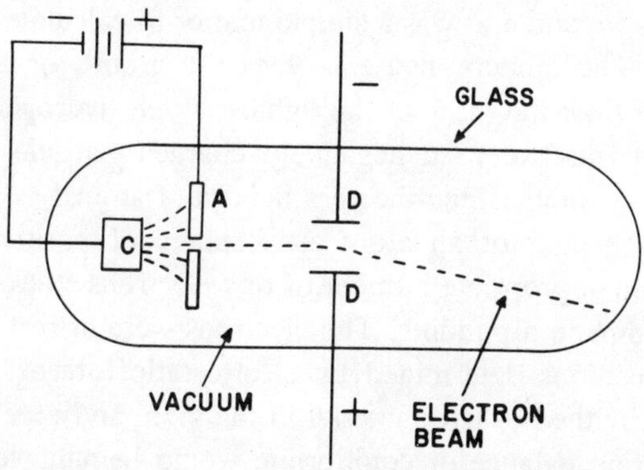

Fig. 144. Thompson's apparatus for measuring *e/m*, the ratio of the charge on an electron to its mass.

an electron, *e*, to its mass, *m*. No matter how fast the electrons were moving, the ratio *e/m* was always the same, 5.27×10^{17} electrostatic units of charge per gram.

Thomson's third measurement made use of Wilson's cloud chamber to determine the charge on an electron. His method was to pass X-rays through the chamber, thereby producing a cloud around the negatively charged particles. The size of the droplets was calculated from their rate of fall. The smaller the droplet, the slower it settles down, and a formula exists which gives the radius of the droplet in terms of the fall velocity and air pressure. He also measured the amount of water that settled to the bottom of the chamber. Dividing the total amount of water by the volume of a droplet gave the number of droplets that had been produced by the X-rays. Each droplet grew around one electron, so he also knew the number of electrons involved. He then measured the total amount of charge in the collected water. By dividing the total charge by the number of electrons, he calculated the charge on one electron. The figure is 4.8×10^{-10} electrostatic units, or 1.6×10^{-19} coulombs. The amount of negative charge on an electron turned out to be precisely the same as the positive charge carried by the hydrogen ion in electrolysis.

Thomson's Model of the Atom

Knowing e/m and e, it was a simple matter to calculate the mass of an electron. The modern figure is 9×10^{-30} grams, or about 1,840 times smaller than the mass of the lightest atom, hydrogen. Here was a magnificent discovery: a negatively charged particle almost two thousand times smaller than the very lightest of atoms!

Thomson thought of an atom as a sphere of positively charged matter with an appropriate number of tiny electrons stuck in here and there like plums in a pudding. The electrons were at rest in the atom in certain positions determined by electrostatic forces. They would be attracted by the positive material in the atom and repelled by each other. A certain balance or equilibrium would be achieved when the electrons found their "right" places.

According to this view, the hydrogen atom consists of one electron and a much larger mass of positive matter having an equal but opposite charge. Other atoms would have additional electrons and corresponding amounts of positive matter. Each atom would have equal amounts of positive and negative charge and would be electrically neutral. Everything was beginning to fit together beautifully. When an atom became excited—that is, when it received additional energy from the outside—electrons were supposed to oscillate back and forth about their equilibrium positions, emitting light waves of various wavelengths. The reasoning was as follows: Because of its charge, an oscillating electron is really an oscillating current of electricity. Maxwell had predicted that such an oscillating charge should emit electromagnetic waves, and Hertz had verified the prediction. There was every reason to believe that an oscillating electron within the atom should obey the same laws as an oscillating current in a Hertzian antenna. Light emitted by an incandescent material, therefore, must obviously be caused by oscillating electrons within the excited atoms. How wonderfully simple! Laborious calculations were then performed in an attempt to find a correspondence between the light wavelengths emitted by specific elements, and the frequency of vibration of the electrons. But strangely enough, all such attempts failed. And worse still, an even more disturbing situation soon became apparent. Incandescent gaseous elements give off light of *specific* wavelengths, rather than light of a continuous range of frequencies. The observed line spectrum is unchanging for the same element. This implies that the electrons are vibrating at specific frequencies within the atom rather than at a continuous range of frequencies. Why should electrons be so selective in their frequencies of vibration? No answer was available at the time and there the problem rested. Thomson's discovery of the electron had presented a problem that appeared to be insoluble within the framework of Maxwell's theory.

The Discovery of Radioactivity

Still other mysteries were beginning to crop up. Early in 1896 the French physicist Henri Becquerel tried a simple experiment to find out whether X-rays can be emitted by fluorescent materials. Such materials were known to fluoresce brilliantly when struck by X-rays, but they also fluoresce when placed in bright sunlight. Perhaps the action of the sunlight would cause the fluorescent material to give off X-rays. To find out, he wrapped a photographic plate in black paper and put it on a window sill. On top of that he placed a crystal of a fluorescent mineral called uranyle, a double sulfate of uranium and potassium. When he developed the plate, after several hours' exposure of the crystal to sunlight, he found a gray spot at the place over which the uranyle had rested. Rays from the uranyle crystal had affected the photographic plate.

Later, quite by accident, Becquerel discovered that the sunlight had not been necessary. Uranyle fogged the photographic plate even when the two were placed together in the desk drawer, completely protected from light! Here was an unbelievable development. The darkening of the plate had nothing to do with exposure of the uranyle to sunlight. A penetrating radiation, perhaps similar to X-rays, was generated in the crystal all by itself, without any outside excitation.

Becquerel surmised that the uranium in the crystal must be generating the rays. He heated the crystal, cooled it down, ground it to a powder, dissolved it in acids, but the intensity of the mysterious radiation remained the same. Becquerel's radioactivity, as it is called, is not connected with the physical or chemical way in which the uranium atoms are arranged in a compound. It is a mysterious process taking place deep within the very heart of the atom.

Polonium and Radium

Following Becquerel's discovery, Marie Curie (1867–1934) a Polish-born chemist and the wife of French physicist Pierre Curie (1859–1906), carried out an extensive series of tests for radioactivity in all chemical elements. Then a little known chemist, she was to become the first scientist ever to receive two Nobel prizes, one for chemistry and one for physics. Working in Paris, she and her husband announced two new elements in 1898, each more powerfully radioactive than uranium. One was called polonium, in honor of her native land, the other, radium. Then followed four years of exhausting manual labor, to distill a speck of the precious radium salts from tons of "worthless" uranium ore from which valuable uranium had previously been extracted.

The radioactivity of radium was incredible. By far the most radioactive substance known, it was some two million times more radioactive than uranium. From the depths of its atoms there issued an endless stream of energy in various forms. It glowed in the dark and maintained its own temperature a little warmer than its surroundings; it was later found to give forth a heavy radioactive gas, radon; but above all, it bore stark witness to a terrible cataclysm taking place unceasingly within the very heart of the atom. Although they could not know it, Becquerel and the Curies had initiated the atomic age.

The Nucleus Is Born

It is largely to the New Zealander, Ernest Rutherford (1871–1937), that we owe our understanding of radioactivity. Working in Canada with his English collaborator, Frederick Soddy (1877–1956), he performed a series of masterful experiments which uncovered the fundamental facts involved. According to Rutherford, radioactive atoms explode and tear themselves apart. In the process, three kinds of rays are given off, dubbed α (alpha) rays, β (beta) rays, and γ (gamma) rays, after the first three letters of the Greek alphabet. The gamma rays turned out to be electromagnetic waves similar to X-rays, but far more penetrating and of much shorter wavelength. The beta rays proved to be streams of electrons. The alpha rays were completely new to science. They were rapidly moving fragments of the exploded radioactive atom. Yet, they were more than mere fragments. They were ions of the inert element helium—ions and not atoms because they lacked the two electrons normally carried by helium atoms. Each alpha particle carries two units of positive charge. The two electron "plums" are missing from the helium atom, leaving an unbalanced positive charge.

Not only are rays given off when a radioactive atom explodes, but the fragment that remains is a different and lighter element. This in turn is radioactive and explodes, as do its residues after it, in a long series of transmutations from element to element. Finally, after millions of years, only ordinary lead is left.

Rutherford saw the fast-moving and heavy alpha particle as an excellent projectile with which to belabor the atom, and possibly wrest from it the secret of atomic structure. In 1911, he bombarded a thin foil of metal with a stream of alpha particles. Those that passed through the foil struck a screen on the other side which produced a scintillation of light at each point of impact. Scintillations appeared all over the screen, indicating that many alpha particles were deflected through various angles in passing through the foil. He also found that some alpha particles were seemingly reflected by the metal foil, some even bouncing back toward the source of radiation.

Rutherford counted the number of light scintillations at various angles and found that the scattering of alpha particles was considerable. With these data, he was able to show that Thomson's plum-pudding picture of the atom could not be correct. According to Thomson's model, the positive and negative charges were almost uniformly distributed throughout the entire body of the atom. If that were true, the internal charges in the atom would tend to neutralize each other. The resulting electrostatic forces would not be strong enough to deflect the positively charged alpha particles through the large angles that Rutherford had observed. The only possible explanation was that the atom's positive charge, and most of its mass, are concentrated in a tiny region at the very center of the atom. Rutherford called the massive, positively charged central core the *atomic nucleus*. Circling around it are the negatively charged electrons, more or less the way our planets circle around the sun. Each electron is prevented from leaving the atom by the force of attraction between its negative charge and the positive charge of the nucleus. This electrostatic force takes the place of the gravitational force that keeps each planet in its orbit around the sun. Thus, Rutherford's atom emerged as a miniature solar system with a tiny but massive nucleus surrounded at a relatively great distance by a swarm of orbiting electrons.

It was later established by two of Rutherford's students, H. Geiger and E. Marsden, that the number of electrons in each elemental atom is equal to the atomic number of that element in Mendeléev's periodic table. The number of units of positive charge in the nucleus is also equal to the atomic number because the amount of positive charge is equal to the amount of negative charge in each atom. Thus, hydrogen has one electron and one unit of positive charge. Carbon has six electrons and six units of positive charge. Oxygen has eight, sodium eleven, sulfur sixteen, and so on. Whereas Mendeléev had used atomic weight to establish the chemical order of the elements, it was now clear that the number of electrons in the atom is the important quantity. It is these electrons which somehow determine the manner in which chemical elements join together to form compounds.

The Ultraviolet Catastrophe

It is now time for the strange little *quantum* to enter our story. The quantum, by now, has become almost an obsession with scientists. It finds its way into almost every important equation, it dictates the direction of research, and it has even led to rather fruitless arguments about free will and the like. Though it is one of the most fundamental concepts of the physical world, it was not recognized until the turn of the present century. Even then, it turned up in a field where its hints were obscure at best. The very concept was contrary to all

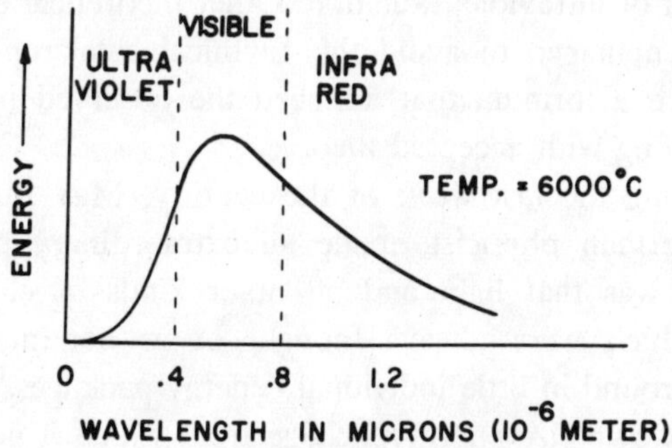

Fig. 145. A graph illustrating the distribution of energy in the various electromagnetic wavelengths coming from a hot body. (One micron is equal to one-millionth of a meter.)

scientific beliefs of the time, and its discovery was an act of scientific desperation.

The quantum made its official entrance to physics in connection with a theoretical dilemma called the "ultraviolet catastrophe." The dilemma arose during the last decade of the nineteenth century. It was well known in those days that the radiation emitted by a glowing solid, such as a red hot poker, is in the form of electromagnetic waves. The length of these waves varies continuously over a wide range, including the visible part of the spectrum as well as the longer infrared wavelengths and the shorter ultraviolet. The *amount* of radiation is not the same at all wavelengths, however, but depends on the temperature of the poker. Figure 145 illustrates how the energy is distributed among the wavelengths at a temperature of about 6000°C. There is a certain wavelength which receives more energy than any of the others and the amount falls off for shorter and longer wavelengths. Increasing the temperature moves the peak to the left, toward the shorter wavelengths; reducing it moves the peak to the right. All of these were experimental facts, known to be true. Then physicists tried to use theoretical ideas to calculate the distribution of energy in the various wavelengths coming from a glowing body. Such attempts developed a formula which implied that all of the energy in all of the world's matter should long since have escaped in a catastrophic burst of ultraviolet radiation! Other theoretical approaches to the problem managed to avoid this mythical catastrophe, but none could produce a formula that matched the observed results. Something was wrong with accepted theory.

Then, during the last week of the century, Max Planck (1858–1947), a German physicist, made an extraordinary proposal. His general idea was that light and all other kinds of electromagnetic radiation—which were always thought of as continuous waves—actually go around in little individual "energy packages." Each wavelength has its own size energy package from which it never deviates, and which is different from that of every other wavelength. Planck called these discrete bundles of energy, *quanta*. He decreed, in effect, that nature never generates one-third of a quantum, or two-sevenths of a quantum; they come only in whole units, and that is that. He showed that the amount of energy in a quantum depends upon the wavelength, or what amounts to the same thing, the frequency of the electromagnetic wave. The higher the frequency (or the shorter the wavelength) the higher the energy per quantum. The formula is simpler in terms of frequency:

$$\text{Energy per quantum} = \text{h} \times \text{frequency}$$
$$\text{E} = \text{hf}$$

where *h* is called *Planck's constant*.

Using his quanta, Planck was able to derive a formula that is in beautiful agreement with the way a glowing body radiates electromagnetic waves. But what a price to pay for the achievement. The very idea was so fantastic that only after the lapse of seventeen years did Planck receive the Nobel prize. The idea of the quantum was like saying a child's swing could only move in arcs of 1 foot, or 2 feet, or 5 feet and so on, but never in an arc of 1½ feet, or any other in-between length. The idea was preposterous, but it did lead to the right answer.

Planck's constant *h* is now recognized as one of the few fundamental constants of nature, equal in importance to *c,* the velocity of light in a vacuum. More than any other number, it is a symbol of the new physics. In a very few years it would give rise to momentous events, yet its magnitude is a mere .000000000000000000000000000000000663 joule seconds, a number more compactly written as 6.63×10^{-34} joule seconds.

An example will emphasize the small size of a quantum of light. Blue light has a frequency of about 6×10^{14} cycles per second. Multiplying that number by Planck's constant, a quantum of blue light is found to have an energy of about 4×10^{-19} joules. This number is so unbelievably small as to defy description. Ordinary incandescent lamps emit light waves that carry off many joules of energy each second, corresponding to countless billions of trillions of quanta. But despite its small size, the quantum has produced a revolution in physical thought unsurpassed since the days of Galileo and Kepler.

The Photoelectric Effect

Planck's invention of the quantum might have been a nice mathematical trick, but many scientists believed that the radiation of glowing bodies would eventually be explained on more conventional grounds. But in 1905, the quantum took a giant step toward scientific respectability. In that year, the great Albert Einstein (1879–1955) applied the idea of light quanta to the explanation of the so-called "photoelectric effect."

The effect was first noticed by Heinrich Hertz during his brilliant researches on radio waves. In 1887, he happened to notice that the passage of a spark in a circuit is facilitated by the passage of another spark in its vicinity. Investigating further, he found that the effect is caused by ultraviolet light emitted by the latter spark. An ordinary spark is rich in invisible, high-frequency light in the ultraviolet portion of the spectrum. When this light falls on the electrodes of a nearby spark gap, it greatly facilitates the passage of a spark.

The following year, Wilhelm Hallwachs discovered that a charge rapidly leaks off a negatively charged plate when it is exposed to ultraviolet light. In 1898, Ernest Rutherford showed that when ultraviolet light falls on a metal plate, the surrounding air becomes filled with negatively charged particles. With Thomson's discovery of the

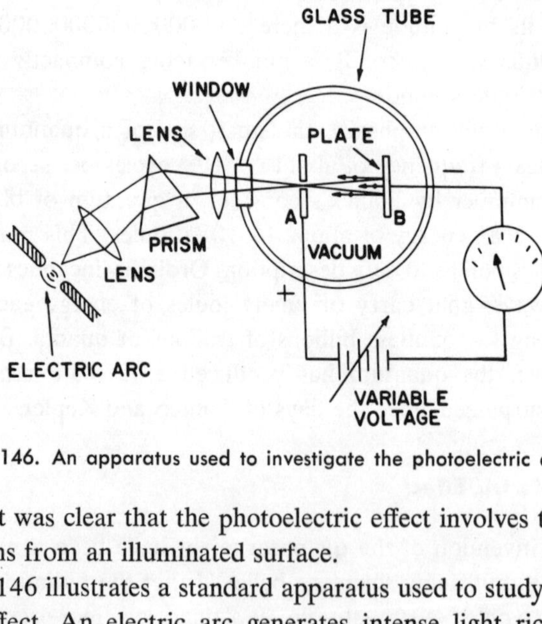

Fig. 146. An apparatus used to investigate the photoelectric effect.

electron, it was clear that the photoelectric effect involves the ejection of electrons from an illuminated surface.

Figure 146 illustrates a standard apparatus used to study the photoelectric effect. An electric arc generates intense light rich in ultraviolet radiation. This invisible light passes through quartz lenses and a prism before entering an evacuated glass tube through a quartz window. Quartz must be used because ordinary glass is opaque to high-frequency ultraviolet light. The lenses and prism are called a "monochromator." Ultraviolet light is spread into a spectrum by the quartz prism, just as ordinary visible light is spread out by a glass prism. By rotating the prism slightly, it is possible to select a desired frequency of ultraviolet light and direct it into the evacuated tube. In this way, it is possible to measure the photoelectric effect over a wide range of frequencies.

Ultraviolet light of the desired frequency passes through a hole in the copper anode, *A*, and strikes the cathode *B*, a plate made of a specific metal to be tested. A variable battery voltage is connected to the anode and cathode, and a sensitive meter measures any current flowing in the circuit. When light of an appropriate frequency strikes the test plate *B*, negatively charged electrons are ejected and move across the vacuum to the positively charged anode *A*. This motion of electrons is an electric current, the remaining portion of the circuit being through the battery, meter and connecting wires. It is characteristic of series circuits of this kind that the amount of current flow is the same everywhere in the circuit. This means that the meter records the same current as that flowing from cathode to anode. In performing a typical experiment, a plate, *B*, is irradiated by ultraviolet light whose frequency is changed successively from one frequency to another. The amount of current flowing is measured for each frequency by reading the meter. This enables the physicist to calculate the number of electrons emitted by the test plate for any frequency of interest. It is also possible to keep the frequency constant, and change the intensity of the light. This tells the physicist how the photoelectric effect depends on the intensity of light.

Using these methods, two laws were discovered that govern the emission of electrons by light. The first comes into play when the frequency is maintained constant and the intensity is varied; the second, when the intensity is kept constant and the frequency is varied.

1. For a constant frequency of light, the velocity, and therefore the kinetic energy of the ejected electrons, does not change. As the intensity of the light varies, the *number* of ejected electrons goes up, but the *kinetic energy* of each is always the same.

2. When the frequency of the incident light is varied, no electrons are emitted until the light frequency is raised to a specific threshold frequency, no matter how intense the light. For higher frequencies, the kinetic energy of the electrons increases in direct proportion to the frequency.

The two laws are illustrated graphically in Figure 147. In part (*a*), the light frequency is held constant as its intensity is varied. The number of photoelectrons increases in direct proportion to the intensity of the light striking the test plate. Part (*b*) shows how the energy of each photoelectron changes as the frequency is increased. For frequencies below that of point *A*, the energy is zero because no electrons are ejected from the metal. As the frequency is increased above that point, the energy increases uniformly with increasing frequency.

The laws of photoelectricity defied all attempts to explain them on the basis of Maxwell's classical theory of electromagnetism. The

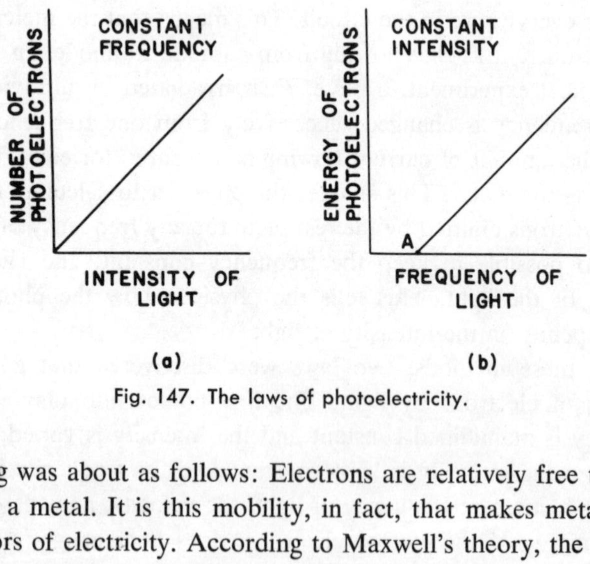

Fig. 147. The laws of photoelectricity.

reasoning was about as follows: Electrons are relatively free to move about in a metal. It is this mobility, in fact, that makes metals good conductors of electricity. According to Maxwell's theory, the oscillating electric field of a light wave exerts a force on such an electron, throwing it out of the metal. As the *intensity* of the light increases, the electric field should also increase in strength. This stronger electric force should throw an electron out with greater velocity and, what amounts to the same thing, with correspondingly greater kinetic energy. But experiment proved this conclusion to be false. Even if the light intensity is increased a thousandfold, photoelectrons keep popping out with exactly the same energy. Only their number increases with increasing intensity.

On the other hand, the energy of a photoelectron definitely *is* connected with the *frequency* of incident light. The higher the frequency, the higher the energy. There is no apparent reason for this interdependence of frequency and energy in the classical electromagnetic theory of light. It was at this stage of the theoretical impasse that Einstein entered the picture.

Einstein pointed out in 1905 that the idea of discrete light quanta, each carrying a definite amount of energy, provides a natural explanation of the photoelectric effect. Planck had already defined the energy in a quantum in terms of the frequency of the light: $E=hf$. Einstein suggested that when a light quantum, or photon as it is also known, strikes the surface of a metal it is either reflected, or absorbed by one of the electrons. If the latter, it must transfer *all* of its energy to one electron, because there can be no energy less than one quantum. We cannot have half a quantum going to one electron and half to another. Increasing the intensity of the incident light merely means more quanta *at the same frequency*. The increased number of quanta eject a greater number of photoelectrons, but their speed and energy is always the same.

Quite a different explanation arises when the frequency is increased. Each light quantum now has more energy because its energy, hf, goes up with increasing frequency. When this increased energy is absorbed by an electron, the latter is ejected with higher velocity. So, increasing the frequency increases the energy of the ejected electron.

Einstein also explained why no photoelectrons are emitted by light below the threshold frequency. In passing through the surface of the metal, a photoelectron loses a certain amount of energy received from the light quantum. This is analogous to the energy lost by a missile on escaping from the earth's gravitational field. If a bullet is shot straight up with sufficient velocity, it eventually leaves the earth's vicinity forever. Its final velocity, however, is considerably lower than the initial velocity. The reduction in velocity represents the energy given up in escaping from the earth's attraction.

The amount of energy given up by an electron in leaving the surface of a metal is known by the inappropriate name, *work function*. Thus, the energy relations of the photoelectric effect are as follows:

Energy of photoelectron = energy of light quantum − work function

$$E = hf - W$$

where *W* stands for the work function of the metal in question. The magnitude of *W* varies from one metal to another. The equation tells us that the energy of a photoelectron is equal to the energy of the incident light quantum minus the energy lost in leaving the metal. If *hf* happens to be less than *W*, no electrons can escape from the metal. As soon as *hf* becomes larger than *W*, photoelectric emission begins. With Einstein's explanation of the photoelectric effect, Planck's idea of the quantum began to assume the status of a physical reality.

The Quantum and the Atom

In 1913, a young Danish student of Rutherford, Niels Bohr (1885–1962), published an improved atomic theory. He began by pointing out a theoretical difficulty connected with Rutherford's atomic model, a small positively charged nucleus surrounded by orbiting electrons. Viewed in the light of classical theory, this picture of the atom could not last longer than a minute fraction of a second. An electron rotating around the nucleus is really an electric oscillator which ought to radiate electromagnetic waves in accordance with classical theory. As energy is given off, there would be less left for the electron and it would slow down rapidly in its orbit. It was easy to prove that orbiting electrons would spiral into the nucleus in less than a millionth of a second. Of course, this really doesn't happen; our atoms have been around for a long time. The paradox was just as difficult to explain as the ultraviolet catastrophe and seemed to warrant equally drastic action.

Bohr decided to *quantize* the atomic orbits. If radiant energy can only exist in certain minimum amounts, why not assume that the kinetic energy of an atomic electron is also limited to specific amounts? This means that an orbiting electron cannot have just *any* orbit, but only one of certain very specific orbits. It can jump from one orbit to another, but it cannot decide to stay in a "forbidden" orbit. Using this approach, Bohr was able to show a remarkable correspondence between his atomic orbits and the light frequencies emitted (or absorbed) by atoms. Each orbit represents a certain level of kinetic energy. If a light quantum of the correct energy (and frequency) comes along, the electron can absorb it and jump to an orbit of higher energy. Later, the electron can return to a lower energy orbit and emit an appropriate light quantum. Thus, an atom can emit only those light frequencies corresponding to specific orbit or energy-level transitions, and it can absorb only those frequencies that it can emit. These frequencies make up the familiar emission and absorption spectra that had long been associated with the various elements. Because the allowed orbits (and energy levels) are different for each chemical element, each has a characteristic spectrum.

The Hydrogen Atom

When Niels Bohr announced his quantum theory of the atom in 1913, there was little doubt that he was on the right track. Planck had conceived the quantum, Einstein had used it to explain photoelectricity, and Bohr had used it to explain atomic structure and the riddle of the spectral lines. We can understand the latter by following Bohr's description of the simplest of all atoms, the atom of hydrogen.

The hydrogen atom consists of a single electron revolving around a nucleus that carries a single positive charge. The hydrogen nucleus is called a *proton* in modern terminology. The proton's mass is about 1,840 times greater than that of the electron, but its charge is equal and opposite in sign to that of the electron. Because the electron's orbit is so much larger than either a proton or an electron, most of the atom is empty space.

When the hydrogen atom is heated to incandescence, it emits visible light that consists of only four lines when viewed through a simple spectroscope—a red line, a blue line and two violet lines. This group of lines is called a *series* by spectroscopists, and each element has its own characteristic series of lines. Of these, the hydrogen series is the most typical and regular. As far back as 1885, a Swiss school teacher named Johann J. Balmer (1825–1898), discovered that the frequencies of the lines in the hydrogen series (now called the Balmer series) can be calculated by a simple equation that he devised:

$$f = R\left(\frac{1}{4} - \frac{1}{n^2}\right)$$

where *R* is a numerical constant and *n* stands for the integers: three, four, five, six, etc. The letter *n* cannot stand for one, because the frequency would then be negative, or for two, because the frequency

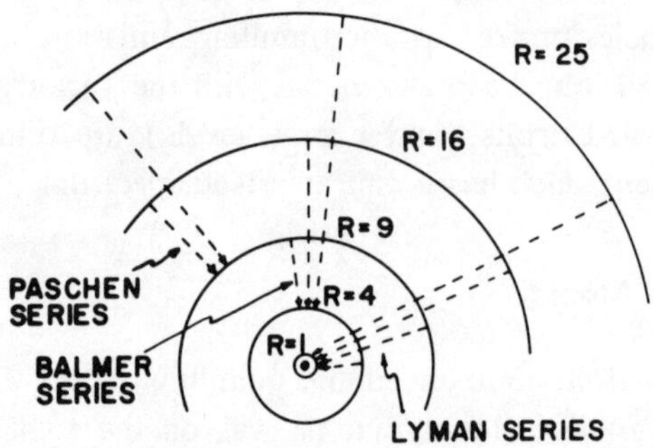

Fig. 148. The possible orbits of hydrogen's electron according to Bohr.

would be zero. Both solutions have no physical meaning as far as light is concerned.

Using Balmer's equation, Bohr was able to show that the allowed energy levels of the electron in a hydrogen atom correspond to the circular orbits shown in Figure 148. The fascinating aspect of the arrangement is the relationship between the radii of the orbits. If the radius of the first orbit is called a unit of length, the second has a radius of four units, the third has a radius of nine units, and so on. These numbers, 1, 4, 9, 16, etc., are the squares of the integers 1, 2, 3, 4, etc. Knowing this fact, Bohr was able to determine which physical quantity is "quantized," that is, the quantity that increases by the same amount from one orbit to the next. It turned out to be a physical quantity called "action," the product of mechanical momentum and the length of the orbit. Furthermore, the change in "action" from one quantum orbit to the next was precisely equal to Planck's constant, *h*!

Bohr was able to show that Balmer's series of visible hydrogen lines correspond to "electron jumps" to the second orbit from the orbits beyond it. Each such jump, or transition as it is called, involves just the energy difference needed to produce a light quantum of the correct frequency. When the transition is from a higher energy level to a lower level, energy is given off as light; in the opposite direction, a light quantum is absorbed.

Bohr also pointed out that transitions to the first orbit from those beyond would involve higher energy and, therefore, higher frequency light quanta—quanta in the ultraviolet region of the electromagnetic spectrum. Similarly, transitions from the outer orbits to the third orbit must give a series of lower frequency light quanta in the infrared region. Both of these series were discovered, the former by Theodore Lyman, an American physicist and the latter by Friedrich Paschen, a German physicist. The ability of Bohr's atomic theory to explain the spectral emission of the hydrogen atom gave strong support to the quantum theory. The quantum had entered physics to stay.

Elliptical Orbits

During the decade following Bohr's publication, his theory achieved great success in explaining the optical spectra and chemical interactions of the elements. The theory was given an important addition in 1915 by Arnold Sommerfeld. His contribution was to add some elliptical orbits to the circular orbits proposed by Bohr. The first orbit had no ellipses added to it. The second orbit had three elliptical orbits added to it. Each of these new orbits has precisely the same energy level as that of the neighboring circular orbit. An electron has the same kinetic energy, therefore, whether circling around the second circular orbit or around one of the neighboring elliptical orbits. To the third circular orbit, Sommerfeld added eight elliptical orbits, and to the circular orbits farther out, he added more and more elliptical orbits. Each circular orbit along with its family of elliptical orbits became known as a "shell." As we shall see, this concept led to an understanding of the way that chemical elements combine. The atom was growing more complicated, but it began to fit better and better with the observed properties of the elements.

The Size of the Atom

As we have seen, the hydrogen atom has one electron and one unit of positive charge in its nucleus. Atoms of the heavier elements contain two, three, four, five and more electrons, the exact number corresponding to the atomic number of the element in the periodic table. The positive charge in the nucleus of each atom is equal to the total negative charge of the orbiting electrons. For these heavier, more complex atoms, the general pattern of quantum orbits is the same as for hydrogen, but the size of these orbits is much reduced. The greater positive charge in the nucleus attracts the electrons with greater force, causing a progressive shrinking of the size of the orbits as we go to elements of higher and higher atomic number.

We might now ask ourselves how the greater number of electrons are distributed among the orbits. According to classical theory, the answer is simple. They all merely fall into the inner orbit and play Follow the Leader around the nucleus. This assumption, if true, would lead us to expect that the size of the atom decreases as we consider heavier and heavier elements. We know that the diameter of the inner orbit is smaller for heavier elements, so such elements should be smaller than hydrogen. But experiment tells us that this does not happen. No matter how many or how few electrons an atom may have, its overall size is approximately the same, about 10^{-8} centimeter (less than a hundred millionth of an inch).

The paradox was explained by the German physicist Wolfgang Pauli (1900–1958). According to Pauli's "exclusion principle," each quantum orbit can hold no more than two electrons. The principle tells us that the first orbit can hold either one electron (hydrogen) or two electrons (helium). If both of these electron vacancies are filled, the next electrons must be accommodated in other orbits. When all of the vacancies in a given shell (or energy level) are filled, electrons begin to occupy orbits in the next shell which correspond to a higher energy level.

If we consider the elements in order of increasing atomic number, the size of the corresponding orbits shrink in size because of the higher charge on the nucleus. But more and more orbits become occupied by electrons. The two effects counteract each other and the size of the atoms remains about the same on the average all the way from the lightest to the heaviest elements. Only minor variations in size take place as we go from one completely full shell to another.

The Atom and the Periodic Table

The electron shells of various atoms in the periodic table are filled according to their increasing energy levels. No electron remains in a high-energy orbit if the atom has a lower-energy orbit that is either empty or contains only one electron. The first shell, having the lowest energy level, is completely filled by the two electrons of helium. Lithium, the next element, has three electrons. According to the exclusion principle, one of these must occupy a place in the second shell. This shell, consisting of four orbits (one circular, three elliptical) can accommodate up to eight electrons. Thus, the first two shells can hold a total of ten electrons, and both are completely filled by neon with its ten electrons.

Study has shown that no atom ever has more than eight electrons in its outermost shell. It is possible, however, for an atom to have more than eight electrons in shells closer to the nucleus.* Calcium, for example, has 20 electrons distributed in the first four shells as follows: 2, 8, 8, 2. The twenty-first electron is accommodated in the third rather than the fourth shell. Electrons fill the various shells in much this way throughout the periodic table—always being careful not to exceed eight in the outermost shell.

The Atom and Valence

We now know that every atom has a positive nucleus containing protons surrounded by negative electrons arranged systematically in successive shells. When the number of electrons is equal to the number

* Beginning with the third shell, all of the orbits in a shell do not have precisely the same energy levels. There is some *overlap*, for example, of orbital energies between some of the third and fourth shell orbits. This is caused by variations in the complexities or geometrical shapes of the orbits.

of protons in the nucleus, the atom is electrically neutral. We might expect such an atom to be completely stable, but this is usually not so. If it were, the elements would not combine to form compounds. Actually, only the inert gases are relative hermits, lacking any real desire to combine with other elements.* And, as one might expect, a clue to their behavior is given by the electron populations in the various shells. The distribution of electrons is given below for the inert gases:

	Atomic number	1st shell	2nd shell	3rd shell	4th shell	5th shell	6th shell
helium	2	2					
neon	10	2	8				
argon	18	2	8	8			
krypton	36	2	8	18	8		
xenon	54	2	8	18	18	8	
radon	86	2	8	18	32	18	8

Each of these exceptionally stable elements has eight electrons in its outer shell. Conversely, *no other element* has an outer shell containing eight electrons. The other chemical elements have between one and seven electrons in their outer shell, and these elements are able to react with each other in various ways to form millions of compounds. The active elements react with each other by gaining, losing, or sharing these electrons.

Long before anything was known about the structure of the atom, the concept of *valence* was conceived to account for the forces that hold atoms together in compounds. We know today that valence forces involve the transfer or sharing of electrons by atoms. These *valence electrons* are primarily those in the outermost or *valence shell.* Only occasionally do other electrons also act as valence electrons.

Active chemical elements react with each other in an attempt to achieve the stable electron configurations of the inert gases. This can be illustrated with the help of Figure 149. The elements from helium

* Recently, some of the "inert" gases have been induced to combine with other elements. The statement given above is still valid, however, as far as the great majority of chemical processes are involved.

to neon—both inert elements—are shown in order of increasing atomic number. Each element listed between the two could acquire the stable structure of helium by shedding the electrons in its outermost shell. Similarly, each could acquire the neon structure by adding enough electrons to build up an 8-electron shell. Both events can take place if appropriate elements are caused to react. For instance, when

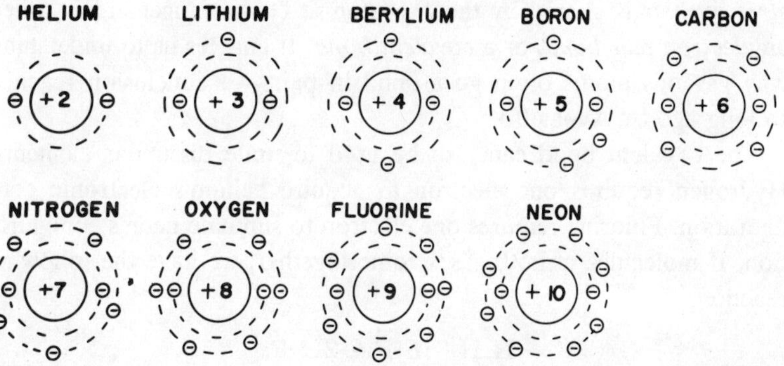

Fig. 149. Electron configurations for atoms from helium to neon in the periodic table.

lithium and fluorine are put together, a violent reaction takes place. Lithium loses one electron leaving it with the stable structure of helium. Fluorine accepts that electron, building its outer shell up to eight electrons, the same as neon. In other words, lithium "wants" to get rid of one electron, and fluorine "wants" to pick up one electron.

The lithium fluoride reaction described above involves *electron transfer*. After the reaction each lithium atom becomes a positive ion because it has a +3 charge on its nucleus and a −2 charge on its orbiting electrons. A net unbalanced charge of +1 exists on the ion as a whole. Similarly, the fluorine ions have a net charge of −1. Because lithium lost one electron in the reaction it is said to have a valence of +1. Fluorine, which gained an electron has a valence of −1. Oxygen, which "likes" to accept 2 electrons, has a valence of −2, and so on. Elements which give up electrons in chemical reactions are called *metals,* and the nonmetals usually accept electrons.

In addition to the transfer of electrons, atoms can get the electrons they seek by *sharing* them with other atoms. Consider fluorine, which needs one additional electron in its outer shell. If two fluorine atoms come together, they can achieve a measure of satisfaction by sharing an electron apiece between them.

$$:\underset{\cdot\cdot}{\ddot{F}}\cdot + :\underset{\cdot\cdot}{\ddot{F}}\cdot \rightarrow :\underset{\cdot\cdot}{\ddot{F}}:\underset{\cdot\cdot}{\ddot{F}}:$$

Each atom contributes one electron to the common purpose, and an *electron pair* is shared by the two atoms. This arrangement is called an *electron pair bond,* or a *covalent bond.* It enables us to understand why gaseous atoms often go around in pairs—a conclusion reached so long ago by Avogadro.

The covalent bond can also be used to unite dissimilar elements. Hydrogen requires one electron to acquire helium's electronic configuration. Fluorine requires one electron to simulate neon's configuration. If molecules of both gases come together, we have the following reaction.

$$H:H + :\underset{\cdot\cdot}{\ddot{F}}:\underset{\cdot\cdot}{\ddot{F}}: \rightarrow 2H:\underset{\cdot\cdot}{\ddot{F}}:$$

Each atom supplies an electron for the covalent bond.

Electron Spin

We know that the planets spin on their axes as they travel around the sun. In 1925 S. Goudsmit and G. E. Uhlenbeck suggested that electrons do likewise as they circle around the nucleus. And because a spinning electron represents a tiny rotational current, it became necessary to think of the spinning electron as a tiny magnet. Instead of a mere point-charge of electricity, we have tiny magnetized charges rotating around the nucleus. We must now take that magnetism into account in our concept of the atom.

An electron can spin in one of two ways in the atom: in the same direction in which it moves in its orbit, or in the opposite direction. It was subsequently determined that the two electrons in the same quantum orbit must spin in directions opposite to each other. This made it possible to state Pauli's exclusion principle in a somewhat different form, and thereby solve another riddle that had long defied solution. The two electrons spinning in opposite directions in the same orbit set up weak magnetic fields that slightly alter each other's orbits. These orbits, although almost identical, are really a little different. Their energy levels, therefore, are also slightly different. This means that electron jumps to these orbits should produce light of slightly different frequencies. And when the spectral lines are examined with great care, it is found that they do consist of multiple frequencies in accordance with theory. Each atomic orbit is thought of, therefore, as a pair of almost identical orbits split apart by weak magnetic fields.

Waves of Matter

In the year 1924, an idea was born that was to have a profound effect on our concept of matter. By that time, light was known to be half particle, half wave. It traveled around in discrete particle-like packages, yet these quanta exhibited all the telltale properties of waves. While such concepts were not easy to accept at first, they do not do great violence to our fundamental or intuitive judgments. Then, a young French physics student, Louis de Broglie (1892–   ), presented to the faculty of the University of Paris a doctoral thesis containing an extraordinary idea. If light, which seems to be a wave, can sometimes behave as a particle, perhaps particles of matter can also sometimes behave as waves.

Louis de Broglie was familiar with Einstein's theory of relativity that had been published in 1905. As a consequence of that theory, physicists have come to regard mass as one of the embodiments of energy. In other words, mass can be converted to energy and *vice versa*. The famous formula $E = Mc^2$ (energy = mass × the square of the velocity of light) states the equivalence of mass and energy in mathematical terms. As we shall see, every explosion—whether nuclear or not—attests to the truth of that formula. Bearing these facts in mind, we can trace a chain of ideas that carries a startling implication:

Particles of matter have mass.
Mass is equivalent to energy.

Energy implies electromagnetic radiation.
Electromagnetic radiation implies frequency.

So, we speculate that particles of matter are involved with waves that have frequency! Proceeding further:

Particles that have a frequency are much like light quanta.
Light quanta are related to light waves.
So material particles should be related to "matter waves."

If all of this seems fantastic, you can sympathize with de Broglie's professors. This young man believed that the motion of material particles is accompanied and guided by certain "pilot waves" that

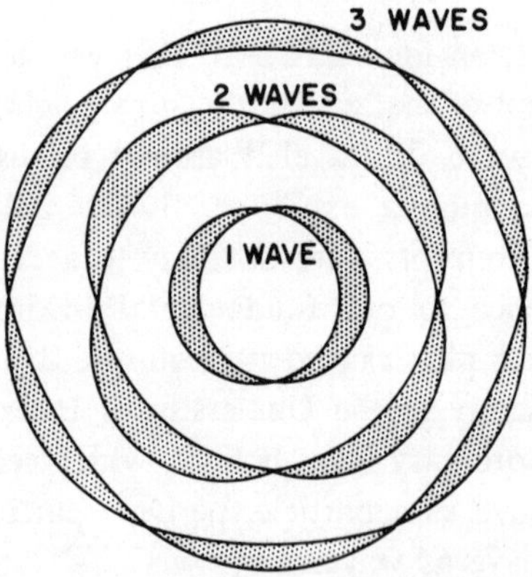

Fig. 150. De Broglie's strange "matter waves" in the atom. Each orbit can contain only an integral number of waves.

move through space with the particles. These waves determine the "allowed" quantum orbits in Bohr's model of the atom. The inner orbit is such that its length is just equal to the wavelength of the electron's pilot wave. This enables the orbiting pilot wave to "catch itself by its tail" as it rotates around the nucleus. The idea is illustrated very schematically in Figure 150. The inner orbit is occupied by a wave consisting of two points of zero amplitude and two points of maximum amplitude, just as would be the case for a wavelength of light or sound. The second orbit contains two of these waves, the third orbit three, and so on. Wherever there is an electron, there is also a "matter wave" that determines its path.

It is clear from de Broglie's theory, that the quantum orbits must have lengths that contain an integral number of these matter waves: one in the first, two in the second, and so on. But we noticed earlier that the length of each orbit multiplied by the momentum (mass times velocity) of the electron is equal to h for the first orbit, $2h$ for the second, $3h$ for the third, and so on. In mathematical terms:

$$\text{Length of 1st orbit} = \frac{h}{mv}$$

$$\text{Length of 2nd orbit} = \frac{2h}{mv}$$

$$\text{Length of 3rd orbit} = \frac{3h}{mv} \qquad \text{etc.}$$

It was apparent to de Broglie that the wavelength of his electron-waves must be equal to:

$$\frac{h}{mv}$$

This would put one wavelength in the first orbit, two in the second, and so on. Orbits of intermediate size would not be possible because they could not accommodate a whole number of wavelengths. With a brilliant stroke of genius, de Broglie gave a kind of form to the strange orbits of Niels Bohr. Particles of matter in motion began to assume characteristics similar to the wave properties of light.

The Electron Microscope

Louis de Broglie's revolutionary proposal was not accepted at first. His ideas sounded plausible enough, but where was the proof? If matter waves really exist, why had they so long eluded the notice of physicists? In short, de Broglie's ideas led to interesting mathematical theories of the atom, but were they physics?

The answer to those questions arrived quite unexpectedly in 1925. In the Bell Telephone Laboratories, in New York City, C. J. Davisson had been experimenting with beams of electrons since 1921, long before de Broglie had conceived of matter waves. He had been bouncing electrons off a very hot piece of nickel in a vacuum. Then came an accident that wrecked his apparatus, broke the vacuum, and let air in to ruin the carefully prepared surface of the nickel. The only way to clean the nickel surface involved prolonged heating. Now aided by Germer, Davisson restored the surface of the nickel and continued with the experiment.

Although the two scientists did not know it at the time, the heat treatment had produced a fundamenal change in the nickel, fusing it into large crystals where before it had consisted of myriad small ones. Each large crystal now contained atoms arranged in regular rows and columns, like soldiers marching on parade. Davisson and Germer proceeded with their electron bombardment of the nickel, and beheld an amazing result. Here before their eyes were the typical X-ray diffraction patterns long known to science. But they had produced no X-rays—only electrons! These apparent X-ray diffraction patterns—so similar to the interference effects of light—were the first direct confirmation of de Broglie's theory. They proved that electrons behave like waves. And more than that, they showed that electrons behave in precisely the way that de Broglie's little wavelength equation, $\lambda = h/mv$, had predicted. The wavelength associated with the electrons could be calculated from the distance between successive peaks and troughs of the diffraction pattern. The electron's wavelength was found to increase and decrease as its velocity was changed, just as de Broglie had predicted. It was later found that other particles behave in the same way, each kind guided by its own particular matter wave as it moves through space. Although de Broglie had been vindicated, the nature of his "pilot waves" was still completely obscure.

It's hard to see what Davisson's experiment had to do with telephones. But it did result in the invention of a new and important optical instrument, illustrating one of the uncanny ways in which science aids humanity. The ability of a microscope to disclose fine detail depends on the smallness of the wavelength of light used. So the most powerful microscopes used ultraviolet light. But the wavelength of moderately fast-moving electrons is thousands of times smaller than that of ultraviolet light. The electron microscope, with its "lenses" of electric and magnetic fields, makes use of these short matter waves to provide magnifications undreamed of before their discovery. This new microscope has become a standard instrument of science and industry.

The New Atom

De Broglie's ideas were extended and put in mathematical form in 1926 by the Austrian physicist Erwin Schrödinger (1897–1961), in the form of the famous "Schrödinger's equation." The new equation contained all of the results of Bohr's quantum theory, and even more. The "old" theory could account for the *frequency* of each of hydrogen's spectral lines, but not its *intensity,* or "brightness" in the spectrum. Schrödinger's equation gave both. In place of circular or elliptical quantum orbits, the atom's structure was described by the so-called ψ (psi) functions which correspond to various kinds of de Broglie waves that can exist in the region surrounding the atomic nucleus. The same results were obtained about the same time by a young German physicist, Werner Heisenberg (1901–   ), but in a more complex mathematical form. Both methods turned out to be identical in principle.

The Uncertainty Principle

The commonly accepted meaning of de Broglie waves, which guide material particles, became clear a few years later when Heisenberg announced the "uncertainty principle." The principle tells us, in essence, that there is a theoretical limit to the precision with which we can ever know the path and velocity of an atomic particle, such as an electron. We can illustrate the concept with a "thought experiment."

It is well known that light exerts pressure. Although the amount is too small to produce observable effects on everyday things, it is quite strong enough to push the "tails" of comets around as a comet moves around the sun. The tail, consisting as it does of dust particles, always faces away from the sun because of the pressure exerted by sunlight. Now imagine a ping-pong ball bouncing back and forth on a table. Light strikes the ball and is reflected to our eyes. The impact of light quanta on the ball produces forces in a real sense, but we would hardly expect the ball to be measurably deflected. The ball is just too massive and the light energy too small. But substitute an electron for the ping-pong ball and the situation is changed completely. A light quantum striking an electron has quite enough energy to knock it off course. Therefore, *the very attempt to observe an electron's position or velocity, produces a change in those quantities.* As successive quanta (or photons) strike the moving electron, it follows a zigzag course because of the impacts. If we reduce the photon's energy—by lowering its frequency—we can reduce the force of impact. But another difficulty arises. The lower the frequency (or the longer the wavelength) the less sharp the image of the electron, because of diffraction. So we are less exact in our knowledge of the electron's position. In short, high frequency photons give us an exact position of the particle, but they interfere with its velocity. Low frequency photons interfere less with its velocity, but define its position less accurately. If we choose a middle ground, we disturb the particle's path only moderately, and still determine its path with reasonable accuracy.

When these ideas are applied to the atomic electron, scientists conclude that they cannot know its orbit with any precision. The amount of uncertainty, in fact, is about as large as the radius of the orbit!

All of this leads us to the conclusion that when describing motions on the atomic scale, we cannot think of the trajectory as an infinitely thin line as we do for the motion of ordinary objects. In the small world of the atom, the motion of a particle is guided by a wave which is thought of as giving a "smeared-out" trajectory. We can only calculate the *probability* that the particle is at any specific place in its "smeared-out" path. It is impossible *in principle* to know both its position and velocity any more accurately.

We can illustrate the idea with an example. Suppose we shoot an electron toward a screen, such as Davisson's lump of nickel. We cannot calculate exactly where it will strike. A second electron will probably arrive at a slightly different place. And so with a third and fourth. When the effects of all the electrons are observed, we see a diffraction pattern similar to that produced by light waves. The wave function, ψ (or rather its square), guides the electron with a degree of uncertainty that obeys laws of probability.

Heisenberg's principle tells us why this uncertainty disappears for ordinary trajectories in our macroscopic world. The amount of the uncertainty depends upon Planck's constant *h* divided by the mass of the particle *m*. Because *h* is so extremely small, a particle as small as one-thousandth of a gram results in an uncertainty that is quite negligible. This means, in effect, that our measuring instruments have no significant effect on the particle's trajectory.

Although the vast majority of scientists accepted the so-called probabilistic interpretation of quantum mechanics, there were a few notable exceptions. Einstein, for one, never accepted the uncertainty principle. To the end of his days he hoped that physics would some day return to the deterministic point of view—the same determinism that we use for large scale phenomena. Similar points of view were held by Schrödinger and, more recently, by de Broglie. The controversy still rages behind the scenes, in the traditional manner of scientific disputes.

Antimatter

According to Einstein's theory of relativity, the mass of a particle should increase as its velocity is raised. The change is negligible for low velocities, but becomes enormous as the velocity approaches that of light. The effect has been observed over and over again in the laboratory and is now considered completely verified. Schrödinger's equation was worked out only for low-velocity "non-relativistic" electrons. In addition, the equation considered the electron as a point, and attempts to apply the equation to a spinning electron all met with failure.

In 1930, Paul A. M. Dirac (1902–) devised a new equation that takes care of both deficiencies. It takes account of relativity, working equally well for slow or fast electrons, and also leads to the conclusion that electrons act like little magnetized spinning balls.

But, as usual, this new theoretical discovery led to a complication.

According to the theory, there ought to be two completely different worlds: one a "positive" world with which we are familiar; and another amazing "negative" world that does just about everything backwards. An object in this strange world would have *negative mass* —placed in our gravitational field it would fall "up" instead of down! Push it one way, and it would move in the opposite direction. Also, the new theory required that an electron should have negative energy levels as well as the familiar positive energy levels. If so, what is to stop an ordinary electron from jumping into one of these negative energy levels? Just imagine the result. In jumping from a positive level to a negative level, it would give up energy in the form of a photon. But it could then jump to another more negative energy level and give up still more energy. This could go on forever, the electron jumping to levels of still greater negative energy, giving off photons in the process. This doesn't happen, of course, so Dirac had to explain why.

Dirac assumed that all of the negative energy levels are filled by the strange electrons from the "negative" world. Therefore, ordinary electrons are prohibited from jumping down by Pauli's exclusion principle. All of space is filled to capacity with these negative-world electrons, but we never perceive them. Fantastic? Well, it's a theory and we can take it or leave it.

Unfortunately, there are further messy complications that will not let us ignore the negative world. Suppose one of these negative energy levels happens to be empty—that is, a negative-world electron is missing. How would we observe the errant negative-world electron? The absence of a negative charge is equivalent to the presence of a positive charge, so we would look for a positive particle. Also, the absence of a negative mass is equivalent to the presence of a positive mass, so our missing negative-world electron ought to show up as a positively charged particle having positive mass. Could such a particle exist?

In 1931, an American physicist, Carl Anderson, discovered a particle having just these properties. Its mass was equal to that of an electron and it had a positive charge equal in magnitude to the charge on an electron. These positive electrons, or *positrons,* obeyed all of

Dirac's predictions. A pair of electrons—one positive, one negative—was produced by the impact of a high energy photon (a gamma ray) against the nucleus of an atom. When a positron happens to meet an electron, the two are annihilated and their mass flies off in the form of a high-energy photon.

Later, physicists discovered the negative proton—or antiproton—a particle analogous to the positron. Its charge is equal and opposite to that of an ordinary proton, and its mass is 1840 times that of the electron, just as expected. Like the lighter particles, the proton and antiproton disappear in a violent explosion when they meet.

The existence of the antiparticles poses perplexing questions for science. Is the matter in our world the predominant kind throughout the vast reaches of space? Or are the different galaxies made up of the two kinds of matter, more or less evenly divided? Are there anti-worlds here and there waiting to explole in a great cataclysm when they meet ordinary worlds? Were the two kinds of matter once together and now separated? There are no answers at present to such questions. Solving the riddle of the atom has merely led to a multitude of equally perplexing questions. Perhaps one day scientists will come to learn all there is to know about matter and energy. But that day is not now.

Nuclear Energy

17

The public had heard little about the atom or its nucleus prior to the first atomic explosion, but physicists had long suspected that the nucleus contained gigantic reserves of energy. Even to scientists, however, the problem of releasing that energy seemed insurmountable. The great Lord Rutherford had publicly ridiculed the possibility of extracting useful energy from the atomic nucleus. Then, almost overnight, the atom reached the headlines of newspapers. The nucleus, which had scarcely been mentioned previously in high school physics courses, suddenly became a household word.

Since then, the general public has had to consider fission bombs, hydrogen bombs, nuclear reactors, radioactive isotopes, atomic warships and radioactive fallout. These momentous developments are a direct result of man's increasing knowledge of the structure of the atomic nucleus. Unfortunately, this new knowledge has thus far been used mostly for destructive ends. We can only hope that our growing knowledge of the nucleus will eventually lead to more constructive uses.

The Puzzle of Nuclear Make-up

With the discovery of the nucleus by Rutherford, the picture of the hydrogen atom began to make sense. Hydrogen had one electron revolving around a positively charged nucleus. Being the lightest of all atoms, its mass was taken as unity. Its simple nucleus was given a special name—the proton—which had a positive charge of one unit.

With hydrogen thus taken care of, physicists tried to use the same ideas for the nuclei of other elements. The trouble began, however, even with the next heavier atom, helium. Helium contains two electrons, so its nucleus must have a positive charge of two units. Clearly, the helium nucleus contains two protons. If helium had been twice as heavy as hyrodgen, that simple explanation would have sufficed. But helium is actually about *four* times as heavy as hydrogen. Although it can contain only two protons, it has a mass number of four.*

Matters grow no better for heavier atoms. Like helium, their nuclei weigh about twice as much as they would if only protons were present. For very heavy elements, the situation is even worse. Uranium, the heaviest natural element, has 92 electrons and weighs about two and one-half times as much as it "ought to."

Attempts were made to explain these discrepancies by assuming that the nucleus contains extra protons. The positive charge on these extra protons was supposed to be cancelled or balanced by an equal number of electrons in the nucleus. Thus, one proton-electron pair would add slightly more than one unit of mass to the nucleus without increasing its net charge. But as more became known about subatomic particles, it was clear that the electron would not "fit" in the tiny nucleus—it is just too large. Like all particles, the electron has a de Broglie wavelength given by the equation $\lambda = \frac{h}{mv}$. For an electron to fit within the nucleus its wavelength would have to be about the same as that of a proton. But the proton's mass *m* is about 2,000 times greater than the electron's mass. To have the same wavelength, the electron's velocity would have to be about 2,000 times greater than that of the proton. Unfortunately for that theory, the electron's energy would then be so high that it could not remain in the nucleus. An explanation of the extra mass of the nucleus would have to be sought elsewhere.

* The mass number of an atom is equal to the number of protons and neutrons in its nucleus. The neutron is a particle slightly heavier than a proton having zero electric charge.

Isotopes

Scientists seeking to understand the nucleus had another perplexing matter to explain. There are cases in which different atoms *of the same chemical element* have different weights. Uranium, for example, comes in two varieties: U^{235} and U^{238}; both atoms are uranium, containing 92 electrons, but their mass numbers differ by three units, 235 and 238. Both kinds of uranium have precisely the same chemical properties. Atoms of the same chemical element that have different atomic weights are called "isotopes," and they occupy the same place in Mendeléev's table. Isotopes are the identical twins of chemistry.

During the mid-twenties, the idea gradually arose that there must be "chargeless protons" in the nucleus—neutral particles weighing about the same as protons. A program was begun at the Cavendish Laboratory to "kick" some of these neutral particles out of the nuclei of light elements. The results were negative. The work was followed up in Germany by Bothe and Becker. By bombarding light elements such as boron and beryllium with alpha-particles emitted by polonium, they discovered a new type of nuclear transformation. The rays coming from the nucleus were much more penetrating than alpha or beta rays, and the scientists concluded (incorrectly) that they must be a form of electromagnetic radiation.

Two years later, in 1932, the experiments were repeated by James Chadwick (1891–   ), a student of Rutherford's, at the Cavendish Laboratory. Studying these mysterious, highly penetrating rays, he proved that they were really streams of neutral particles with a mass about equal to that of a proton. Thus a new nuclear particle, the *neutron*, was born in Rutherford's laboratory.

Because it has no charge, the neutron's passage through material bodies is not affected by the powerful electric fields within the atom. Thus neutrons are able to pass easily through several inches of lead, whereas protons produced in the usual radioactive processes can be stopped by a thin sheet of lead. The most powerful radiations known in the 1930's were gamma rays, and neutrons were more penetrating even than these. We now know that a neutron is *not* a combination of a proton and an electron as was originally thought; it is actually a particle existing in its own right—a particle that constitutes a basic component of the nucleus.

The Neutron and Nuclear Structure

Once the existence of the neutron was established, the answer to the vexing question of nuclear structure became a bit more clear. The excess weight of all nuclei heavier than hydrogen was simply accounted for by assuming the nuclei to contain an appropriate number of neutrons. Helium, the second lightest atom, obviously contains two protons and two neutrons in its nucleus, which is surrounded by two orbiting electrons. The mass number of helium would be four, but the charge on the nucleus (the atomic number) would be just two. Similarly, oxygen, with a mass number of sixteen and an atomic number of eight, would have a nucleus containing eight neutrons and eight protons surrounded by eight eletcrons. And, providentially, the neutron provides no size problem, for its mass is about equal to that of the proton. Its de Broglie wavelength should be of the right length to "fit" into the nucleus.

Although electrons fly through the vast open spaces at the outer reaches of the atom, the neutrons and protons are packed as tightly as olives in a jar within the tiny nucleus. The constitution of several atomic nuclei are shown very schematically in Figure 151, in terms of the neutrons and protons they contain.

The nucleus is thought of as an incompressible substance, much as we regard a drop of water. The particles are held together by great cohesive forces which more than overcome the electrostatic repulsion of the positively charged protons. For these reasons, the nucleus becomes larger as more particles are added, and its density, or weight per unit volume, remains relatively constant. The density of this nuclear "fluid" is greater than that of water by a factor of 10^{14}, or:

$$100{,}000{,}000{,}000{,}000$$

and a cubic inch of it would weigh almost two billion tons!

The neutron succeeded in clearing up another mystery of nuclear structure—that of isotopes. Chlorine, for example, has two kinds of atoms, some of mass number 35 and others, 37. Clearly, both isotopes contain 17 protons in the nucleus, one for each orbiting electron. But, in addition to these, Cl^{35} has 18 neutrons in its nucleus, while Cl^{37}

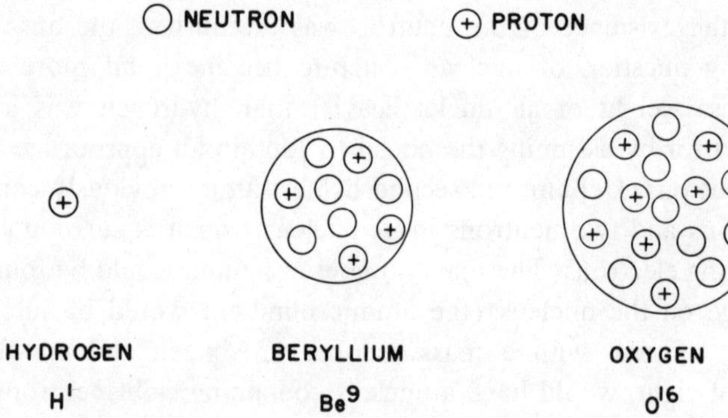

Fig. 151. A schematic representation of the nuclei of hydrogen, beryllium and oxygen showing the presence of protons and neutrons.

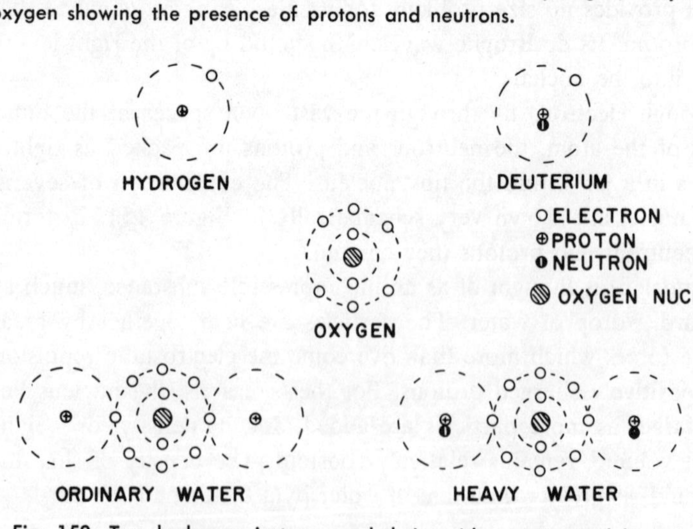

Fig. 152. Two hydrogen isotopes and their oxides, water and heavy water.

contains 20 neutrons. Similarly, U^{238} contains 146 neutrons compared with 143 for U^{235}. Actually, most elements have several naturally occurring isotopes. Those for hydrogen are shown in Figure 152. The heavy isotope of hydrogen is called *deuterium*, and water made of deuterium is called heavy water. Its chemical symbol is D_2O instead of H_2O. About one molecule out of three thousand of ordinary water contains one atom of deuterium.

Radioactive Families

Increased knowledge of the nucleus led to a better understanding of radioactivity. Rutherford had shown in 1899 that there are three kinds of radiation coming from the atom:

1. Alpha (α) rays, which proved to be helium nuclei—and which can be stopped by a thin sheet of paper.
2. Beta (β) rays, which turned out to be streams of fast-moving electrons, and which can pass through thin aluminum foil a few thousandths of an inch thick.
3. Gamma (γ) rays, similar to X-rays and light, but of much shorter wavelength, which can pass through lead shields many inches thick.

Figure 153 shows how these radiations are affected in passing through a magnetic field. The (positive) alpha particles are deflected one way, the (negative) beta particles the opposite way, and the gamma rays, being electromagnetic waves, remain undeflected.

The phenomenon of radioactivity is really the spontaneous transformation of one chemical element into another. When an alpha particle is emitted by an atom, its nucleus loses two protons and two neutrons. Its charge is reduced by two units of positive charge, producing a new element which is located two steps to the left in Mendeléev's periodic system. The atomic weight of the new element is four units lighter than before the transformation. The emission of a beta particle (a negative electron) moves the element one step to the right in the periodic system and does not change its atomic mass number.* The emission of gamma rays is merely the result of a disturbance within the atom caused by the ejection of a material particle.

* The emitted electron is not one of the orbit electrons. It comes from the transformation of a neutron into a proton and electron. The additional proton increases the atomic number by 1, adding one unit of charge to the nucleus.

In the course of radioactive decay, the unstable radioactive atoms reduce their atomic number and mass until they finally reach a stable state which is always an atom of lead. An alpha decay (ejection of an alpha particle), results in a reduction in atomic mass of 4 units, but

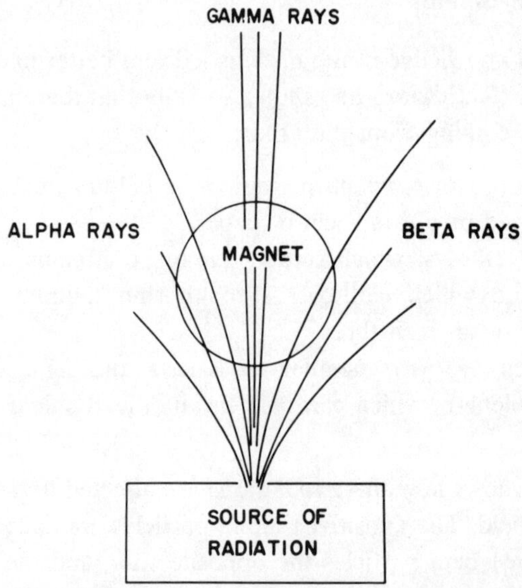

Fig. 153. Deflection of alpha, beta and gamma rays by a magnetic field.

a beta decay (ejection of an electron), has no effect on the mass number. In theory, therefore, there are 4 possible families of radioactive elements, each moving down the periodic table from the heavier to the lighter elements. Three such families can be found in nature, those beginning with uranium (238), thorium (232), and protactinium (231). The fourth family can be produced artificially in nuclear reactors.

The uranium family starts with uranium (238) and, after eight alpha and six beta transformations, becomes stable lead (206). The intermediate elements are unstable isotopes of such elements as uranium, radium, radon, and polonium. Similar patterns of radioactive decay exist for the other radioactive families.

Radioactive Lifetimes

The life expectancies of everyday things, such as automobiles or puppies, vary somewhat depending on the nature of the item and its environment, or wear. Some live a little longer, others a little less. If we imagine that a large number of puppies are born on a certain day, we can draw a graph, as shown on the left of Figure 154, that

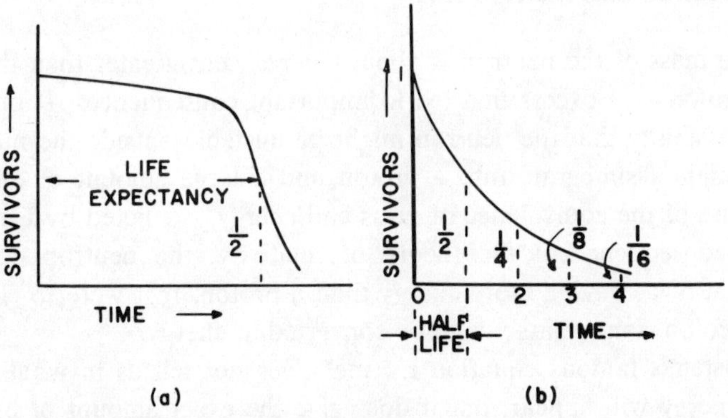

Fig. 154. Graphs illustrating the half-lives of animals and ordinary things *(a)* and radioactive elements *(b)*.

gives the number still alive after the passage of various lengths of time. The graph shows that there is a certain life "expectancy" or half-life of the population, at which time one-half of the individuals are still alive. The half-life is about seventy-five years for a man, perhaps ten years for a car, and twelve years or so for a dog. The curve shows that any one individual has a relatively small chance of dying much before the half-life point, and an equally small chance of living considerably beyond that point.

The life expectancy of a radioactive atom is quite another matter. The instant at which a given radioactive atom will decay is independent of the period of time that has elapsed since it was "born." This sort of life expectancy leads to the curve shown on the right of the diagram, a curve that mathematicians call an "exponential curve." For this situation, the half-life is the period of time during which one-half of the atoms will decay. There is no way of predicting which individual atom will decay first.

The half-lives for different radioactive elements vary from extremely long to extremely short periods of time. For uranium, the half-life period is 4.5 billion years, for radium 1,590 years, but it is only one ten-thousandth of a second for a certain isotope of radium, RaC!

The Neutron and the Neutrino

The mass of the neutron is about 0.1 per cent greater than that of the proton—an excess that led to important consequences. It opened the possibility that the neutron might be unstable outside the nucleus and might disintegrate into a proton and a large amount of energy. Because of the equivalence of mass and energy, predicted by Einstein as a consequence of the theory of relativity, the neutron can be thought of as having more energy than a proton. If it were to change to a proton, some mass would be converted to energy.

Einstein's famous equation $E=mc^2$ does not tell us in what form that energy will appear, but it does give the exact amount of energy to expect. In addition to the energy, an electron would almost certainly be released in the transformation. The law of conservation of charge would then be obeyed, the proton and the electron having equal and opposite electric charge.

When the experiment was performed, it was found that a neutron is unstable outside the nucleus and does indeed disintegrate into a proton and an electron. But the results led to another important discovery. The law of conservation of momentum (momentum is mass times velocity) leads to the conclusion that the proton and electron should be emitted in opposite directions after the "explosion" of the neutron. But they were found to go off in directions that were not exactly opposite. To satisfy the law, scientists postulated the existence of a mysterious third particle that carried off some of the energy. This particle, having zero mass and zero charge, was named the "neutrino" or "little neutron" by Enrico Fermi. With its help, all of the energy of the neutron disintegration could be accounted for, provided that allowance was made for the energy carried away by the neutrino. Furthermore, the half-life of a neutron outside the atom was found to be about ten minutes, while neutrons exist indefinitely within the nucleus.

The experimental verification of neutron-decay led to another perplexing question. How can an electron emerge from a neutron, which is certainly far too small to contain an electron? We run into the same problem in many other examples of radioactive decay: electrons pop out of nuclei in which, as we have already seen, no electron could possibly have been present. The neutron's structure must indeed be more complex than was originally supposed. Electrons must be produced, somehow, from the mass or energy of the neutron during the process of disintegration. Despite such questions, Fermi's theory of neutron-proton transformation, with an electron and neutrino as by-products, is in excellent agreement with experimental fact.

Mesons and Hyperons

In the early thirties, physicists were happy enough with their fundamental particles, the proton, electron, and neutron. The neutrino was an embarrassment, of course, but no one foresaw the great proliferation of sub-atomic particles that was to come. In 1936, the Japanese physicist, H. Yukawa (1907–), predicted a new particle on purely theoretical grounds, in order to explain the enormous forces within nuclei. These *mesons*, of mass about one-seventh that of the proton, are continually exchanged between proton and neutron within the nucleus. When a meson becomes separated from a nuclear particle, it disintegrates almost immediately into an electron and a neutrino.

Mesons were later discovered experimentally by Carl Anderson in the cosmic rays showering down to earth from the upper atmosphere. Cosmic rays consist of high energy particles, mostly protons, that arrive from outer space, causing nuclear transformations in atoms with which they collide.

Later, several other mesons were discovered, some having positive charges, others negative charges, and one with no electric charge at all. The so-called π meson (or "pion") has a slightly greater mass than the μ meson (or "muon"). Studies have shown that a positive or negative pion decays into a (positive or negative) muon and one neutrino.

In the years following, particle after particle made its appearance in the world of nuclear physics. There was the K meson with a mass of 965 electrons, the Tauon, with a mass of 966.5 electrons, and several particles heavier than neutrons that were called "hyperons." To date, well over eighty different particles have been observed, and no one knows when the end the will be reached. Although physicists are learning a great deal about the characteristics of these particles, there is no accepted theory that leads to a real understanding of them.

Nuclear Fission

By the time the neutron was thirteen years old, it had played its part in the uncontrolled release of nuclear energy. The subsequent production of useful nuclear power, and the wide availability of radioactive isotopes for medicine and industry, are practical results of the nuclear chain reaction, made possible by neutrons.

The nuclear chain reaction is a self-sustaining process that can release enormous amounts of energy. It is based on a nuclear transformation possible only in heavy nuclei, a process called *fission.* It is because neutrons can cause fission in heavy nuclei that useful energy can be abstracted from the nucleus.

Figure 155 illustrates how the fission of U^{235} can be thought of as occurring. A neutron is absorbed by the nucleus, which becomes unstable and begins to vibrate, passing through a series of elongated shapes. The cohesive forces tend to restore the nucleus to a spherical shape, but the protons' electric forces of repulsion tend to increase the elongation. A cleavage eventually develops and the nucleus breaks neatly into two almost equal parts. But most important of all, the fission of U^{235} is accompanied by the ejection of two or three (more exactly, 2.5 on the average) neutrons which, in turn, may hit other nuclei causing them to break up. Thus, a chain reaction may develop which rapidly causes fission of many atoms in a piece of uranium with the liberation of enormous amounts of energy.

The energy produced by a nuclear reactor or atomic bomb comes from the conversion of mass to energy. In the aggregate, the new lighter elements produced by the fission of uranium weigh less than the uranium from which they were formed. The difference shows up as energy in accordance with Einstein's equation, $E=mc^2$. It should also be pointed out that the same equation is responsible for the production of all energy, whether by chemical or mechanical means. Whenever energy is liberated by a process, the mass of the products is less than the mass of the initial ingredients by an amount equivalent

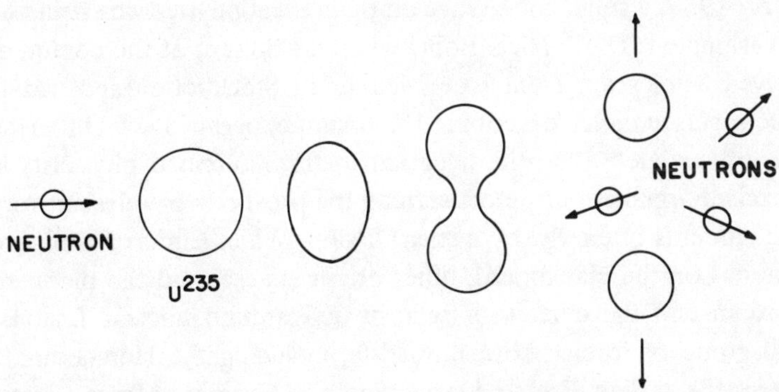

Fig. 155. Nuclear fission. A neutron enters the Uranium-235 nucleus, causing it to break into smaller fragments which include two or three secondary neutrons.

to the energy released. The gases released by a TNT explosion weigh less than the original sticks of TNT. Similarly, a glass of water weighs more when hot than when cold.

In all of these ordinary examples, the amount of weight lost is extremely small, much too small to be detected by the most sensitive balance. Only in the case of nuclear reactions is the amount of converted mass great enough to be detected by the delicate methods of nuclear experimentation. Although the weighty caloric of an earlier scientific era has passed away, science has finally shown that energy is not the simple weightless property it was once thought to be. Energy is on a par with matter as a fundamental substance of the universe.

Nuclear fission was first detected in 1938 by two German scientists, Otto Hahn and Fritz Strassman. They had been bombarding various elements with neutrons, just as many other physicists were doing all over the world. They discovered that while most nuclei change only slightly after swallowing a neutron, the uranium nucleus undergoes a dramatic change. After the bombardment, they discovered traces of barium in their previously pure uranium sample, an element half-way down the periodic table. Uranium had been transformed, in part, into a much lighter element, barium.

Otto Hahn wrote a letter to a woman physicist, Lise Meitner, describing the unexpected results, and by a series of coincidences, the information found its way to physicists in the United States. On January 27, 1939, a small conference on theoretical physics was being held in Washington, D. C. Niels Bohr, who was present at the conference, received a telegram from Lise Meitner in Stockholm (she had fled Hitler's persecution) describing the fission experiment of Otto Hahn. Bohr immediately read the telegram to the assembled physicists and an excited argument arose concerning the prospects of releasing large-scale amounts of energy by nuclear fission. While Enrico Fermi wrote equations on the blackboard, other physicists escorted the press from the room and the curtain of security fell around nuclear fission. A small group of foreign-born physicists, including the Hungarian Leo Szilard, the Italian Enrico Fermi, and the German Albert Einstein, used their influence to induce a skeptical American government to undertake a program of nuclear research aimed toward developing the atomic bomb.

It was clear from the start that it is not the more abundant isotope of uranium, U^{238}, that is capable of fission, but the extremely rare isotope U^{235} which is present in the small amount of 0.7 per cent in natural uranium. Even more discouraging is the fact that "useless" U^{238}—which makes up the remaining 99.3 per cent—captures neutrons at such a fast rate that it prevents any chain reaction from getting under way in the admixture of U^{235} atoms. There were only two ways around the problem: either to separate the U^{235} from the neutron-hungry U^{238} atoms, or find a way to reduce the appetite of the U^{238} atoms for neutrons.

The first method—isotope separation—was tried at a secret plant at Oak Ridge, Tennessee. Production of U^{235} was achieved by the diffusion method, based on the fact that lighter compounds containing U^{235} diffuse through membranes somewhat faster than compounds containing the heavier isotope.

Enrico Fermi took the other approach, and managed to produce the first nuclear chain reaction in 1942 at the University of Chicago. It was found that heavy uranium has a great ability to capture fast neutrons while the lighter isotope prefers very slow ones. The energetic neutrons produced in the fission process have very high velocities, however, so a process called moderation was used to slow them down to a level acceptable to U^{235} atoms. This was achieved by mixing ordinary uranium metal ($U^{238}+U^{235}$) with a so-called "moderator,"—a substance whose molecules reduce the velocity of neutrons by the collision process without absorbing them. The two most efficient moderators are deuterium (heavy hydrogen) and carbon. Modern nuclear reactors use one or the other of these moderators. When deuterium is used, it is in the form of heavy water.

Fermi's reactor was built under the grandstand of the University of Chicago using a carbon moderator in the form of graphite bricks. In operating the "atomic pile" it was found that while U^{235} atoms underwent fission, some of the U^{238} atoms also swallowed up neutrons. This led to a most important byproduct of the reactor. When U^{238} absorbs a neutron, it is transformed temporarily into neptunium (Np^{239}) and finally into Plutonium (Pu^{239}). These new radioactive elements do not occur in nature, neptunium having the atomic number 93, and plutonium 94. They are two of the man-made *transuranium elements,* elements heavier than the heaviest natural element, uranium. In all, eleven such elements have thus far been made in the laboratory.

Neptunium is merely a transitory stage in the process, but plutonium is long lasting and capable of nuclear fission along the same lines as U^{235}. In fact, plutonium is a better fuel for nuclear reactors than U^{235}. It breaks up more easily and produces a greater number of secondary neutrons in the fission process (about 2.9). But plutonium has an even greater advantage. Being a different chemical element from uranium, it can easily be separated chemically from the remaining uranium after it is produced in the reactor. Using this principle, the so-called "breeder reactor" can actually produce more fuel than it uses up. The reactor's fuel is plutonium combined with ordinary uranium. During the nuclear reaction, some of the uranium is converted to plutonium and this more than makes up for the amount of original plutonium that is consumed.

In the generation of useful power, all reactors produce heat which is used to drive an engine, as aboard ship, or an electrical generating machine.

Fission Bombs

The only additional concept needed to produce a nuclear fission bomb is that of *critical size*. To illustrate, imagine a small quantity of pure U^{235} or Pu^{239} resting on a table. Fission takes place slowly and harmlessly in atoms here and there in the sample, but no chain reaction takes place. The atom, after all, is mostly empty space, and there is little likelihood that a neutron from one of the exploding nuclei will strike another nucleus and cause fission. On the average, a neutron must travel about 4 inches before it runs into another nucleus. So if the size of the sample is much less than that, most of the neutrons that are given off spontaneously will escape before they can cause new transformations. If the sample is too small, therefore, no chain reaction can take place. As we go to larger and larger samples, each of these neutrons has a greater chance to produce another fission by striking a nucleus. When a *critical size* is reached, many secondary neutrons produce fission in the metal and the process rapidly builds up in intensity. Because plutonium produces a greater number of secondary neutrons than U^{235}, its critical size is somewhat smaller. In any event, it is clear that a relatively small amount of pure fissionable material is capable of producing a violent explosion.

To make a bomb, of course, the fissionable material must be kept in several small, subcritical pieces until the instant of detonation. When that instant arrives, the pieces can be shot into one supercritical mass at high speed, so that the chain reaction will not get under way appreciably before the entire lump is formed. There are undoubtedly more clever, but secret, ways of accomplishing the same result.

Fusion

Fusion is another type of chain reaction that holds great promise for the future, but which presents formidable difficulties for scientists. It is based on the nuclear process of *fusion* which is just the opposite of fission. It involves the fusion—or joining together—of light nuclei, rather than the disruption of heavy nuclei. The major difficulty lies in the nature of the particles involved. Fission takes place readily because uncharged neutrons can enter a nucleus with little difficulty. In contrast, the process of fusion involves the joining together of light nuclei, such as hydrogen and lithium, which contain positive charges. The great forces of repulsion make it difficult for such nuclei to join together, or fuse.

In practice, the atoms of a gas, such as hydrogen, must be raised to a temperature of millions of degrees in order to give then extremely great velocities. This enables the nuclei to touch in spite of the enormous electrical forces of repulsion. Only at such great temperatures are the nuclei moving fast enough to combine and release energy by fusion. The process is called a "thermonuclear" reaction because it is caused by heat or thermal energy.

Once again, the energy released by fusion is derived from the conversion of mass to energy. The new heavier elements formed by fusion have less mass than the sum of the elements that went into their formation. The difference shows up as energy. This energy causes the temperature to rise still higher, and the nuclei to fuse more rapidly, providing the essential conditions for a chain reaction.

The difficulty in producing a controlled fusion reaction lies in the temperatures involved. No material container such as steel or concrete can withstand the necessary temperatures even for an instant. At present, scientists are attempting to contain the hot ionized gases by using magnetic fields.

Unfortunately, the fusion chain reaction is easily used to produce an uncontrolled reaction—the so-called hydrogen bomb. The elements used are the hydrogen isotopes, deuterium, D^2, and a still heavier isotope called tritium, T^3. Although deuterium is found in ordinary water, tritium is an unstable isotope and must be manu factured in nuclear reactors. Because of its great expense, tritium is used only in small amounts in a bomb, as a kind of nuclear "kindling" to initiate the nuclear reaction in inexpensive deuterium.

The first hydrogen bomb was exploded on November 1, 1952 at Elugelab, a small coral island in the Pacific Ocean. The heavy hydrogen isotopes were compressed and heated to hundreds of millions of degrees by the explosion of a "small" fission bomb of plutonium or uranium. These conditions produced the fusion of the deuterium and tritium to produce helium isotopes along with the release of enormous amounts of energy. The complete disappearance of the island attested to the unimaginable power of the fusion reaction. While there is a practical limit to the size in which fission bombs can be made, there appears to be no such limitation when it comes to fusion bombs. Fissionable material, such as U^{235}, must be kept in small, separate pieces prior to detonation, but the hydrogen isotopes impose no such requirement. This explains why hydrogen bombs come in much larger "sizes" than fission bombs.

For thousands of years, the human mind has grappled with the seemingly unending task of unearthing the secrets of man's environment. His first great success, the calendar, led to the science of astronomy which can tell us the chemical composition of stars incredibly distant from the earth. His mastery of the concept of number made astronomy and all of the other exact sciences possible. His discovery of lenses led to optics and the realization that human senses can be supplemented usefully in the observation of natural occurrences. With the mastery of the laws of motion came dependable clocks and the ability to navigate routinely over the face of the earth. Light, a great mystery through the ages, provided the connecting link with electricity that led to radio and the electronic marvels of our day. Chemistry, the science of the very matter of our planet, provided the countless new materials upon which our civilization depends. Heat, once thought to be a chemical element, gave us the engines through which man's efforts are magnified a million-fold in his daily work.

These accomplishments, great as they are, are merely stepping stones to a more promising future. But today, the atomic nucleus threatens to disrupt the entire process of human enlightenment and progress.

Of all the scientific knowledge amassed during man's lengthy stay on earth, we can safely say that most was acquired during the present century. This is by no means a reflection on the great minds of the past, but the point is this: the rate of increase in scientific research has been fantastic in recent decades. Properly exploited, the gains of the next half century can eliminate poverty, hunger and disease as the scourges of mankind. Historically, scientific achievement has led inevitably to a richer and more fruitful human existence. In our time, man's political immaturity and the nuclear chain reaction raise grave doubts about the future course of human progress. And yet, at century's end, we may well look back in disbelief at the alternatives offered the human race: a great new world made possible by intelligent exploitation of scientific research; or a mad scrambling between men of different regions to acquire a disproportionate share of the good things of life.

Bibliography

FOR FURTHER READING

ANDRADE, E. N. da C. *Sir Isaac Newton.* New York: Doubleday & Company, Inc., 1958.

———. *An Approach to Modern Physics.* New York: Doubleday & Company, Inc., 1957.

———. *Rutherford and the Nature of the Atom.* New York: Doubleday & Company, Inc., 1964.

BITTER, FRANCIS. *Magnets.* New York: Doubleday & Company, Inc., 1959.

BROWN, SANBORN C. *Count Rumford, Physicist Extraordinary.* New York: Doubleday & Company, Inc., 1962.

DRAKE, STILLMAN. *Discoveries and Opinions of Galileo.* New York: Doubleday & Company, Inc., 1957.

DUQUESNE, MAURICE. *Matter and Antimatter.* New York: Harper & Row, Publishers, Incorporated, 1960.

FARADAY, MICHAEL. *Experimental Researches in Electricity.* New York: E. P. Dutton & Co., Inc., 1951. (Originally 3 Vol., 1839–1855.)

FINDLAY, ALEXANDER. *Chemistry in the Service of Man.* New York and London: Longmans, Green and Co., 1957.

GALILEI, GALILEO. *Dialogues Concerning Two New Sciences.* New York: Dover Publications, Inc. (Based on the 1914 translation.)

GAMOW, GEORGE. *Biography of Physics.* New York: Harper & Row, 1961.

GARDNER, MARTIN. *Great Essays in Science.* New York: Pocket Books, Inc., 1957.

HERTZ, HEINRICH. *Electric Waves.* New York: Dover Publications, Inc., 1962. (Republication of 1893 translation.)

HOGBEN, L. *Mathematics for the Million.* New York: W. W. Norton & Co., Inc., 1937.

HUGHES, DONALD J. *The Neutron Story.* New York: Doubleday & Company, Inc., 1959.

JAFFE, BERNARD. *Michelson and the Speed of Light.* New York: Doubleday & Company, Inc., 1960.

———. *Chemistry Creates a New World.* New York: Pyramid Publications, Inc., 1962.

———. *Crucibles: The Story of Chemistry.* New York: Premier Books, 1960.

KLINE, MORRIS. *Mathematics in Western Culture.* New York: Oxford University Press, 1953.

LESSING, LAWRENCE P. *Understanding Chemistry.* New York: New American Library of World Literature, Inc., 1959.

MACDONALD, D. K. C. *Faraday, Maxwell and Kelvin.* New York: Doubleday & Company, Inc., 1964.

MASSEY, H. M. *The New Age in Physics.* New York: Harper & Row, Publishers, Incorporated, 1960.

NEUGEBAUER, O. *The Exact Sciences in Antiquity.* New York: Harper & Row, Publishers, Incorporated, 1962.

NEWTON, SIR ISAAC. *Opticks.* New York: Dover Publications, Inc., 1954. (Based on the 1730 edition.)

PANNEKOEK, A. *A History of Astronomy.* New York: Interscience Publishers, Inc., 1961.

PIERCE, JOHN R. *Electrons, Waves and Messages.* New York: Hanover House, 1956.

———. *Symbols, Signals and Noise.* New York: Harper & Row, Publishers, Incorporated, 1961.

PLEDGE, H. T. *Science Since 1500.* New York: Harper & Row, Publishers, Incorporated, 1959.

READ, JOHN. *A Direct Entry to Orangic Chemistry.* New York: Harper & Row, Publishers, Incorporated, 1960.

ROMER, ALFRED. *The Restless Atom.* New York: Doubleday & Company, Inc., 1960.

SANDFORT, J. F. *Heat Engines.* New York: Doubleday & Company, Inc., 1962.

SARTON, GEORGE. *Ancient Science and Modern Civilization.* New York: Harper & Row, Publishers, Incorporated, 1959.

SINGER, CHARLES. *A Short History of Scientific Ideas to 1900.* New York and London: Oxford University Press, 1959.

SINGER, CHARLES; HOLMYARD, E. J. and A. R. HALL. *A History of Technology.* New York and London: Oxford University Press, Inc., 1954.

TATON, RENÉ. *Ancient and Medieval Science.* New York: Basic Books Inc., Publishers, 1963.

TAYLOR, LLOYD W. *Physics, The Pioneer Science.* New York: Dover Publications, Inc., 1959.

THIRRING, H. *Energy for Man.* New York: Harper & Row, Publishers, Incorporated, 1962.

VAN BERGEIJK, W. A. *Waves and the Ear.* New York: Doubleday & Company, Inc., 1960.

VAN MELSEN, A. G. *From Atomos to Atom*. New York: Harper & Row, Publishers, Incorporated, 1960.

WHITTAKER, SIR EDMUND. *A History of the Theories of Aether and Electricity*. New York: Harper & Row, Publishers, Incorporated, 1960.

WOLF, A. *A History of Science, Technology, and Philosophy in the 16th and 17th Century*. New York: Harper & Row, Publishers, Incorporated, 1959.

———. *A History of Science, Technology, and Philosophy in the 18th Century*. New York: Harper & Row, Publishers, Incorporated, 1961.

Index